Bridging Traditions

Habent sua fata libelli

Early Modern Studies Series

Bridging Traditions

Alchemy, Chemistry, and Paracelsian Practices in the Early Modern Era

Edited by
Karen Hunger Parshall,
Michael T. Walton, and
Bruce T. Moran

Early Modern Studies 15
Truman State University Press
Kirksville, Missouri

tsup.truman.edu

Cover art: "The master and assistant distilling alcohol, a primitive form of reflex condenser," from Conrad Gesner, *The newe jewell of health*, translated by George Baker (London: H. Denham, 1576). M0012934, Wellcome Library, London.

Cover design: Teresa Wheeler

Library of Congress Cataloging-in-Publication Data

Bridging traditions : alchemy, chemistry, and Paracelsian practices in the early modern era : essays in honor of Allen G. Debus / edited by Karen Hunger Parshall, Michael T. Walton, and Bruce T. Moran.

pages cm.—(Early modern studies ; vol. 15)

Includes bibliographical references and index.

ISBN 978-1-61248-134-0 (library binding : alk. paper)—ISBN 978-1-61248-135-7 (e-book)

1. Chemistry--History. 2. Debus, Allen G. 3. Paracelsus, 1493-1541. I. Parshall, Karen Hunger, 1955- II. Walton, Michael Thomson, 1945-2013 III. Moran, Bruce T. IV. Title: Alchemy, chemistry, and Paracelsian practices in the early modern era.

QD14.B84 2015

540.9'031--dc23

2014016934

The paper in this publication meets or exceeds the minimum requirements of the American National Standard for Information Sciences—Permanence of Paper for Printed Library Materials, ANSI Z39.48–1992.

To the memory of Allen G. Debus and
Michael T. Walton.

Contents

PART TWO: Regional Contexts and Communities of Texts

PART THREE: Evaluations and Perceptions

Illustrations

Chapter 2

Chapter 3

Chapter 10

Chapter 11

Introduction

Twenty men crossing a bridge,
Into a village,
Are twenty men crossing twenty bridges,
Into twenty villages,
Or one man
Crossing a single bridge . . .
—Wallace Stevens, "Metaphors of a Magnifico"

Reference to Wallace Stevens's early poem "Metaphors of a Magnifico" (1918) may seem an odd way to introduce a collection of essays related to early modern science and medicine and honoring Allen Debus (1926–2009). Yet there is a link between the poem and Debus's historical research, namely, an awareness that in daily—as well as in historical—life, human beings are sometimes both in and out of the spheres in which others may perceive them. Debus was particularly concerned about making this distinction in regard to the traditions of knowing that shaped the contours of natural knowledge in the early modern era. Against those who sought to describe a single direction in the march toward modern science, he argued that those who seemed, from a historical distance, to be crossing the metaphorical bridge of the Scientific Revolution were not in lockstep. As actors in a scene yet to be written, they held differing views of what they were involved in and glimpsed various outlines of what they approached.

"[T]he Scientific Revolution was not," Debus declared, "simply the forward march of a new experimental method coupled with the powerful tool of mathematical abstraction. For some the two were incompatible."[1] Nevertheless, despite the differing intellectual traditions and ways of knowing that distinguished Renaissance thinking, Debus identified a certain intellectual posture

1. Debus, *Chemical Dream of the Renaissance*, 32.

in regard to uncovering nature's secrets that emerged from an interplay of practices described as mystical, artisanal, and experimental. Mixtures of method were essential to the process of inspiring new learning, and Debus had neither apprehension nor reluctance in joining together what some historians considered separate domains. In this way, he emphasized points of connection between apparently distinctive zones of comprehension and experience: magic and experiment, alchemy and mechanics, practical mathematics and geometrical mysticism, things earthly and heavenly, and especially, although it seems unnecessary from today's perspective, science and medicine. Such unions of apparent opposites came together for him in what he called the chemical philosophy, the chemical-Paracelsian approach to nature in which chemistry and chemical processes gained "divine significance," while also leading, through analytical methods and material experience, to the "fundamentals of nature." The book of creation was a chemistry book, and Debus seemed to share something of the religious awe of chemical philosophers when reading it. At the end of one of his earliest works, an essay published as the short pamphlet entitled *The Chemical Dream of the Renaissance* (1968), he revealed what may have been a private thought: "I would close by saying that I do believe that there was a chemical dream in the Renaissance—it was a search for our Creator through his created work by chemical investigations and analogies."[2]

Most of the essays collected in this volume introduce individual perspectives relating to the chemical and/or Paracelsian understanding of nature in the sixteenth and seventeenth centuries. Karen Parshall introduces Allen Debus's intellectual place within the history of early modern science highlighting, in the context of an intellectual biography, the historiographic circumstances for the emergence of a vision of Renaissance science in which the esoteric and the practical, the organic and the mechanical, the philosophical and the religious combined to characterize a quest for natural knowledge. Debus's vision and his approach to historical sources were not uncontroversial, especially as both problematized the "grand narrative" of the Scientific Revolution. His emphasis upon printed as opposed to manuscript sources was criticized as doing too little. At the same time, his contentions for the relevance of vernacular works were castigated as claiming too much. Debus did more than scratch the surface, and

2. Ibid., 32–33.

his major work, *The Chemical Philosophy*,[3] although introductory, was groundbreaking as a result. Yet even here there were detractors, and one openly hostile review rebuked a perceived attempt to distance a Latinate elite from the social and intellectual settings of Renaissance alchemy and natural philosophy. Debus himself stepped both in and out of vernacular and Latinate traditions, and combining the two by means of renewed attention to manuscript and printed sources has become a major current among contemporary methods in the history of science and medicine, and underscores many of the approaches reflected in papers collected for this volume.

The title of this volume refers also to combining subjects that otherwise might be held distinct, and here too contemporary approaches continue the direction explored in introductory, but nevertheless pioneering, ways in Debus's accounts of "chemical philosophy." In that regard, the essays in this volume's first section, "Curious Practices and Practices of Curiosity," introduce an array of practices—some material, some textual, some public, some personal, some emerging from institutions—that reflect the variety of approaches and processes that gave shape to a knowledge of nature and the body in the early modern world.

Debus had referred to an "Elizabethan compromise" as a way of describing how certain altered features of Paracelsian theory and pharmaceutical practice came to be incorporated within a philosophically broadened Galenic system of medicine. Jole Shackelford expands and further targets this notion by focusing upon examples of physicians who, rather than compromising original Paracelsian ideas, remodeled and restructured them as practical chymical therapies. Bringing an abundance of new historical material to light from vernacular and Latin sources, he explores, in particular, the development of chemical uroscopy in the medical practice of the seventeenth-century German physician Johann Hayne (fl. 1620). While some viewed uroscopy as a disadvantage to blending chemical and Galenic practice, others presented it as a positive and practical therapy and diagnostic tool. In bringing together his own training and experience as a practitioner with traditional elements of medical theory and Paracelsian notions of astral disease and chemical diagnosis, Hayne transformed an apparently disordered and numinous system (based in Paracelsian notions) into therapies that influenced later therapeutic practices accommodating materialist and even corpuscularian views of matter.

3. Debus, *Chemical Philosophy*.

In his essay, Bruce Moran focuses on Andreas Libavius (ca. 1555–1616), whose emphasis upon the mutual dependency of artisanal skill and philosophical understanding led to the reconstruction of the art of *chymia* as a self-sufficient and self-limiting intellectual discipline. Beyond practices linked to material and textual worlds, however, establishing *chymia* as an art within a public domain required, in the Libavian view, the articulation of other practices that were essentially social and communal. Libavius was a learned Aristotelian chemist who condemned Paracelsian physicians and drugmakers not only for their secretive language and techniques, but also for their ignorance of medieval alchemical traditions. True chymists observed civic and domestic responsibilities, pursued procedural openness, and, most of all, invited qualified witnesses to verify and repeat techniques and discoveries. In correspondence, particularly with the Jena professor of medicine Zacharius Brendel (the elder) (1553–1638), Libavius advanced the vision of combining social and artisanal practices with philosophical precepts within a university discipline that would both reform and refine the art of *chymia*.

Reflecting further upon the material and social practices at play in defining the purview of *chymia*, Margaret Garber, in a richly textured study of chemical curiosities published in the journal *Curiosi* of the Leopoldina Academy, notes occasions in which the Academy's physician members sought status as adepts not only through material practices related to making chemical medicines but also through participation in projects of metallic transmutation. The Academy, and especially its journal, operated as a vehicle of social legitimacy among physician members who pronounced authoritatively upon medical matters. At the same time, these physicians transmitted curiosities in the journal's pages based upon a knowledge of alchemical procedures. Through witnessed histories of transmutation and the description of material practices, as well as by means of discussing theoretical and practical principles related to working with and understanding metals and minerals, the journal brought alchemical traditions and experience to bear upon a certain variety of medical self-fashioning. The secretive subject of *chymia* thus gained public attention in large part due to physicians engaged, on the one hand, in projects related to pharmaceuticals and, on the other, in processes pertaining to the material practices of *chrysopoeia*.

By the early eighteenth century, both the vitalism of Georg Ernst Stahl (1659–1734) and the mechanistic descriptions of Hermann Boerhaave (1669–1738) had rejected the broad theoretical claims of Paracelsian and Helmontian chemical medicine. Debus emphasized the divergent positions of these two theorists in his study of chemistry's complex relationship to medicine. As he saw it, chemical physicians (iatrochemists) and those inspired by the mechanical philosophy (medical mechanists or iatrophysicists) battled one another over the proper understanding of the functions of the body.[4] Traditions of vitalism and mechanism made room for practical chemical pharmacy, but the theoretical commitments of each kept the other at arm's length. As Kevin Chang shows in his study of the design and publication of Stahl's "inflammability principle," however, there was room within vitalism itself for an altered understanding of the nature of metals and for a means of drawing together nature's kingdoms of minerals, animals, and plants. The significance of Stahl's notion of phlogiston, Chang argues, is best understood contextually as it evolves within a tradition of criticism aimed at refashioning the chemical principles of Paracelsus, namely, the *tria prima* of sulfur, salt, and mercury. Chang thus extends Debus's analysis of chemical philosophical thought with a finely focused study on chemical principles grounded in an eighteenth-century analytical context.

Domestic, regional, and gendered perspectives relating to chemistry, medicine, and the body orient the focus of the volume's second section, "Regional Contexts and Communities of Texts." Here practices related to gender, commerce, and patronage accentuate a shifting social landscape linked to alchemy and chemical medicine, while textual practices related to spiritual alchemy, *magia*, and kabbalah bridge religious communities and inspire a specific, chemical understanding of creation and the processes of nature.

One study draws attention to regional practices in Spain and argues for a revision of claims, some of them advanced by Allen Debus, concerning the early modern status of chemical knowledge and the reception of Paracelsian ideas within the Spanish domain. In recent historiography, serious efforts have been made to bring the Iberian experience and the role of the Spanish Empire into the general discussion of the Scientific Revolution.[5] Individual studies have, in this regard, brought to light long-overlooked details of Spanish involvement in cosmography, natural history, chemistry, and Paracelsian medicine during the early modern era.[6] Representing this historiographic direction, Mar Rey Bueno situates *chymia* within Spanish everyday life, highlighting the chemical resources

4. Debus, *Chemistry and Medical Debate: von Helmont to Boerhaave*, 135–74.

5. See, for example, Brotóns and Eamon, *Beyond the Black Legend*.

6. See, for example, Portuondo, *Secret Science*; Osorio, *Experiencing Nature*; and Bleichmar, *Visible Empire*. For chemistry in Spain, see especially López Pérez, "Novatores or Alchemists?"

and avenues of expertise that were not only available within commercial and domestic settings but were also often promoted by the patronage interests of provincial aristocrats. Here domestic pharmacy, books of secrets, and agricultural manuals converged to create local chemical arenas in which the instruction and participation of women, particularly in processes of distillation, played a prominent role. Reversing a narrative that influenced Debus's perception of Spanish science, Rey Bueno convincingly argues that a resurgence of chemistry in Spain was *not* connected with a delayed reception of Paracelsian doctrines and, in terms of practice, entered into numerous dimensions of urban, religious, and domestic life, existing far beyond the influence, and aristocratic mimicry, of court interests.

In her essay, Anke Timmermann shifts the regional focus to the Hapsburg territories of the Austrian archduchy and brings together elements of regionalism, craft experience, collecting, and experimentation in a focused study of the practices of two physicians, Wolfgang Kappler (1493–1567) and Nicolaus Pol (ca. 1470–1532). As writers, readers, and pharmaceutical practitioners representing different social surroundings, Kappler and Pol characterized conditions and practices that defined local relationships between alchemy and medicine. Emphasizing conditions of trade, industry, and practical crafts, the essay brings to light details of individual interests and experiences, and builds a broader perspective from which to view the various purposes—intellectual, social, and practical—served by alchemical interests nurtured within private and professional worlds in central Europe.

Parts of the German intellectual world, specifically the works of Oswald Croll (ca. 1563–1609) and Heinrich Khunrath (ca. 1560–1605), are also the focus of Michael Walton's study of alchemy as both a spiritual and material art. Crossing boundaries between the textual word-magic of kabbalah and empirical investigations of nature, particularly within the milieu of Paracelsian practice, the essay traces the origins of a chemical kabbalah connecting the laboratory with the oratory. Such a tradition emerged from texts that defined an earlier kabbalah with a Christian focus and emphasized creation by means of words and letters. The merging of kabbalah with traditional alchemy became, on this account, most pronounced in the coincidently intriguing and confounding text, *Voarchadumia contra alchimiam* by Giovanni Agostino Pantheus (fl. 1518), who signified aspects of alchemy and nature by means of Hebrew letters. This was an art that influenced others, including Heinrich Khunrath and Oswald Croll. In their works, the magic of the word fused with the practical endeavors of the chemist to define a form of kabbalah that combined practical chemistry and medicine with a biblical account of creation.

In a final section, "Evaluations and Perceptions," historiographical, etymological, and visual themes predominate. One of the immediate influences upon the work of Allen Debus was the scholarship of Walter Pagel (1898–1983), who interpreted Paracelsus within the context of Renaissance philosophy while extracting him from the nationalistic framework of German folk hero. In his essay, Dane Daniel reexamines the debate between Pagel and another leading twentieth-century Paracelsian scholar, Kurt Goldammer (1916–97). Pagel tied Paracelsus to Renaissance Neoplatonism and gnosticism, underscoring links to humanist philosophers, especially Pico della Mirandola (1463–94) and Marsilio Ficino (1433–99). Goldammer, on the other hand, thought that the religious context of Reformation Germany was more important in fashioning Paracelsus's natural philosophy and medical cosmology. Emphasizing Paracelsus's notions regarding relationships between the microcosm (man) and the macrocosm (the universe), Daniel finds a stable platform to argue for a *via media* between philosophical and religious contexts as sources for Paracelsus's ideas.

Confounding the grand narrative of the Scientific Revolution led Debus to focus his attention upon the works of two English mystical authors, John Dee (1527–1608/9) and Robert Fludd (1574–1637). Dee represented a bridging of chemical and mathematical traditions in the Renaissance, and Fludd characterized a synthesis of scripture, mystical alchemy, mechanical argument, and medicine. Paying attention to historiographic treatments of Dee, Nicholas Clulee draws attention to continued questions that affect evaluations of Dee's writings and the role of the occult in the formation of natural knowledge. Clulee tracks the different positions regarding Dee, sometimes by authors who make use of the same texts, especially in relation to evaluating Dee's occult activities. By following the trail of historiographic interpretation, he underscores recent discoveries concerning Dee's fortunes and maneuverings among factions at the Elizabethan court. Concerning the question of whether a unified perception of Dee is possible, Clulee offers his own perception of Dee's intellectual biography, one based on shifting social contexts and the specific intellectual problems upon which he focused.

Closing the collection, Heinz Schott methodologically shifts the approach of the preceding essays toward emblematics and visual culture. Departing from the now well-worn paths treating Paracelsian alchemy and natural magic of Debus and others, this essay analyzes variations in the imagery of Natura, especially as a feminine topos. Images of Natura (clothed and unclothed, terrestrial and celestial), Schott argues, resolved the realms of heaven and earth and influenced early modern natural philosophy, alchemy, and medicine by means of a figurative language of imitation and an "erotic tension" aroused by gendered depictions of natural artistry and divine wisdom. Allegories of the female figure combined art and nature, while in the writings of Agrippa von Nettesheim (1486–1535) nature herself decidedly favored women, since the female body as well as the female intellect were both created only by God, without physical interference from the stars.

The idea for this volume, and most of the essays in it, followed as a consequence of what can appropriately be called Michael Walton's labor of love. Michael, who died in August 2013, had been a student of Allen Debus and the two maintained a close personal and intellectual friendship throughout their remaining lives. The two collaborated in the production of *Reading the Book of Nature* (1998), a collection of essays that emphasized current research in the history of Renaissance and early modern chemistry and medicine, and emphasized the evaluation of historical authors within political, cultural, religious, social, and intellectual spheres.[7] Michael recognized the significant role Allen had played in blending the history of medicine with the history of science and in stressing the importance of chemical traditions in the Renaissance. Allen prized Michael's work as well, including, in his collection of significant papers from *Ambix*, the journal of the Society for the History of Alchemy, not one, but two Walton contributions.[8] In his life and work, Michael bridged several cultural-religious traditions and social roles, taking significant part in the world of scholarship while rising each morning to a "day job" as a successful entrepreneur. He was always a scholar, and in the evenings, like Machiavelli, he returned to his house and metaphorically took off the clothes of the day, dressed himself in the robes of learning and erudition, and dined in his study on the intellectual food for which he was born. There, he read Renaissance texts and pursued interests in the biblical account of creation, Paracelsian and early modern medicine, and the history of chemistry.

7. Debus and Walton, *Reading the Book of Nature*.

8. Walton, "John Dee's *Monas Hieroglyphica*" and "Boyle and Newton on the Transmutation of Water and Air."

The fruit of some of that labor appeared most recently as *Genesis and the Chemical Philosophy* (2011) and continues into the present volume.

The essays here illustrate recent and new directions in the history of science and medicine, as well as in early modern cultural history. They also represent a depth of analysis attainable only through engagement with manuscript and printed sources previously left unread and consequently unexamined or through broad historiographic expertise. Some essays relate to well-known figures and texts and argue for influence in novel ways. Others bring to light evidence for the presence of alchemy and chemical medicine in unfamiliar social sites and professional locations. The stage of alchemy, chemistry, and Paracelsian medicine has considerably expanded since Allen Debus began to recognize its importance for both the history of science and the history of medicine. It should be no surprise that the actors who now perform on that stage are no longer limited to a few well-known figures, but amount to a cast of thousands.

Contributors to this volume have made individual attempts to reengage and recombine the crosscurrents of natural knowledge in the early modern era. In that regard, many participate in examining parts of the "chemical philosophy" articulated by Allen Debus a half century ago. At the same time, however, like Wallace's twenty men on a bridge, they step in and out (sometimes critically, sometimes by pressing further) of that "chemical philosophy" as a category of historical interpretation. Taken collectively, these essays both widen and deepen our understanding of a type of natural knowledge that embraces the macrocosm as well as the microcosm and that intersects at the boundaries of alchemy, chemistry, and Paracelsian medical philosophy. Exploring varieties of that knowledge continues to thrive with new scholarship, and it is a tribute to the work of Allen Debus that his own endeavors opened passages to the persistent refinement and understanding of subjects that were once snubbed as suitable only to the refuse heap of the history of science.

WORKS CITED

Bleichmar, Daniela. *Visible Empire: Botanical Expedition and Visible Culture in the Hispanic Enlightenment.* Chicago: University of Chicago Press, 2012.

Brotóns, Victor Navarro, and William Eamon eds. *Beyond the Black Legend: Spain and the Scientific Revolution.* Valencia: Instituto de Historia de la Ciencia y Documentacion, 2007.

Debus, Allen G. *The Chemical Dream of the Renaissance*. Cambridge: Heffer, 1968.

———. *The Chemical Philosophy: Paracelsian Science and Medicine in the Sixteenth and Seventeenth Centuries*. 2 vols. New York: Science History Publications, 1977. Reprint, Mineola, NY: Dover Publications, 2002.

———. *Chemistry and Medical Debate: van Helomnt to Boerhaave*. Canton, MA: Science History Publications, 2001.

Debus, Allen G., and Michael T. Walton, eds. *Reading the Book of Nature: The Other Side of the Scientific Revolution*. Kirksville, MO: Thomas Jefferson University Press, 1998.

López Pérez, Miguel. "Novatores or Alchemists? A Spanish Historiographical Problem." In *Chymia: Science and Nature in Medieval and Early Modern Europe*, edited by Miguel López-Pérez, Didier Kahn, and Mar Rey Bueno, 331–36. Newcastle upon Tyne: Cambridge Scholars Publishing, 2010.

Osorio, Antonio Barrera. *Experiencing Nature: The Spanish Empire and the Early Scientific Revolution*. Austin: University of Texas Press, 2010.

Portuondo, Maria M. *Secret Science: Spanish Cosmography and the New World*. Chicago: University of Chicago Press, 2009.

Walton, Michael T. "Boyle and Newton on the Transmutation of Water and Air." In *Alchemy and Early Modern Chemistry: Papers from Ambix*, edited by Allen G. Debus, 477–84. [London]: Jeremy Mills Publishing, 2004.

———. "John Dee's *Monas Hieroglyphica*." In *Alchemy and Early Modern Chemistry: Papers from Ambix*, edited by Allen G. Debus, 178–85. [S.I.]: Jeremy Mills Publishing, 2004.

Chapter 1

Crafting the Chemical Interpretation of Nature

The Work of Allen G. Debus

Karen Hunger Parshall

Mathematization and mechanization: these were among the keywords in the vocabulary that had come—by the middle of the twentieth century—to describe the period in the history of science known as the Scientific Revolution. An epoch traditionally understood as one marked by the overthrow of the ancient, Aristotelian cosmos by the modern, Copernican worldview, the Scientific Revolution was thus viewed by its early to mid-twentieth-century interpreters—scholars such as Edwin Burtt, Eduard Dijksterhuis, Alexandre Koyré, Herbert Butterfield, and A. Rupert Hall—as having been driven by developments in mathematics and the physical sciences, particularly astronomy and physics.[1] This was the picture of the Scientific Revolution that Allen Debus confronted as a graduate student in the history of science at Harvard in the mid-1950s.

1. See, for example, Burtt, *Metaphysical Foundations*; Dijksterhuis, *Val en worp* and *De mechanisering van het* (and the latter's English translation *Mechanization of the World Picture*); Koyré, *Études galiéennes* and *From the Closed World*; Butterfield, *Origins of Modern Science*; and Hall, *Scientific Revolution 1500–1800*.

Debus, like so many others of his generation, had come to the history of science from science. After taking an undergraduate degree in chemistry at Northwestern in 1947, Debus joined his former professor, the early modern British historian John Murray, at Indiana University to pursue a master's degree in history.[2] Murray, by actively encouraging Debus to use his scientific expertise to inform his historical research, oversaw his student's earliest forays into the history of early modern English chemistry. The master's thesis, "Robert Boyle and Chemistry in England, 1660–1700," that resulted from this work in 1949 also generated Debus's first publication in the history of science, a short paper titled "Robert Boyle and His Sceptical Chymist."[3]

Following George Sarton's stricture that historians of science prepare themselves for the doctorate by first earning two master's degrees—one in history and one in a science—Debus applied and was admitted to the master's program in chemistry at Indiana.[4] There, in addition to pursuing his coursework, he actively engaged in laboratory research on the electrical conductivity of salts at very low concentrations. In 1951 and one course shy of completing his degree, he left the program on the occasion of his marriage to fellow chemistry graduate student Brunilda López Rodríguez.

For five years the couple worked as practicing chemists at Abbott Laboratories[5] before they decided to leave industry in order for Debus to pursue a PhD in the history of science at Harvard University. In courses with I. Bernard Cohen on the medieval background of Galileo's thought, Charles Taylor on the intellectual debates of the Middle Ages, and Wilbur Jordan on Tudor and Stuart England, among others, Debus was exposed to what has been termed the Great Tradition in the historiography of the Scientific Revolution.[6] Cohen, in particular, "emphasized the development of the physical sciences in the period from Copernicus

2. Biographical information on Debus's life and career has been drawn from an autobiographical memoir entitled "From the Sciences to History." For more on Debus's life and interests (and in particular his strong side interests in antique cars and early recordings), see Parshall and Chang, "Éloge: Allen George Debus."

3. Debus, "Robert Boyle and His Sceptical Chymist."

4. See Debus, "From the Sciences to History," 240. Debus mentioned having read this stricture in one of Sarton's editorials in *Isis*. Compare Sarton, "Preface to Volume 37"; "Third Preface to Volume Forty"; and *Guide to the History of Science*, 61. In none of these places, however, was Sarton as specific as Debus recalled on the issue of actual degrees.

5. Allen Debus ultimately held five patents based on his research, one of which was for an improved and more economical technique for synthesizing Novocaine. See the select bibliography (to 1993) of his works in Theerman and Parshall, *Experiencing Nature*, 281–82.

6. Cohen, *Scientific Revolution*, 21–150.

through Newton" as he guided his students not only through the key primary sources but also through then-recent work of scholars such as Koyré.[7]

Debus read Koyré with interest but found his approach at once internalistic and positivistic in its insistence on the "'glorious progress' of the evolution of scientific ideas."[8] Moreover, neither Cohen nor Koyré ventured far into the history of sciences other than the physical, and both approached the field from the point of view of a history of ideas as opposed to a history in which the development of science was grounded in some broader social context. It was the then-neglected history of chemistry and this kind of broader contextualization that Debus, the trained chemist, resolved to pursue in his doctoral research. He was not, however, encouraged in this objective. In his words, his "fellow students in the graduate program warned [him] against this, pointing to the fact that the study of the Scientific Revolution was firmly anchored in the physics of local motion and the acceptance of the heliocentric system" and that "[a]ny study of chemistry would necessarily be of lesser interest."[9] Undaunted, Debus launched what would become a lifelong research program with a paper in Jordan's Tudor-Stuart seminar on the English Paracelsians and shortly thereafter decided to embark on a "dissertation on the work of the Paracelsians and the rise of chemical medicine in the sixteenth and seventeenth centuries."[10]

A key turning point in Debus's intellectual and scholarly journey came in 1959 when he was awarded a Fulbright fellowship for study in England. Just prior to his departure, he had read the work of the German-born, England-based pathologist and historian of medicine Walter Pagel, and particularly Pagel's six-part essay, "Religious Motives in the Medical Biology of the XVIIth Century," and his newly published book, *Paracelsus: An Introduction to Philosophical Medicine in the Era of the Renaissance.*[11] In these writings, as he later acknowledged, Debus encountered an approach that "seemed closer to [his] own than that of any other author [he] had encountered."[12] Pagel's work not only highlighted the intricacies of the history of sciences other than the mathematical sciences, but it also situated that history in the wider context of early modern religious,

7. Debus, "From the Sciences to History," 243.

8. Ibid.

9. Ibid., 244.

10. Ibid., 243.

11. Ibid. See Pagel, "Religious Motives in the Medical Biology"; and Pagel, *Paracelsus.*

12. Debus, "From the Sciences to History," 244.

philosophical, and social concerns. He practiced and advocated an approach to historical texts that considered and interpreted them in full, that refused to cherry-pick them for those ideas that seemed most relevant to modern science, that eschewed the positivism of scholars in the so-called Great Tradition. As Pagel put it in the provocatively entitled essay "The Vindication of Rubbish," written in 1945, interpretations "based on the selection of material from the modern point of view, may endanger the presentation of historical truth."[13] He thus strove to analyze the writings and to understand the philosophy of Paracelsus on its own terms and in its cultural context. If positivists like Burtt, Dijksterhuis, Koyré, and Hall had seemingly consigned many of the texts of the (al) chemists to the rubbish bin,[14] Pagel sought to vindicate them by restoring them to what he deemed their rightful place in a fuller discourse on the Scientific Revolution.

This methodology struck a chord in Debus, who arranged to meet Pagel shortly after his arrival in England. The two men thus entered into an intellectual dialog that would end only with Pagel's death in 1983. It is no exaggeration to say that, although he learned and benefited much from Bernard Cohen, Debus was intellectually a disciple not of Cohen but of Walter Pagel.

Debus spent the Fulbright year in England on the research that would result first in his 1961 dissertation and then in his 1965 (first) book, *The English Paracelsians*.[15] He justified his analysis of the thought of such sixteenth- and seventeenth-century English chemists as Richard Bostocke,[16] Thomas Moffett, Thomas Tymme, William Turner, Edward Jorden, and Robert Fludd in Pagelian terms. "If," he said,

> the work of Paracelsus and other major chemists of the sixteenth and seventeenth centuries is characterized either as unworthy of discussion or as acting as a force which retarded the general growth of science, a true understanding of the period is made difficult if not impossible. There is a vast body of

13. Walter Pagel, "The Vindication of Rubbish," *Middlesex Hospital Journal* (Autumn 1945), 1, as quoted in Debus, "Chemists, Physicians, and Changing Perspectives," 70.

14. Hall, in particular, wrote that while "many have searched" the alchemical literature "for the beginnings of a chemical attitude," "[t]here the grain of real knowledge is concealed in a vast deal of esoteric chaff"; *Scientific Revolution 1500–1800*, 309.

15. Debus, *English Paracelsians*.

16. The name of this figure has been a source of considerable confusion and debate. Debus tended to refer to him simply as "R. Bostocke"; others referred to him as "Robert Bostocke"; still others as "Richard Bostocke." More recent research suggests that he was in fact "Richard Bostocke," otherwise spelled as "Rychard Bostok." See Harley, "Rychard Bostok of Tandridge, Surrey."

> Renaissance chemical literature, and a large portion of it was written not by gold-seeking alchemists, but rather, by scholars who honestly felt that the true key to nature's secrets was to be found in the study of chemistry. Today we can say that many of their explanations were incorrect, but to their contemporaries their work often seemed a stimulating force, moving man ever closer to a true understanding of nature.[17]

In bringing back to light the ideas of the English Paracelsians, Debus demonstrated the extent to which they incorporated Paracelsian ideas into their chemical and medical views and practice. He argued that there was, in fact, an "Elizabethan compromise" by which "the occult aspects of Paracelsian thought were rejected while the new remedies were eagerly adopted, provided they proved their worth."[18] This compromise, moreover, was negotiated at what we might see as the interface between chemistry and medicine. It underscored not only the interconnectedness of these two domains but also the Elizabethan reaction to the confrontation of traditional Galenic and revived Hermetic and Neoplatonic thought as fashioned in the writings and teachings of Paracelsus. Debus's first book-length foray into this relatively virgin territory "attempt[ed]," in his words, "to define some of the major problems of concern to English iatrochemists prior to 1660." It did not aim, however, to provide "a complete discussion of all phases of English iatrochemical thought in the sixteenth and seventeenth centuries."[19]

While some reviewers appreciated the book on its own terms,[20] others criticized it for not doing more. Writing in the pages of the *British Journal for the History of Science,* Conrad Josten, former curator of the Museum of the History of Science and a scholar of the early history of chemistry and astronomy, faulted Debus for his reliance on published primary—as opposed to manuscript—sources. While acknowledging that Debus had explored "a relatively uncharted field of study" and, in so doing, "had to restrict his approach to that of a pioneer," Josten nonetheless branded the book "a preliminary survey" from which "superficial judgements and half-truths" had been derived such as, precisely, the claim that the English Paracelsians had rejected "the occult aspects of Paracelsian thought."[21]

17. Debus, *English Paracelsians,* 14.
18. Ibid., 175.
19. Ibid., 9.
20. See, for example, Leicester, "Chemistry in England, 1557–1640."
21. See Josten, "Book Review, *The English Paracelsians,* by Allen G. Debus."

In Josten's view, "[a]s anyone familiar with manuscript collections of the period knows, the sympathetic and 'magnetic' cures recommended by Paracelsus gained at least some popularity in England, as did his teachings on *mumia*, magic sigils, conjurations, astrological influences, and the transmutation of metals."[22] In closing, Josten allowed that "Professor Debus is at his best in describing the influence of Paracelsus on practical medicine," but sniped that "[i]n that respect, more than in any other, his book will serve as a useful guide to those who, as is to be hoped, will undertake more detailed research in this rewarding field."[23]

Debus himself had been clear about the scope of his first book, and it was he who immediately undertook the next phase of "more detailed research in this rewarding field" during a yearlong sabbatical at Churchill College, Cambridge, in 1966/67. As he conceived it, his next project would focus on what he—like the Paracelsians themselves—understood as a "chemical philosophy" that had held sway in certain quarters in the sixteenth and seventeenth centuries and that had coexisted and "competed powerfully" with an emergent mechanical philosophy for "explaining the operations of nature."[24] It would, however, be ten years before *The Chemical Philosophy: Paracelsian Science and Medicine in the Sixteenth and Seventeenth Centuries* ultimately appeared.[25]

Those ten years were extremely busy and productive for Debus on several levels. He built and firmly institutionalized the program in the history of science at the University of Chicago; he brought out the *World's Who's Who in Science: A Biographical Dictionary of Notable Scientists from Antiquity to the Present* that, for the first time, incorporated the chemical philosophers meaningfully alongside the traditional "big names" in the history of science;[26] he produced two *Festschriften*, one in honor of Walter Pagel in 1972 and one in memory of the noted historian of medicine C. Donald O'Malley in 1974;[27] he undertook a project to bring out a series of primary sources relating both to the evolution of a chemical philosophy of nature and to recapturing a fuller sense of scientific practice in the sixteenth and seventeenth centuries;[28] and he produced numerous research arti-

22. Ibid. (his emphasis).

23. Ibid., 297.

24. Cohen, *Scientific Revolution*, 111.

25. Debus, *Chemical Philosophy*. The manuscript was actually completed in 1973; the original preface was dated 1 January 1974. Various developments ultimately resulted in the delay of its publication until 1977.

26. Debus, *Who's Who in Science*.

27. Debus, *Science, Medicine and Society in the Renaissance* and *Medicine in Seventeenth Century England*.

28. Among these are Elias Ashmole, *Theatrum chemicum britannicum . . . : A Reprint of the London Edition*

cles that may be seen as case studies leading to the synthesis he would present in *The Chemical Philosophy* in 1977. As he put it, "I believed that the study of the works of the chemical philosophers would give us a deeper understanding of the rise of modern science—not because their works contain anything equivalent to Newtonian mechanics, but because they offered an alternative system that became a center of debate beginning in the late sixteenth century. This was a conflict that was to go beyond internal medical and scientific questions to touch on education, religion, politics, and philosophy—a wide spectrum of intellectual and cultural issues."[29]

The chemical philosophy that Debus detailed hinged on the work of Paracelsus and on the interpretations and utilizations of that work in the sixteenth and seventeenth centuries. First and foremost, the chemical philosophy was a reaction against the (pagan) Aristotelian and Galenic worldviews that had persisted in various forms into the Renaissance and that dominated the curriculum of the medieval and early modern universities. Chemical philosophers—at once Christian, Neoplatonic, and Hermetic—sought to understand nature in terms of fresh observations and experiments, ever guided by their reading of the Bible and the divine revelations into nature and its mysteries that it would inspire. Chemistry, not the logic or mathematics of the schools, would lead the way to this new understanding, since the macrocosm of nature and the intimately interconnected microcosm of man could both be explained in the same (al)chemical terms. Creation had been a vast chemical unfolding of nature, although since nowhere in the scriptures was mention made of the creation of fire, it could not be one of the elements out of which all of nature was fashioned. Natural phenomena such as lightning and the growth of metals could be explained in chemical terms. Diseases in man mirrored (al)chemical processes in nature and so could be treated effectively by alchemical means.

This chemical philosophy confronted and challenged traditional Aristotelian and Galenic thought most directly in the medical faculties of the universities. If fire was not an element, then the Aristotelian and Galenic philosophies

of 1652, with a New Introduction by Allen G. Debus (New York: Johnson Reprint, 1967); Debus, *Science and Education in the Seventeenth Century: The Webster-Ward Debate* (London: Macdonald; New York: American Elsevier, 1970); and John Dee, *The Mathematicall Praeface to the Elements of Geometrie of Euclid of Megara (1570)* (New York: Science History Publications, 1975). See also *Robert Fludd and His Philosophicall Key: Being a Transcription of the Manuscript at Trinity College, Cambridge, with an Introduction by Allen G. Debus* (New York: Science History Publications, 1979).

29. Debus, "Chemists, Physicians, and Changing Perspectives," 70–71.

crafted as they were in terms of the four elements—earth, air, water, and fire—the two associated pairs of primary opposites—hot and cold, wet and dry—and the four humors—blood, phlegm, yellow bile, and black bile—collapsed. Because Galenism prevailed in the curriculum, the challenge of the chemical philosophers sparked spirited debate in medical circles that often spilled over into broader intellectual discourse.

In *The Chemical Philosophy,* Debus followed the evolution of this approach to understanding nature from the initial debates in England and on the Continent over interpretations of Paracelsus's radical position, to the even more radical and mystical stance exemplified in the writings and pronouncements of Robert Fludd, through the re-formation and demystification of a vitalistic, chemical, and medical paradigm in the work of Jean Baptiste van Helmont that led to serious discussions of actual educational reform. In the process, Debus highlighted, as he stated in his preface, "some of the major concepts of interest to the chemical philosophers as well as the conflicts that they engaged in with their contemporaries" over a chronological period extending "roughly from the death of Paracelsus in 1541 to the publication of Robert Boyle's *Sceptical Chymist* one hundred and twenty years later."[30] Debus was also clear as to the limitations of any synthesis. "In an attempt to make a very large subject manageable," he explained, "it has been necessary to be selective. Thus, although a great number of chemical philosophers are referred to in these pages, there are many others with an equal claim for inclusion who are not here. I hope," he added, "that the most important themes have been mentioned."[31]

As was the case with *The English Paracelsians,* at least one key reviewer, this time no less an authority than Charles Webster, in an essay review in the fourth number of *Isis* in 1979, took issue with the consciously synthetic nature of the work and faulted Debus's methodological choices. He forewarned that "[t]he reader will . . . be disappointed with the small proportion of new material contained in the two volumes. That material tends to be confined to introductory passages and interpolations of minor importance inserted for the sake of completeness. For the most part," he continued, "the digests of older material included compare unfavorably with the papers as they were originally published."[32] In his view, moreover, Debus has disregarded the extensive surviving

30. Debus, *Chemical Philosophy* [Dover edition], xxi.

31. Ibid.

32. Webster, "Essay Review of *Chemical Philosophy, Man and Nature in the Renaissance,* and *Der sächsische*

body of Renaissance alchemical literature that was circulated in manuscript form, and he has not taken account of the fact that the educated elite possessed a ready knowledge of Latin, so the absence of vernacular alchemical or natural philosophical works was no barrier to their acquaintance with the various traditions of Renaissance philosophy.[33]

Webster ended his biting critique of Debus's work with the statement that "[t]he basic weakness in *The Chemical Philosophy* lies in the consequences of its avowed 'internalist' methodology (p. xii). Any analysis of cultural trends carried out in terms of self-contained summaries of a small selection from the relevant literature entails the risk of misrepresentation. It will be apparent," he continued, ". . . that 'socio-economic interaction,' which Professor Debus regards as a dispensable dressing to his work, is fundamental to the generation of any meaningful historical perspective."[34]

Webster's essay review also sharply criticized Debus's undergraduate textbook, *Man and Nature in the Renaissance*, which had appeared in 1978.[35] Not surprisingly, Debus's goal in that text had been to complicate the narrative of the Great Tradition by infusing and intertwining it with a discussion of the chemical philosophers who worked in tandem with—and who were sometimes one and the same as (think Johannes Kepler and even Isaac Newton)—natural philosophers working in the emergent mathematical and mechanical paradigms of the seventeenth century. For Webster, this effort resulted negatively in "a hybrid between *The Chemical Revolution* and the familiar *Scientific Revolution*" with "[t]he reader being misled" in the process "by overexposure to minor English scientific writers, especially Robert Fludd."[36] "Thus," he closed derisively, "although representing a fashionable line of interpretation, which will guarantee some ephemeral applause, in the longer term this book does disservice to the movement working for a more genuine historical appreciation of Renaissance science."[37] Ironically perhaps, given this negative assessment, *Man and Nature in the Renaissance* remains in print over thirty years after its initial publication and has been translated into almost a half dozen languages.

Paracelsist Georg Forberger," 588.

33. Ibid., 590.

34. Ibid., 591.

35. Debus, *Man and Nature in the Renaissance.*

36. Webster, "Essay Review of *Chemical Philosophy, Man and Nature in the Renaissance,* and *Der sächsische Paracelsist Georg Forberger,"* 592 (his emphasis).

37. Ibid.

It is perhaps also ironic that the *first* number of the volume of *Isis*, in which Webster's highly critical reviews later appeared, carried the citation for the Pfizer Award for the best book published in 1977, an award that went to Allen Debus for *The Chemical Philosophy*. In making the presentation, Pfizer Committee chair Nicholas Steneck noted that

> [f]or nearly two decades [Debus] has been endeavoring to chart in ever-greater detail the terrain of Paracelsian science and medicine that Walter Pagel and other pioneers in our discipline first brought to light and to link this seemingly aberrant form of science to the mainstream of the scientific revolution. His latest entry on the scene, *The Chemical Philosophy: Paracelsian Science and Medicine in the Sixteenth and Seventeenth Centuries*, is an impressive summation of this work, bringing together in one place the theories, the controversies, and the concerns that animated the chemical-philosophical tradition and made it a part of the turmoil of the early modern era.[38]

With the publication, essentially in tandem, of *The Chemical Philosophy* and *Man and Nature in the Renaissance*, Debus "was not sure . . . what major project to pursue."[39] Would he push forward chronologically into the seventeenth and eighteenth centuries? Would he return to focus on the Renaissance? Ultimately, he decided to do for France what he had done in his dissertation and first book for England, namely, a study focused on the reception of and debates on Paracelsian chemistry and medicine within the French academic and medical establishment. This research took him from the sixteenth into the early decades of the eighteenth century as he once again explored in a particular national context "a Galenic medical establishment challenged by a group of chemical physicians who saw their new preparations as the basis for a sweeping program of chemical reform."[40] Perhaps not surprisingly, the analysis again centered on debates, this time within the medical faculty at the University of Paris, between the medical faculties in Paris and Montpelier, on the floor of the Académie des Sciences, and on the pages of periodicals such as the *Journal des sçavans*. This national context and this time period, however, witnessed the increasing delineation from medicine of an actual *science* of chemistry at the same time that it attested to the persistence of a vitalist tradition of thought into the eighteenth century. In his review of *The French Paracelsians* in *Isis*, Bruce

38. Kohlstedt, "News of the Profession," 148–49.
39. Debus, "From the Sciences to History," 274.
40. Debus, *French Paracelsians*, xv.

Moran agreed with Debus that "an awareness of the vitalist tradition 'is essential for understanding both the chemical revolution and the Romantic reaction against the mechanical science of the philosophers at the end of the eighteenth century.'"[41] Debus's work on Paracelsianism in the French context thus served as a bridge from the history of Renaissance and early modern chemistry both to the eighteenth-century developments in chemistry associated with the names of Lavoisier, and others, and to the Romantic movement of the late eighteenth and early nineteenth centuries.

Debus's final, original, book-length study, *Chemistry and Medical Debate: van Helmont to Boerhaave,* continued the theme of debates at the interface of chemistry and medicine, but this time he explored the vitalist points of view of the chemical physicians in contrast to the mathematical and mechanistic points of view of the so-called iatrophysicists.[42] This new group of scholars "recognized," in Debus's words, "the important role played by mathematics and physics in the development of the new science and hoped to apply the same methods to medicine."[43] The iatrophysicists, then, brought the "new science"—in the sense of the literature of the Great Tradition—to bear on medicine, which had previously been, as Debus had so cogently argued, a battleground between the Galenists and the Paracelsians. In the late seventeenth and early eighteenth centuries, party lines had come to be drawn between practitioners of the new mechanistic and the older animistic medicine, while die-hard followers of Galen and particularly of Hippocrates persisted. Were the body and disease to be explained and understood in chemical or in mechanical terms? And if in mechanical terms, what, if any, role was chemistry to play in medicine? Finally, if it was to play no role or a lesser role, then where did chemistry stand? Did it represent a field quite distinct from medicine? These were among the questions that shaped the debates that Debus chronicled with characteristic depth and richness of detail. Seymour Mauskopf singled out precisely this aspect of Debus's work in his review in the *Journal of the History of Medicine* in this way. *Chemistry and Medical Debate,* he wrote, "contains a wealth of information about an aspect of the Scientific Revolution that had been neglected until the research of Debus himself. This research has inspired others to take

41. Moran, Review of *The French Paracelsians,* 576, quoting Debus, *French Paracelsians,* 208.

42. Debus, *Chemistry and Medical Debate.*

43. Ibid., xiv.

up the study of sixteenth- and seventeenth-century chemistry and alchemy, indeed to the point of producing what is now quite an industry."[44]

The Hermeticist reinterpretation of the Scientific Revolution, according to Floris Cohen in *The Scientific Revolution: A Historiographical Inquiry,*

> came down to an ongoing demonstration that esoteric notions not nowadays considered scientific persisted until a far later date into the 17th century than had previously been taken for granted. Thus the historian of chemistry Allen Debus, in a large number of books and articles, called particular attention to the huge interest evoked in the 17th century by debates in chemistry, in particular involving the Paracelsians and the critique of Paracelsus by van Helmont—subjects written about before but only now seeming to have found the right, Hermetic or Rosicrucian, framework, in which they began to make proper sense.[45]

To be sure, Debus did not affect this "righting" of the historical record alone. His work takes its place alongside that of other key scholars such as Pagel, Frances Yates, Paolo Rossi, and Charles Webster in shaping a new approach to the Scientific Revolution that challenged and fundamentally altered the understanding of the period that had emerged from the Great Tradition. Their work "began to give," in Cohen's words, "a quite distinct, new coloring to our conception of what early modern science itself stood for."[46] Nicholas Clulee perhaps captured it best in the citation he read when the Sarton Medal, the highest honor in the history of science, was awarded to Allen Debus in 1994. "No sea change, historiographic or otherwise," he wrote, "is the work of one person, but Allen Debus can claim a place in this one as an early and constant proponent of the need to give the sciences of organic nature a place of importance in our histories alongside the mathematical and mechanistic study of physical nature."[47]

It is fair to say that Allen Debus did not consciously strive to produce an "industry" in the history of early modern chemistry and alchemy. As a graduate student, he had encountered a historical narrative of the Scientific Revolution—as crystallized in, for example, Butterfield's *The Origins of Modern*

44. Mauskopf, Review of *Chemistry and Medical Debate,* 96.
45. Cohen, *Scientific Revolution,* 174.
46. Ibid., 170.
47. Clulee, "News of the Profession," 284–85.

Science—that seemingly had no place in it for the history of chemistry and medicine. In fact, he had found in Butterfield and elsewhere outright ridicule of pre-Boylean chemistry. "It is difficult for us to imagine," Butterfield wrote in 1949 just a few years before Debus would read his book in a graduate class at Harvard, "the state of chemical enquiry before the days of Boyle, or to comprehend on the one hand the mystifications and mysticisms, on the other hand the anarchical condition of things, amongst the alchemists in general. . . . Concerning alchemy it is more difficult to discover the actual state of things, in that historians who specialize in this field seem sometimes to be under the wrath of God themselves; . . . they seem to become tinctured with the kind of lunacy they set out to describe."[48] Butterfield had become famous for his critique of what he termed a "Whiggish" interpretation of history and had taken his fellow historians to task for failing often to recognize, first, that "[t]he value of history lies in the richness of its recovery of the concrete life of the past" and, second, that "if history can do anything it is to remind us of those complications that undermine our certainties, and to show us that all our judgements are merely relative to time and circumstance."[49] Yet, as Debus pointed out not without a certain amount of irony, Butterfield "would have produced a more balanced account if he had followed his own advice when he prepared . . . *The Origins of Modern Science*."[50] It was that "more balanced account" that Debus sought to effect from his first publication in the obscure *Indiana Journal for Bookmen* in 1949 to his final publications just before his death in March of 2009. His life work aimed to illuminate the history of his own particular science, chemistry, a history, in his view, that was central to "any discussion of the Scientific Revolution" and that was "key . . . for an understanding of the fundamental changes of that period—no less so than the work of Copernicus, Galileo and Newton."[51] In this goal, he succeeded, as the ongoing research and reevaluation reflected in the chapters that follow attest.

48. Butterfield, *Origins of Modern Science*, 98, as quoted in Debus, "Some Thoughts on Butterfield's *Origins of Modern Science*," 45.

49. Butterfield, *Whig Interpretation of History*, 74–76, as quoted in Debus, "Some Thoughts on Butterfield's *Origins of Modern Science*," 45.

50. Debus, "Some Thoughts on Butterfield's *Origins of Modern Science*," 45.

51. Debus, "Preface," in *Chemical Promise*, xix.

WORKS CITED

Burtt, Edwin A. *The Metaphysical Foundations of the Modern Physical Science: A Historical and Critical Essay*. London: K. Paul, Trench, and Trubner, 1924.

Butterfield, Herbert. *The Origins of Modern Science, 1300–1800*. New York: Macmillan, 1949.

———. *The Whig Interpretation of History*. London: G. Bell, 1931. Reprint, New York: Norton, 1965.

Clulee, Nicholas. "News of the Profession: Sarton Medal Citation." *Isis* 86, no. 2 (1995): 284–85.

Cohen, Floris H. *The Scientific Revolution: A Historiographical Inquiry*. Chicago: University of Chicago Press, 1994.

Debus, Allen G. *The Chemical Philosophy: Paracelsian Science and Medicine in the Sixteenth and Seventeenth Centuries*. 2 vols. New York: Science History Publications, 1977. Reprint, Mineola: Dover Publications, 2002.

———. *The Chemical Promise: Experiment and Mysticism in the Chemical Philosophy 1550–1800: Selected Essays of Allen G. Debus*. Sagamore Beach, MA: Science History Publications, 2006.

———. *Chemistry and Medical Debate: van Helmont to Boerhaave*. Canton, MA: Science History Publications, 2001.

———. "Chemists, Physicians, and Changing Perspectives on the Scientific Revolution." *Isis* 89, no. 1 (1998): 66–81.

———. *The English Paracelsians*. London: Oldbourne, 1965.

———. *The French Paracelsians: The Chemical Challenge to Medical and Scientific Tradition in Early Modern France*. Cambridge: Cambridge University Press, 1991.

———. "From the Sciences to History." In *Experiencing Nature: Proceedings of a Conference in Honor of Allen G. Debus*, edited by Paul H. Theerman and Karen Hunger Parshall, 237–80. Dordrecht: Kluwer Academic Publishers, 1997.

———. *Man and Nature in the Renaissance*. Cambridge: Cambridge University Press, 1978.

———. "Robert Boyle and His Sceptical Chymist." *Indiana Journal for Bookmen* 5 (1949): 39–47.

———. "Some Thoughts on Butterfield's *Origins of Modern Science*." *Historically Speaking* 8 (April 2007): 44–45.

———, ed. *Medicine in Seventeenth Century England: The C. D. O'Malley International Symposium*. Berkeley: University of California Press, 1974.

———, ed. *Science, Medicine and Society in the Renaissance: Essays to Honor Walter Pagel.* 2 vols. New York: Science History Publications; London: Heinemann, 1972.

———, ed. *World's Who's Who in Science: A Biographical Dictionary of Notable Scientists from Antiquity to the Present.* Chicago: Marquis Who's Who, 1968.

Dijksterhuis, E. J. *De mechanisering van het wereldbeeld.* Amsterdam: Meulenhoff, 1950. Translated as *The Mechanization of the World Picture* (Oxford: Oxford University Press, 1960).

———. *Val en worp: Een bijdrage tot de geschiedenis der mechanica van Aristoteles tot Newton.* Groningen: Noordhoff, 1924.

Hall, A. Rupert. *The Scientific Revolution 1500–1800: The Formation of the Modern Scientific Attitude.* Boston: Beacon Press, 1962.

Harley, David. "Rychard Bostok of Tandridge, Surrey (c. 1530–1605), M.P., Paracelsian Protagonist and Friend of John Dee." *Ambix* 47 (March 2000): 29–36.

Josten, Conrad H. "Book Review, *The English Paracelsians,* by Allen G. Debus (London: Oldbourne Press, 1965)." *British Journal for the History of Science* 3, no. 3 (1967): 296–97.

Kohlstedt, Sally. "News of the Profession: Annual Meeting of the History of Science Society, 1978." *Isis* 70 (1979): 146–51.

Koyré, Alexandre. *Études galiéennes.* Paris: Hermann, 1939–40.

———. *From the Closed World to the Infinite Universe.* Baltimore: Johns Hopkins University Press, 1957.

Leicester, Henry M. "Chemistry in England, 1557–1640," review of *The English Paracelsians,* by Allen G. Debus. *Science* 154 (11 November 1966): 758.

Mauskopf, Seymour. Review of *Chemistry and Medical Debate: van Helmont to Boerhaave,* by Allen G. Debus. *Journal of the History of Medicine* 58 (2003): 94–96.

Moran, Bruce. Review of *The French Paracelsians: The Chemical Challenge to Medical and Scientific Tradition in Early Modern France,* by Allen G. Debus. *Isis* 84, no. 3 (1993): 575–76.

Pagel, Walter. *Paracelsus: An Introduction to Philosophical Medicine in the Era of the Renaissance.* Basel: S. Karger, 1958.

———. "Religious Motives in the Medical Biology of the XVIIth Century." *Bulletin of the Institute of the History of Medicine* 3 (1935): 97–128, 213–31, 265–312.

Parshall, Karen Hunger, and Ku-ming (Kevin) Chang. "Éloge: Allen George Debus, 16 August 1926–6 March 2009." *Isis* 101, no. 1 (March 2010): 159–62.

Sarton, George. *A Guide to the History of Science.* Waltham, MA: Chronica Botanica Co., 1952.

———. “Preface to Volume 37: Qualifications of Teachers of the History of Science.” *Isis* 37, no. 1 (1947): 5–7.

———. “Third Preface to Volume Forty: Qualifications of Teachers of the History of Science (Second Article).” *Isis* 40, no. 4 (1949): 311–13.

Theerman, Paul H., and Karen Hunger Parshall, eds. *Experiencing Nature: Proceedings of a Conference in Honor of Allen G. Debus*. Boston: Kluwer, 1997.

Webster, Charles. Essay Review of *The Chemical Philosophy: Paracelsian Science and Medicine in the Sixteenth and Seventeenth Centuries* by Allen G. Debus; *Man and Nature in the Renaissance* by Allen G. Debus; *Der sächsische Paracelsist Georg Forberger. Mit bibliographischen Beitragen zu Paracelsus, Alexander von Suchten, Denys Zacaire, Bernardus Trevirensis, Paolo Giovio, Francesco Guicciardini und Natale Conti* by Rudolph Zaunick, Hans-Heinz Eulner, and Kurt Goldammer. *Isis* 70, no. 4 (December 1979): 588–92.

Part One

Curious Practices and Practices of Curiosity

Chapter 2
Johann Hayne and Paracelsian Praxis

Chemical Physiology as a Link between Semeiotics and Therapeutics

Jole Shackelford

Early in his career as a Paracelsian scholar, Allen Debus developed the concept of an "Elizabethan compromise" to describe what he observed as the assimilation of Paracelsian drugs and select elements of Paracelsus's chemical philosophy into an eclectic medicine that was fundamentally Galenic, but beginning to incorporate corpuscular ideas about matter. He found this assimilation to be especially characteristic of English medical publications in the late sixteenth and early seventeenth centuries, which were the subject of his first book, *The English Paracelsians*.[1] That chemical drugs, widely associated with Paracelsus and

1. Debus, *English Paracelsians*. "Elizabethan Compromise" is the title of chapter two and reflects the author's use of this term in the final sentence of his earlier publication, "Paracelsian Compromise in Elizabethan England." The concept earlier referred to Queen Elizabeth's efforts to forge a religious compromise, as for example in Ross, "George Herbert and the Humanist Tradition," 169, and Debus's reference is clearly meant to place the reception of Paracelsian medicine into the cultural context of the English Reformation. Although Debus's characterization has drawn criticism for oversimplifying English engagement with Paracelsian philosophy and

his followers, were taken into a Galenic polypharmacy is plain from the study of printed pharmacopoeias and physicians' records.[2] Debus's work also made it clear that Paracelsian theory was indeed compromised in salient published treatises of seventeenth-century European medicine, notably by the Helmontians. But the degree to which Paracelsian praxis—a theoretical system of diagnostics, prognostics, and therapeutics—was integrated into actual healing practices has remained obscure. It is perhaps a fitting tribute to Debus's leadership in this field to broaden his vision of the chemical philosophy in early modern science and medicine by illuminating those physicians who did not so much compromise the master's teaching as rework it and bring it into clearer connection with practical therapeutics.[3]

A number of published treatises suggest that such a praxis existed, but few have been studied extensively. Heinrich Nolle's *Hermetical Physick* is recognized as Paracelsian but has not received careful attention as a medical book,[4] and Friedrich Zobell's *Chymische Medicinische Perle*[5] and Johann Hayne's *Drey unterschiedliche newe Tractätlein*[6] are for the most part unknown to historians of science and medicine. Here, I will use Hayne's *Three Diverse New Treatises*, ostensibly separate tracts, but published together in a single pagination, as a tool for explicating the assimilation into a coherent approach to healing of what at first glance appears to be three diverse elements of Paracelsus's work: (1) his

religion and been to some extent revised, Debus's main idea that chemical drugs were incorporated into medical practice and that many of Paracelsus's philosophical and religious ideas were relegated to the intellectual backwater remains cogent, especially when applied to the age of Newton and the Enlightenment. A side effect of this historiography has been that historians have emphasized the assimilation of Paracelsus's ideas rather than investigating continuities in how his medicine was applied as a systematic therapeutic practice.

2. Pumfrey, "Spagyric Art," esp. 23–27, offers a thoughtful discussion of what considerations are useful for defining Paracelsians. He categorizes Paracelsians into three kinds: conscious followers of Paracelsus, promoters of chemical philosophy, and iatrochemists. He notes that the latter category includes medical writers and practitioners who adopt chemical remedies into an eclectic practice, without a commitment to a guiding chemical-philosophical pharmacology, with the result that Paracelsians and Galenists are lumped together. The eclectic nature of iatrochemistry in the seventeenth and eighteenth centuries is a consequence of the adaptation of chemical drugs to multiple kinds of therapeutic systems and the creation of a niche for them in printed pharmacopoeias.

3. The term "chemical philosophy" was not new with Debus's scholarship, which nevertheless did much to bring the term to the modern reader, but was an actor's concept or category in the early modern period.

4. Nolle, *Systema medicinæ hermeticæ generale*. Henry Vaugh translated Nolle's book into English, which I use for convenience, called *Hermetical Physick*. Here, as elsewhere, all spellings reflect the original sources.

5. Zobell, *Chymische Medicinische Perle*.

6. Hayne, *Drey unterschiedliche newe Tractätlein*. The book was published in three later editions—1663, 1683, and 1700—the last two with the title *Trifolium medicum* . . .

insistence on the spiritual origin of many of the most intractable diseases of this age, (2) the chemical and astral nature of these spirits and their manifestations, and (3) his adaptation of traditional medieval and Renaissance uroscopy and physiology to fit his chemical philosophy. Hayne's treatises can in this way serve as an introduction to how Paracelsian medicine, which seems chaotic and dysfunctional to the modern eye, functioned as a practical therapeutic system that incorporated salient features of traditional practices. This would have made his chemical medicine appealing to both patients and healers, even as it exposed them to novel ideas about chemical composition and analysis.[7] Although Hayne is ignored in modern historiography, his book was published in four editions in the seventeenth century and was a major source for the eclectic medical practice of the colonial Pennsylvania physician and religious leader George de Benneville in the eighteenth century. De Benneville is well known to American colonial historians as one of the founders of Unitarian Universalism, but his role as an immigrant colonial healer has only recently received scholarly attention. Given its prominent place in de Benneville's medical manual, Hayne's medicine may have been a more widely known and influential early modern application of Paracelsian chemical physiology to medical practice than the near absence of modern references implies.[8]

The 1663 and later printings of Hayne's *Three Diverse New Treatises* bear a title page and *ad lectorem* written by the well-known seventeenth-century chemical pharmacist and editor Johann Schröder, who identified Hayne's text as a German translation of a Latin original that was undertaken by the editor of the first edition, Georg Faber.[9] This first edition was printed in Frankfurt am

7. The chaotic nature of Paracelsus's writings is commonplace to his readers and attributed in part to the irregular nature of the sources and the contexts in which they were written and preserved. However, an assumption among historians of medicine is that providers of medicine often appeal to their patients on the basis of theory or explanation (rationalization)—this is one of the great strengths of the Hippocratic-Galenic system—and the connection between Paracelsus's chemical philosophy and his medical practice has not seemed sufficiently elaborated to me. The findings reported here are part of an attempt to reconstruct such a system, by looking at how Paracelsus's followers construed his ideas and practices.

8. I claim in "Paracelsian Uroscopy," 32, that Hayne's treatise and its use by George de Benneville are evidence for the existence of a practical Paracelsian therapeutics in the German-speaking world of the seventeenth and eighteenth centuries.

9. This is the account given by Schröder, the editor of the second edition, in the *ad lectorem*, last two pages (unpaginated): "Diese mit andern ihren Kranckheiten / neben auch einem sonderbahren *Iudicio Vrinæ*, hatt vor etliche Jahren unternommen zubeschreiben und in Lateinischer Sprache aussgehen zulassen / der Vornehme *D. Iohan Hayne & c.* welches *Tractæ*tlein also angenehm gewest / dass es folgents *D. Faber* in die Teutsche Sprache zuübersetzen / und in Truck zu geben bewogen hatt."

Figure 2.1: The frontispiece illustration to the third edition of Johann Hayne's book, to which the title *Trifolium medicum* was first prefixed, depicts the chemical physician's use of astrology and uroscopy for diagnosis. Johann Hayne, *Trifolium medicum, oder: Drey höchst-nützliche Tractätlein deren Erstes von astralischen Kranckheiten… Das Andere, von tartarischen Kranckheiten… Das Dritte vom rechten Fundament und Grund wie man die Urinen… erkennen möge* (Frankfurt am Main: George Heinrich Oehrling, 1683). Courtesy of the Wangensteen Historical Library of Biology and Medicine, University of Minnesota.

Main in 1620, supplying a definite *terminus ante quem* for Hayne's composition of these texts. However, Faber makes it clear that Hayne was dead by the time he obtained the treatises, and surviving correspondence suggests that Hayne was active as a physician in the 1570s and 1580s, making this a more likely period of composition. Latin originals for the *Three Diverse New Treatises* that Faber brought together for the book have not yet been identified. In his introduction (*Dedicatoria*) to the first edition, Faber made no mention of Latin texts and noted that these treatises were not published in Hayne's lifetime, but came to him as *manuscripta* through a good friend, who got them from Hayne's "private students."[10] Schröder may have inferred from Faber's use of the Latin word *manuscripta* in the mainly German *Dedicatoria* that Hayne's treatises were written in Latin, or he may simply have asserted a Latin pedigree to enhance the authority of the text. Lacking any convincing evidence to the contrary, it is reasonable to assume that the three *Tractätlein* were written originally in German.

Certainly, there is ample precedent for German physicians—Paracelsus among them—writing in the vernacular at that time. For example, the flamboyant late sixteenth-century chemical physician and printer Leonhard Thurneisser composed and published several elaborate treatises on chemical analysis and medicine, two of which describe diagnostic procedures based on distillation and examination of patients' urine.[11] These books and their author are well known to historians of Paracelsian medicine. On the one hand, Thurneisser was an early proponent of chemical analytical methods, for which he has been regarded as a pioneer of scientific urinalysis. On the other, the impressive physical appearance of his books and their picturesque portrayals of the spatial correspondence between deposits of urine in the analytical flask and the locations of diseases in the body were impressive, even if the latter are quaint and comical to the modern scientific mind.[12] Thurneisser gave these books macronic titles, mixing German

10. Hayne, *Drey unterschiedliche newe Tractätlein (1620)*, fol. iv^r: "Aber doch nicht bey seinen Lebzeiten in öffentlichen Truck kommen / sondern bey seinen Privat Discipulen verbleiben zu lassen gemeint gewesen: Und aber mir unter andern durch einen guten Freund nachfolgende *manuscripta*."

11. Thurneisser's two books on uroscopy are *Προκατάληψις* [*Prokatalepsis*] and *Bebaiosis agonismou*. His book on chemical analysis is *Pison*.

12. Pagel, *Das medizinische Weltbild des Paracelsus*, 19–20n3, mentioned that Thurneisser was given credit for establishing chemical uroscopy by all seventeenth-century authors, that his ideas were given form already in his 1571 treatise on the subject, and that Dorn took the idea of fractional distillation and the representation of the human body by a cucurbit from Thurneisser's 1576 treatise and used it in his own publication, *Aurora Thesaurusque*, in 1577. An illustration showing the human-cucurbit analogy from these last two books was also used in the Huser edition of Paracelsus's works, which clearly associated this somewhat fanciful idea with Para-

and Latin and featuring the occasional Greek word to give them humanistic credentials, but they were basically written in German. Thurneisser published them in Berlin and in nearby Frankfurt on the Oder River, which runs north through Görlitz and Frankfurt to empty into the Baltic at Stettin. It is in this region of Brandenburg and Pomerania that we find traces of Johann Hayne too, and that was also an area inhabited by known Paracelsian enthusiasts. Görlitz itself was a center of Paracelsian medicine and home to the Paracelsian-inspired Lutheran lay theologian Jacob Boehme.[13]

Testing the Water in the Oder River Basin

Thurneisser dedicated his 1571 treatise on chemical uroscopy, *Præoccupatio, Durch zwölff verscheidenlicher Tractaten, gemachter Harm Proben,* "to the learned and esteemed Mr. Johann Hayne, master of the free arts and appointed physician of Stettin in Pomerania, his benevolent, dear, and good friend."[14] This reference sustains the claim made on the title page of Hayne's book that Hayne was physician to the Pomeranian court in Stettin and also identifies him as municipal physician in that Hansa town by the date of Thurneisser's dedication, 6 August 1571. Thurneisser continues, saying that the twelve cases he will discuss in his book, all of which derive from his practice of medicine in and around Berlin and Frankfurt an der Oder, will be well known to Hayne without identifying the patients by name. That Thurneisser expected Hayne to recognize patients by their diagnostic signs strongly suggests that he and his friend Hayne had studied and perhaps practiced medicine in the same social space of Brandenburg, before Hayne migrated downstream seeking employment in Stettin. It also suggests

celsus's medicine. See also Pagel, *Paracelsus,* 192–94, 365. According to Bleker, "Chemiatrische Vorstellungen und Analogiedenken," 71, Thurneisser regarded the visual, qualitative inspection of urine, the primary method of medieval uroscopy, to be of limited usefulness, owing to the weakness and unreliability of sight. One ought rather to base a scientific assessment on quantitative measurement of urine and its fractions. On Thurneisser's Paracelsian conception of medicine and the role of uroscopy in diagnosis, see Morys, *Medizin und Pharmazie in der Kosmologie,* 65–78.

13. Jacob Boehme's biographers identify intellectual and personal links between the lay religious author and Paracelsians in Lusatia. The longtime mayor of Görlitz, Bartholomæus Scultetus, was personally engaged in copying Paracelsus's manuscripts and edited at least one. See Hansen, *Jacob Böhme,* 22, 231–32; and Weeks, *Boehme.*

14. "Dem Wolgelehrten und Achtbarn Herrn Johann Heynen, der freyen Künsten Magistro, und bestelten Physico zu Stettin in Pommern, seinem günstigen lieben und guten freundt" (concise trans. by Frankie Shackelford). Thurneisser, *Præoccupatio,* Aiij^r.

that uroscopy was still a common practice in late sixteenth-century German medicine. In several surviving letters that he wrote to Thurneisser, Hayne called himself Thurneisser's disciple and referred to him as his teacher, suggesting a close personal mentorship, if not a formal academic relationship.[15] The acquaintance of these two physicians, who each professed a Paracelsian medical practice and wrote treatises on the diagnosis of disease on the basis of Paracelsus's chemical interpretation of uroscopy, points to the Frankfurt/Berlin area as an important locus for the implementation of Paracelsian medicine as practice.

Frankfurt was the residence of another, at least regionally, well-known uroscopic practitioner and author in the middle of the sixteenth century, Jodocus Willichius. Born in Rössel in Ermland, Willichius began his studies at the University of Frankfurt in 1516 and taught there until his death in 1552. Willichius published much on classical rhetoric, arithmetic, and other subjects, but aside from his 1544 anatomical commentary, diverse treatises on plague, and an early study of medical physiognomy, nothing printed in his lifetime speaks to his medical teaching or practice, despite the fact that he was appointed personal physician to Elector Johann II of Brandenburg in 1542. It was only posthumously—in 1560—that his treatise on uroscopy was published.[16] In 1582, just a decade after Thurneisser published his first book on chemical analysis and diagnosis of urine, Hieronymus Reusner published in Basel an elaborated edition of Willichius's uroscopy manual with extensive commentary.[17]

Reusner's book is in the traditional genre of scholastic commentary, presenting text attributed to Willichius a chapter at a time, followed by much longer, detailed scholia composed by Reusner. He apparently intended the book for use by practitioners, because he included images of urine flasks, called *matulæ* in Latin, along with descriptions of urine contents and colors. These images were

15. Berlin Staadsbibliothek, Ms. germ. fol. 420a, fol. 420b, and fol. 426 contain records of correspondence. I have not seen these letters, but rely on notes sent to me by Michael Stolberg, whom I thank for drawing them to my attention and sharing this information. The last of these, Ms. germ. fol. 426, 253r–257r, was written in 1583 and indicates that Hayne had resigned from his court appointment, but had been unable wholly to free himself from his lord's service.

16. Willichius, *Ivditia Vrinarvm Doctoris.* Four other medically related treatises by Willichius are listed in WorldCat: *Physiognomonica; Commentarivs Anatomicus; Wie man sich in einer Stadt für der Pestilenz behüten soll und möchte;* and *Wie man sich vorhalten und bewaren sol in den Heuseren.* Variations on these last two appear to have been republished on later occasions. Also, a medical dissertation, *Propositiones de indicationibus curativis ad disputandum propositae,* appears to have been defended by his son (de quibus . . . respondebit . . . Iodocus Willichius Iodoci f. Non: Septembris anno 1573) at Basel University.

17. Reusner, *Vrinarum Probationes.*

printed in black and white and could be appropriately colored by the owner or an illuminator as a visual guide, as was done earlier in medieval uroscopic manuscripts.[18] Hand-colored urine trees or wheel-charts are also evident in other guides to uroscopy, for example, Ulrich Pindar's *Epiphanie medicorum. Speculum videndi urinas hominum. Clavis aperiendi portas pulsuum. Berillus discernendi causas & differentias febrium* (1506) and, of course, in the various editions of Thomas Ketham's *Fasciculus medicinæ*, which helped to codify the manuscript tradition.

Three points need to be made about Reusner's book. First, Willichius's text is a prime example of uroscopy, as it had been elaborated in medieval Latin and Byzantine sources, drawing heavily on the late development of this tradition by the Byzantine physician Johannes Zacharius Actuarius, and called by one modern scholar "probably the high point of this type of work."[19] Actuarius's treatise was itself a late synthesis of the long medieval manuscript tradition of uroscopies dating back to late antiquity.[20] Second, the coupling of Reusner's scholia with Willichius's uroscopy shows how medieval uroscopy was studied by the medical humanists of the sixteenth century. Far from being an obsolete diagnostic science, uroscopy constituted a living part of medical teaching and practice, certainly as late as 1582, when the book was published. Third, we glimpse in Reusner's book the incorporation of Thurneisser's chemical uroscopy into the traditional semeiological literature in the section of his commentary called "On the Spagyricists' utterly new examination of urine; which is done by means of

18. Wellcome, *Evolution of Urine Analysis*, 35, notes that this was also done for the 1540 epitome of de Worde's "Judycyall of Uryns." The copy of Reusner's commentary on Willichius that is in the Hagströmer Bibliothek in Stockholm is exemplary. It contains ninety-six beautifully hand-colored woodcut prints of *matulæ*, illustrating the different kinds of urine samples the uroscopist might encounter. My thanks to Ove Hagelin for bringing this book to my attention.

19. Kiefer, "Uroscopy: Clinical Laboratory of the Past," 164.

20. Johannes Zacharius, usually referred to by his title, Actuarius, was senior physician at the imperial Byzantine court in Constantinople in the fourteenth century. He synthesized the methods of Theophilus, Isaac, and their Greek and Arabic commentators to produce his *peri ouron* (Περὶ Οὔρων), or *De urinis* in Latin, which displaced reliance on the uroscopies and commentaries of the *Articella* among medical humanists and became a standard source for late medieval Latin uroscopy. On Actuarius, see Sarton, *Introduction to the History of Science*, 3.1:889–92. Desnos, *History of Urology*, 126, gives ca. 1130 as a date for Johannes Actuarius, stating that his writings were translated into Latin in the fifteenth century. More recent scholarship by Hohlweg, "John Actuarius' *De methodo medendi*," 122, places him in the fourteenth century, dating the dedication of his *De methodo medendi* to about 1330 and claiming that his uroscopy was his first publication, therefore somewhat earlier, perhaps in the 1320s. Hohlweg (p. 128) notes that Actuarius's text was published in Latin in a 1519 Venice edition and apparently there was a 1531 partial edition in German. On the earlier medieval history of uroscopy, see Wallis, "Signs and Senses."

the separation and resolution of Mercury, Sulphur, and Salt."[21] Taken together, these points suggest that uroscopy was a dynamic feature of late sixteenth-century medicine and that Reusner's work warrants a closer look.

Reusner relates how the chemical uroscopists associated the parts of the urine in the *matula* with corresponding parts of the body and their chemical identities: the *nubes* or cloud forming in the upper part of the *matula* indicates a mercurial or "animal" disease in the head (referring to the Galenic animal spirit); variations in the body of the urine, called here the *enæorema,* indicate a sulphurous or spiritual disease, afflicting the middle part of the body; and a hypostatis or sediment on the bottom of the *matula* signifies a salty or corporeal disease in the abdominal area. It is not hard to recognize here a spatially defined diagnostic scheme that reflects the three basic divisions of the human body into head, thorax, and abdomen, which was commonplace in medieval medicine and goes back to Plato's *Timæus.* These divisions of the urine in the *matula* were also commonplace in medieval uroscopy.[22] In Reusner's text, however, they are associated with the three Paracelsian principles of mercury, sulphur, and salt, which the author aligned with the animal, vital, and vegetative or nutritive spirits of the body.

This relatively un-nuanced semeiological scheme is complicated by Thurneisser's analytical method: correct diagnosis requires careful separation and isolation of these three fractions, each into its own receptacle, which is then weighed with precision "so that in separation we might claim to know very

21. "De spagiricorum nova prorsus urinæ probatione: quæ fit per separationem & resolutionem Mercurii, Sulphuris & Salis"; Reusner, *Vrinarum Probationes,* 286. Bleker, "Chemiatrische Vorstellungen," 73, argued that Thurneisser took over the idea of associating the *matula* with the human body from Actuarius and that Thurniesser's subdivision of the cucurbit into twenty-four zones was a doubling of Actuarius's twelve-zone *matula.* This does not account for the application of color semeiology in chemical uroscopy, however, since Thurneisser believed such direct qualitative observations to be undependable.

22. These terms do have slightly different meanings in medieval uroscopies, perhaps owing to early mistranslations or misunderstandings of the Greek treatises. Actuarius applied the term *enæorema,* which has its roots in "floating" or "swimming" (plural *enæoremata*) to the entire bulk of the urine sample in the *matula,* which is then subdivided into an upper portion, where *nubes* (clouds) can be observed, a middle portion (*sublimia*), and a lower *hypostasis,* where heavier portions of the fluid and any solids will settle. The body of the urine is bounded by its upper surface, which is the crown or circle, and the bottom of the *matula,* called simply the *fundus.* An illustration of these divisions from Actuarius is reproduced in Desnos, *History of Urology,* 131. Wallis, "Signs and Senses," 273, identifies the division of the *matula* into three horizontal zones as beginning with the early medieval text *Liber medicinæ orinalibus,* attributed to the second-century medical writer, Hermogenes. What may have been a post-medieval development is the literal association of the spatial hierarchies in the glass with the spatial arrangement in the body. Wellcome, *Evolution of Urine Analysis,* 47, mentions this in Hamand, 1656, which I take as a reference to Hamand, *Ourography or Speculations on the excrements of urine.*

exactly which element exceeds another, at least in quantity and magnitude, but even in measure and weight."[23] Following Thurneisser further, Reusner noted the weight of the whole urine of a healthy man (he must have meant specific gravity) and argued that if the urine exceeds this figure by much, it indicates a very large quantity of dissolved tartar, which can be separated by distillation. From the distilled saline fraction, predictions about certain kinds of diseases are possible, namely, tartarous diseases such as paralysis, spasms, uterine pains, dyspnoea, scabies, and scurvy, to name a few. Mercurial and sulphurous diseases can be diagnosed from the other two fractions, also by examining the colors of the distilled urine and the places in the cucurbit (distillation flask) where mercurial and sulphurous impurities adhere. Fumes occupying the highest region of the cucurbit (the *capitellus*) signify corrosive chemicals rising into the head, "whence they generate acute diseases of the mind, confuse the memory, debilitate the intellect, [and] excite a sudden hotness and agitation in the entire body," causing exanthemata and "flying heats," especially on the face. "Not infrequently they destroy the man from within by apoplexy."[24] Reusner reported a couple more diagnostic details, but failed to give a full chemical uroscopic semeiology here: "For it is not our intention to write copiously about these experiments of the chemists, since in most places in Europe this manner of judging either is never seen or is rejected by the general consensus." Instead, he referred the reader to Thurneisser's *Præoccupatio*.[25]

If Willichius's careful study of traditional uroscopy is any indication of the prominent place of this form of diagnosis in local medical practice in the Frankfurt area, then Thurneisser's development of the semeiotic art as an element of chemical medicine makes sense from both an intellectual and a marketing perspective. That Thurneisser—who was trained as a goldsmith and was a largely self-made metallurgist and alchemist—turned his attention to

23. "ut in separatione exactissime cognoscamus: quodnam elementum, alterum non duntaxat quantitate & multitudine, sed etiam mensura & pondere excedat"; Reusner, *Vrinarum Probationes*, 287.

24. "Unde cerebri morbos acutos generant: memoriam turbant: intellectum debilitant: subitum calorem & æstum in toto corpore, maxime autem facie excitant, quem nos *die fliegende hitze* nuncupamus: εξανθηματα [exanthemata] item illa producunt, quæ nobis sunt *die Finnen, und ein Maltzig angesicht*: denique non raro apoplexia, hominem e medio tollunt"; ibid., 288.

25. "Non enim nobis animus est, uberrime de hisce probationibus Spagiricorum scribere: cum, quia in plærisque Europæ locis modus hic iudicandi, vel nunquam visus, vel communi fato explosus sit: tum vero, quod materiam hanc penicillo adumbravit Thurnheuserus, in sua πρακαταληψει seu Præoccupatione, contra Uroscopos vulgares"; ibid., 288. One is tempted to read into Reusner's words the conclusion that Paracelsian chemical uroscopy was local to the northeastern German area and slow to diffuse outward from this center.

publishing on chemical uroscopy soon after arriving in Frankfurt an der Oder and his appointment as the Elector's physician, suggests the demand he perceived for medical application of his chemical knowledge in Brandenberg.[26] Reusner's humanist accommodation of Willichius's traditional semeiotics to the chemical analysis introduced by Thurneisser also shows how readily Paracelsus's chemical medicine was translated into practice. The continuation of this practice, at least in northern Germany, is suggested by a 1658 book, *The Galeno-Spagyric Anatomy of the Urine*, published by the Danzig physician Heinrich Martinius.

The full title of Martinius's book identifies his uroscopy as based on the doctrine of Hippocrates and Galen as well as recent authors, chiefly Paracelsus and Thurneisser, but also on the writings of other key chemical philosophers.[27] Martinius's discussion of urine is highly philosophical and academic, defining the material, efficient, and instrumental causes of urine, how it is generated in the body, what its ideal characteristics are, and various ways that urine can depart from this ideal. Chapters 9 through 13 take up pathology, presenting the various parts of the urine and its contents and how these are altered in sick persons. Chapter 14 treats Thurneisser's chemical analysis in particular, and it appears from the typography at the end of this chapter that it was meant to be the end of the book. However, there is a fifteenth chapter, so identified in the table of contents, which deals with the practice of diagnostic uroscopy, "the art of declaring from the urine," which Martinius examines from rational, mechanical, and experimental points of view.[28] What he calls "mechanical" are the analytical methods that he adopted from Thurneisser's 1571 *Præoccupatio contra uroscopos vulgares*, noting that Hieronymus Reusner had earlier "briefly comprehended a general method" from the opinions of Thurneisser and Paracelsus.[29] In fact what

26. Court interest in practical applications of chemical knowledge, including chemical drugs, has been subject to study in recent years by historians, for example, Smith, *Business of Alchemy*; and Nummedal, *Alchemy and Authority in the Holy Roman Empire*, as well as by archaeologists, for example, Martinón-Torres, "Tools of the Chymist"; and Soukup, "Crucibles, Cupels, Cucurbits." The latter article specifically notes that analysis of chemical residues indicates that the laboratory was used for the production of Paracelsian-style chemical medicines.

27. Martinius, *Anatomia Urinæ Galeno-Spagyrica*.

28. "Caput XV Exercitatio ex urinis triplex: Rationalis, Mechanica & Experimentalis; artem pronuntiandi ex urina breviter adumbrat. Et operis conclusio"; ibid., 267.

29. "Mechanicam ex urinis pronuntiandi rationem primus repræsentavit Thurnheiserus, eamque aliquot luculentis exemplis expressit in sua Procatalepsi sive Præoccupatione contra Uroscopos Vulgares. Ex cujus & Paracelsi sententia Hier. Reusnerus universam rationem breviter ita complexus est"; ibid., 285–86.

follows are, as he credited in the margin, several paragraphs copied wholesale from the very end of Reusner's commentary on Jodocus Willichius.[30]

Taken together as three data points on a century-long trajectory of development, the treatises of Willichius, Reusner, and Martinius point to the durability of uroscopy as a diagnostic tool in early modern medicine and to its modification to accommodate the Paracelsian understanding of the body as a chemical laboratory. What these books document is the development of a practical chemical uroscopy that was elaborated by Thurneisser. Hayne's *Three Diverse New Treatises* make it clear that he also developed his practical medicine around a chemical interpretation of uroscopy, but that, unlike Thurneisser, he did not reject the traditional qualitative elements of the art. In this respect, Hayne much more closely followed the lead of Paracelsus, who, despite historians' claims, did not eschew traditional uroscopy as vain superstition, but rather adapted it to his chemical philosophy in a way that made the new theory accessible and diagnostically useful to the reader who was familiar with the medieval tradition.[31] This is significant because it calls for a reassessment of the current historiography of uroscopy in the medicine of the period associated with the Scientific Revolution and for a further consideration of how Paracelsus adapted traditional medicine to the articulation of his chemical philosophy.

Three Diverse New Treatises: Uroscopy, Astrological Medicine, and the Historiography of Medicine

The individual *Tractätlein* that constitute Hayne's *Three Diverse New Treatises* appear at first sight to be separate treatises that were bound together by the

30. Compare the text in Martinius, *Anatomia Urinæ Galeno-Spagyrica*, 286–89, with Reusner, *Vrinarum Probationes*, 286–88.

31. Describing Leonard Thurneisser's chemical analysis of urine, Pagel wrote in *Das medizinische Weltbild*, 19, "Auch hier finden sich nur allgemeine kritische Worte gegen die traditionelle Uroskopie in Paracelsischen Schriften" (Here, too, we find only the general criticism against the traditional uroscopy in the Paracelsian writings). This supports his earlier finding that "Paracelsus opposes 'uroscopy' on the ancient lines. No information, he says, can be obtained from the urine short of its examination by 'extraction,' coagulation and distillation ('ebullition'), i.e., by chemical methods." See Pagel, *Paracelsus*, 190. In *History of Urology*, 124, Desnos toes this same line: "During the Renaissance, this method of uroscopy [that is, traditional inspection] was challenged and discredited by Paracelsus."

publisher and provided with continuous pagination and a common index for the reader's convenience. But on closer inspection they constitute three components of a unified approach to medicine that, owing to its grounding in Paracelsus's writings, can be identified as Paracelsian praxis. Their individual subjects—astral diseases, tartar diseases, and chemical uroscopy—are mutually dependent concepts that make sense when understood as an integrated complex of medical ætiology, diagnostic semeiotics, and guide to therapeutics. They appear this way in Hayne's text and were from this source evidently taken as a system into George de Benneville's medical handbook, *Medicina Pensylvania*.[32] This medical *praxis* unified and codified elements of Paracelsus's medicine that seem unrelated and even superstitious to the modern reader and reveal that Paracelsus's followers in the late sixteenth and seventeenth centuries understood his medicine to be theoretically sound and practicable and that they set about explicating and elaborating it as such. That Hayne actually put his medicine to use with patients is suggested by a letter he wrote to Thurneisser, complaining that his lord and patron insisted that he alone treat his wife for "astralische Läufe" (effects of the planets' "operations") and that the elimination of outside consultants puts the physician at special risk in the case the patient should fail.[33] Thurneisser's *Præoccupatio* is likewise portrayed as drawn from personal practice. Nor was this kind of Paracelsian practice limited to these two. We find in the *Hermetic Medicine* of Heinrich Nolle (1613), for example, an explanation of how astral causes of disease and tartars fit together in a general conception of disease etiology.[34] Somewhat later, in 1636, Friedrich Zobell devoted a medical treatise to the connection between the etiology of tartar diseases and chemical uroscopy, without discussing astral diseases.[35] And, as noted above, Heinrich Martinius's 1658 book focused on incorporating Paracelsian chemical uroscopy as a diagnostic system and presenting it to the empirically minded reader of the mid-seventeenth-century Scientific Revolution. Each of these books can be seen as

32. I have sketched out de Benneville's use of Hayne's text. See note 8 above. De Benneville's manuscript has been digitized and can be viewed at http://contentdm1.accesspa.org/cdm/ref/collection/cppmss/id/801 (accessed 12/20/2014).

33. Berlin Staadsbibliothek, Ms. germ. fol. 426, 253r–257r.

34. See note 4 above. Nolle's elaboration of astral causation within his concept of tartar diseases sheds light on how the Paracelsians integrated celestial and terrestrial systems of causation. I discuss this briefly as preparation for understanding Hayne's book in Shackelford, "Paracelsian Uroscopy . . . in *Medicina Pensylvania*," 15–21.

35. See note 5 above. Zobell's dedication of this book is dated Gottorp, 5 February 1636, but for whatever reason, it does not appear to have been published until Pauli obtained it and published it in Dresden in 1686. WorldCat shows a later edition as Dresden: Winckler, 1701.

its author's attempt to work out a medical system that applies Paracelsus's chemical philosophy to explain traditional practice. Indeed, the diagnostic systems embedded in astrology and uroscopy were two of the most salient features of medieval medical practice. By the late Middle Ages and Renaissance, astrology was understood to be an important element of the physician's practice, expected by the patient,[36] and uroscopy was so ubiquitous in medieval medicine as to be emblematic of the art as a whole.[37]

What the contemporary reader likely apprehended, although a point obscure to the modern reader, is how Paracelsus's medicine readily integrated astrological medicine, which was based on the cosmic connection between the planets and human bodies, with uroscopy, which assumed a material and semeiotic relationship between human bodies and the *matula*. Georg Faber's introduction made this very clear. What traditional medieval medicine regarded as the physician's chief concerns when diagnosing and treating the patient—namely, the six Galenic nonnaturals—are subsumed under the Paracelsian category *Ens naturali,* or the natural being. But, he continued, there are four other *entia* that must be taken into consideration: *Ens astrale, Ens venenale, Ens pagoycum,* and *Ens deale.*[38] These are clearly the same categories as those identified by Paracelsus,[39] as Faber made explicit: "The bright Light of Nature has now taught us plainly about all these, and Theophrastus Paracelsus wrote on this subject well

36. The literature on medical astrology is substantial. Siraisi, *Medieval and Early Renaissance Medicine,* gives a lucid, brief introduction. A more detailed overview of the function of astrology in medieval medicine is French, "Astrology in Medical Practice." Jacquart, "Theory, Everyday Practice, and Three Fifteenth-Century Physicians," argues for the importance of astrology to educated medical practitioners in the sixteenth century, even when they were agnostic about its validity. Stolberg, "Decline of Uroscopy in Early Modern Learned Medicine (1500–1650)," offers a similar argument for the persistence of uroscopy among learned practitioners in an age increasingly insistent on a more scientific basis; patient expectations required physicians to have expertise in uroscopy in order to compete with lay healers.

37. Siraisi calls "a urine flask, the universal symbol of a physician in late medieval iconography" and notes that "[t]he ubiquitous presence of the urine flask as a convenient symbol of the medical practitioner in medieval art is a convention, but it is one that reflects reality." See her *Medieval and Early Renaissance Medicine,* 28 and 125, respectively. In *Canonical Medicine,* French opined that "[t]he most characteristic image of the medieval doctor is of him examining a raised flask of urine," 148. In *Medieval Medical Miniatures,* Jones (p. 56) stated that "[i]f the physician of the Middle Ages had a badge of office, it was surely the jordan, or urine glass. He is portrayed so often in miniatures holding the jordan up to the light as to be instantly recognisable by what he carries." In *History of Urology,* Desnos (131, 133) noted an image of a man holding a *matula,* carved in a wooden plank, that advertised one medieval physician's practice.

38. Hayne, *Drey unterschiedliche newe Tractätlein, Dedicatoria,* ii^{v}–iii^{v}.

39. Clearly, these are the same five disease "entities" identified by Paracelsus in his early *Volumen medicinæ paramirum,* but where what Faber calls *Ens pagoycum* is labeled *Ens spirituale.* Pagel referred to this poisonous entity as *Ens substantiæ.* See his *Paracelsus,* 141.

and thoroughly in the first volume of his Books."[40] Of these, *Ens astrale* is subject to the influences of celestial bodies[41] and *Ens venenale* comprehends a broad class of diseases caused by toxic residues of food and drink that Paracelsus called tartars.[42]

Behind Hayne's *Three Diverse New Treatises* lay Paracelsus's ideas about both of these related subjects: astral causation of diseases and tartar diseases arising from the failure of the inner alchemist. He devoted his third treatise specifically to chemical uroscopy as a means of diagnosing both astral and tartar diseases. The book can thus be regarded as one medical student's successful attempt to assimilate ideas that Paracelsus had composed under varying circumstances and which, for the most part, circulated in manuscript or were published posthumously by various editors and printers.[43] Through Hayne's treatises, we can see how Paracelsus's effort to harness traditional uroscopy and astrological medicine to chemical philosophy—knitting together his worldview—can be viewed as a natural response to the standards of the physician's professional training and the expectations of early modern German medical culture. These three books exhibit a common commitment to translating Paracelsus's medical ideas into a functional practice *without* discarding or compromising his speculations about the nature of the macrocosm and microcosm.

Astral Medicine

Hayne's first tract is on astral diseases, that is, on how knowledge of the astral causes of diseases and their effects can be harnessed for both prevention and

40. "Solches alles lernet uns nun klärlichen das helle Liecht der Nature / unnd hat auch hiervon *Theophrastus Paracelsus* im ersten *tomo* seiner Bücher sehr wol und ausführlichen geschrieben"; Hayne, *Drey unterschiedliche newe Tractätlein, Dedicatoria,* iii^{v}.

41. "Das *Ens Astrale,* wie nemblich die Kranckheiten durch Influentz und Wirckung dess Gestirns entspringen und verursacht werden"; ibid., fol. iii^{r}.

42. "Das *Ens Venenale,* welcher massen fast alle Schwachheiten auss dem Gifft und Unreinigkeit / so in Speiss und Tranck verborgen liegt / (sonst *Tartarus* genant) mögen geboren werden"; ibid.

43. The authorship and publication of Paracelsus's treatises have been studied and debated in the scholarly literature for over a century. The standard edition remains Karl Sudhoff's *Theophrast von Hohenheim gennant Paracelsus,* but the textual tradition has come under recent scrutiny by Weeks, *Paracelsus: Speculative Theory,* 35–43, who has also produced some excellent translations of key doctrinal texts—*Opus paramirum, Opus paragranum, De morborum, De matrice,* and *De causis morborum invisibilium*—in Weeks, *Paracelsus . . . Essential Theoretical Writings.*

cure.[44] Superficially, this was the role of astrological medicine in medieval practice, too, and the similarity between Hayne's ideas and traditional ones bears scrutiny. The treatise begins with a general distinction that defines astral diseases as categorically distinct from the closely related tartar diseases that he discusses in the second treatise. The human being has a twofold nature, "syderisch und corporalisch" (sidereal and corporeal), and it is his sidereal aspect that is subject to the stars, which act on him spiritually, not corporeally. These cause disease via elemental intermediaries, air and fire, the defining elements of the superior globe.[45] "Stars" (chiefly the planets) can be either benevolent or malevolent, as was common in medieval astrology, and the malevolent ones cause diseases by means of a spiritual poison, which cannot be expelled from the body by purging or bloodletting, but can be overcome with powerful strengthening and diaphoretic medicines.[46] Individual planets are the causes of various kinds of disease, depending on their disposition. The Sun is a benevolent planet and does not itself cause disease, unless it is complicated or infected by other planets, and then can cause lethargy and weakness of the eyes. The planet Venus governs fertility and thus, when infected by the malevolent planets Saturn and Mars, causes gonorrhea and diseases of the uterus, kidneys, and the organs of reproduction as well as all the varieties and *sequellæ* of the French disease. It was, in fact, on the basis of Venus's cycle that Johannes Regiomontanus prognosticated the coming of this new disease.[47] The Moon, the lowest of the planets, causes diseases in the brain that manifest as feebleness, spasms, twisting of the mouth (probably the result of a stroke), and falling sickness (epilepsy) as well as disease of the stomach, constipation, *schlag,* catarrh, colic, and menstruation.[48] Hayne's account of the planets is otherwise cursory and wholly in keeping with tradition: Saturn and Mars are malevolent, Jupiter magnificent, and Mercury variable.

Hayne devoted the fifth chapter of this treatise to the "other" stars, the fixed stars of the firmament, not addressing them directly, but by way of considering

44. "Von Astralischen Kranckheiten / wie nemlichen dieselbe von den *Astris* oder Gestirnen verursacht / und wie sie beneben ihren viel und mannigfaltigen Zufällen erkand werden / auch wie man sich darfür nechst Göttlichen Beystand nicht allein præserviren und bewahren / sondern auch dieselben mit guter *Diæt* und andern Mitteln widerumb abwenden möge"; Hayne, *Drey unterschiedliche newe Tractätlein,* title page.

45. Ibid., 1. Paracelsus and his followers—for example, Petrus Severinus—redefined Aristotle's elemental spaces by eliminating the celestial quintessence and distinguishing an inner, lower elemental sphere of earth and water from an outer, superior sphere of air and fire or firmament, which embraced the planets and stars.

46. Ibid., 2.

47. Ibid., 3.

48. Ibid., 3–4.

the virtues that they manifest in specific plants and insects (for example, Spanish flies). Some specifics have powerful and even fatal astral virtues, such as *napellus* (thistle), *basilicus* (basil), and white hellebore, but can be used to treat epilepsy and spasm. Yet, if too strong, these same medicines can cause fatal epilepsy and spasms.[49] Colocinth is a strong purgative, as are turbith, esula, tithymalus, and similar drugs, which work as poisons in the body and function as more or less powerful stomach and intestinal purges.[50]

Tartar Diseases

Hayne's second treatise takes up the subject of tartar diseases—that is, dysfunctions that are caused by mixtures of stones and salt—and how they can be diagnosed by the inspection of urine and then fully cured.[51] This is a class of diseases and a theoretical approach to pathology that is familiar to historians of Paracelsian medicine, but warrants more careful scrutiny.[52] Tartar diseases are exemplary of Paracelsus's ideas about the operation of the *archeus*, or inner alchemist, and the failures of chemical process that result in diseases. From the beginning, Hayne acknowledged Paracelsus as a source of the theory of tartar buildup in the body, caused by the body's failure to separate out and excrete tartar that is consumed in various foods and drink, resulting in an accumulation of tartar obstructions in the forms of *viscus*, *bolus*, *arena*, and *calculus*. In fact, the reader is referred multiple times throughout this treatise to the Latin edition of Paracelsus's *De tartaro* that was published in Basel.[53] According to this theory, slimy foods such as peas, fruits, roots, and herbs contain tar, mucus, and glue, which are coagulated in the body by the spirit of salt and from tartar deposits. Such

49. Ibid., 4. We see here Paracelsus's principle that medicines are poisons and vice versa, depending on dosage and application.

50. Ibid., 5.

51. "Von Tartarischen Kranckheiten / das ist / Von den gründlichen Ursachen aller Schwachheiten Menschliches Leibes / so vom *Tartaro*, Steinen und Saltz vermischt / herkommen / auch was für peinliche Schmertzen und Wehetagen / sampt andern vielen gefährlichen Zufällen mehr hierauss erwachsen / und wie solche auss dem Urin eygentlich erkand / und vollkömlich curirt werden sollen"; ibid., title page.

52. Pagel, *Paracelsus*, 153–65, briefly introduced Paracelsus's theory of tartar diseases and its importance for van Helmont's conception of specific diseases, while Debus, *Chemical Philosophy*, 362–65, elaborated on van Helmont's conception, but neither treated the general importance of the concept of tartars for the broader reception of Paracelsus's ideas. This warrants further research.

53. A Latin edition, *Libri quinque de causis, signis, & curationibus morborum ex tartaro utilissimi*, was published by Perna in Basel, 1563, who also published an edition, *De tartaro libri septem*, in 1570.

tartar commonly appears in the urine as a viscous, whitish slime (*viscus*). Tartar contained in fish, meats, and milk products when coagulated yields *bolus* or *tartar bolaris,* which has a reddish color that is imparted by the organs in which it is formed. The practicing physician should chemically separate these components in the urine in order to recognize them and help the patient.[54]

Hayne described how these tartars, which are more prevalent in drinks derived from fruit and grain (that is, wine and beer) than in foods, are coagulated and compacted in various organs of the body by the spirit of salt, not by the spirits of sulphur or mercury.[55] Such coagulations could happen in various places in the body, but principally in the stomach, liver, and blood vessels, where the three primary stages of digestion take place.[56] Each of these digestions has a natural outlet for wastes that are separated, expelling them by defecation, urination, and diaphoreses through the body's pores. These outlets, generally referred to in medieval Latin medical literature as *emunctoria,* can become blocked by coagulated tartars, causing inflammations, pain, paroxysms, and other symptoms.

After general consideration of the chemical etiology of tartar diseases, Hayne treated them piecemeal by location, basically following the order of the digestions that sequentially refine the body's nutriment: tartars forming in the mouth and stomach (chap. 4), intestines (chap. 5), mesaraic veins (chap. 6), liver (chap. 7), kidneys (chap. 8), bladder (chap. 9), lungs (chap. 10), brain (chap. 11), heart (chap. 13), gallbladder (chap. 14), spleen (chap. 15), blood vessels (chap. 16), flesh (chap. 17), marrow (chap. 18), male and female reproductive organs (chap. 19), and finally tartars in the bones and joints of the arms, hands, legs, and feet, principally evident as forms of gout (chap. 20). Hayne offered for each general location and corresponding type of tartar buildup both uroscopic signs and therapeutic recipes, sometimes drawing on Paracelsus, and sometimes taking personal credit (for example, "von meinem *Arcano*"),[57] but often just recommending anonymous mixtures. For example, in chapter 6, for tartar diseases of the mesaraic veins—the veins of the mesentery were at that time

54. Hayne, *Drey unterschiedliche newe Tractätlein,* 48. Hayne's words suggest that he advocated distillation and chemical analysis of urine, such as Thurneisser described in his books, but he did not elaborate on this here.

55. Ibid., 48–52.

56. These digestions are a chemical version of the corresponding Galenic coctions, which are required in the making of blood from ingested foods and drink. On the relationship between the three coctions and uroscopy in medieval uroscopies, see Wallis, "Inventing Diagnosis," 35.

57. Hayne, *Drey unterschiedliche newe Tractätlein,* 85.

thought to convey chyle from the intestines to the liver—he recommended a potion made from *carduus benedictus*, betony, chicory leaf, "Odermennigkraut," angelica, chicory root, rhubarb, and ginger.[58] For blockage of the liver, which he said the Galenists labeled incurable, he noted that Paracelsus taught in the Latin edition of his *Liber de tartaro* that it could be cured by a medication made from the calcined dregs of wine (that is, from Weinstein by *similia similibus*).[59] However, he also offered a potion made from a decoction of chicory leaf, *carduus benedictus*, agrimony, centaury, chicory root, polypody, rhubarb, and *passula minor* (a kind of currant), mixed and sweetened with sugar, to be given each morning. The urine in this kind of disease is very red and thick, but when the liver is fully plugged up and becomes hard, the urine will appear as clear and pale as well water and without sediment, since the liver is no longer able to separate out impurities that would be seen in the urine glass.[60]

In addition to entering the body as impurities in foods and drink, tartars could come into the lungs with damp air and "chaos,"[61] especially troublesome in the case of the arsenical diseases. Such diseases, which Hayne said were not described by Galen and others, are particularly dangerous and difficult to cure, inasmuch as the arsenical tartar can become even more toxic as the result of the influence of the stars. Nonetheless, Hayne offered recipes and refers the reader to Paracelsus's volume on miners' diseases.[62]

Tartars in several specific locations were connected with familiar diseases and warranted Hayne's extended commentary. Following chapter 11 on tartar in the brain, he devoted chapter 12 to diseases that are characterized by catarrhs and fluxes mixed with tartar, digressing on the curing of fluxes with dissolved tartar, including a cure described by Paracelsus in chapter 1 of the second book of *De vita longa*.[63] Chapter 16, ostensibly on diseases of the vessels, marrow, and

58. Ibid., 68. The recipe is given mainly in German.

59. Ibid., 74–75. Weinstein was a manifestation of tartar, making this a clear example of the Paracelsian principle like-cures-like.

60. Ibid., 75. This recipe is given mainly in Latin, with uroscopic observations in German. The fact that some recipes are given in Latin, others in German, and some macaronic suggests Hayne's incorporation of medicines from diverse—perhaps a mixture of textual and verbal or experimental—sources.

61. "das solcher Tartar vornemlich auss fauler dumpfiger Lufft / oder dickem trübem *Chaos* . . ."; ibid., 87. The addition of this characteristically Paracelsian term for airborne elements that are finer than vapors ("gases") to chemistry is commonly credited to van Helmont. See Pagel, *Paracelsus*, 358, for consideration of the relationship between van Helmont's use of the word "gas" and Paracelsus's use of "chaos."

62. Hayne, *Drey unterschiedliche newe Tractätlein*, 86.

63. Ibid., 106–8. Paracelsus, *Sämtliche Werke I*, 3:249–92, reproduces the texts of *De vita longa libri*

flesh, constitutes the bulk of the treatise on tartar diseases. While most chapters occupy two to seven pages, and the second most lengthy takes up fourteen, chapter 16 fills fifty-two pages—more than a third of the treatise—and even so relegates tartars of the marrow and flesh to chapters 17 and 18, respectively. The greater number of pages devoted to tartars of the blood in the vessels reflects both the prevalence and high visibility of diseases associated with the blood during the early modern era, notably scurvy, fevers, and rashes.[64] For Hayne, diseases arising in the blood challenge the traditional humoral etiological categories of choleric, melancholic, and phlegmatic blood, since it is evident that blood contains fluid with resolved tartar and other salts.[65] According to Paracelsian theory, these salts are resolved and then coagulated again by spirit of salt and form granules ("Körnlein und Grän") that lodge in the pores and can result in stones, making consideration of the chemically evident tartars more significant than the elusive Galenic humors when it comes to diagnosing and treating these kinds of diseases. Hayne offered recipes for various powders, potions, baths, gargles, and inhalants as well as dietary advice to treat various manifestations of scurvy, in particular.

Turning to febrile diseases that arise in the blood from tartar—which are characterized by excessive heat (that is, "fever" in the modern use) and thirst—Hayne provided information both about how to diagnose these through inspection of the urine and about various treatments. The causes of all fevers are not to be found in a single tartar, but rather in various tartars that are resolved in the stomach, liver, and kidneys, and yet it is the tartar in the blood vessels that produces the fever.[66] The diagnostic process therefore must be sensitive to the complexity of the etiology. We see this in the first treatment Hayne presented in this section, namely, purgation with an herbal decoction to be administered to the febrile patient on "good" days, when the feverishness and paroxysms are absent. But, he added, if this potion does not help, and the patient's urine is very evi-

quinque. A Latin edition of *Libri quatuor De vita longa* was published by Bodenstein in Basel, 1560, by Perna in 1562, and by Gohory in Paris, 1567, any of which Hayne might have used.

64. "Capitel XVI. Von Kranckheiten dess Bluts in Adern / Marck und Fleisch vom Schorbock / Fiebern und Flechten / auch Unreinigkeit und Räudigkeit etc. auss *Sale Tartari*"; Hayne, *Drey unterschiedliche newe Tractätlein,* 124. It is worth noting that vessels, marrow, and flesh are not organs according to traditional medieval medicine, but rather what Galen called homoeomerous parts. The relative shift evident here from diseases of organs to diseases of the homoeomerous parts may well signal an important aspect of Paracelsian ætiology as distinct from scholastic pathology.

65. Ibid., 124–25.

66. Ibid., 160.

dently light red, translucent, and clear of particulates ("hell und klar"), or if the patient complains of bitterness in the mouth, then a paroxysm in the stomach is implicated, and one should, each morning until the patient is better, administer water in which lesser centaury has been soaking. But, if there is too much blood in the veins (plethora, a common medieval diagnosis) and the blood has no air in it, then Hayne advised venesection of the *vena hepatica*. If none of this works, he resorted to a powder made from *carduus benedictus* seed and root, chicory, *chelidony major*, gentian, and burned hart's horn (*cornu cervi*), administered with a drink that has some drops of vitriol added.[67] Many other variants follow, depending on the nature of the fever, the qualities of the urine, etc. Poisonous acute and continuous fevers, for example, result from poisons affected by astral influences, which are not easily observed in the urine at first, but upon distillation become so. They may also be known from the undistilled urine, which will have a blue, lime color at the meniscus ("oben am Circkel") in the urine flask.[68] Pestilential fevers are another case in which astral influences are important, and Hayne devoted several pages to these, also referring the reader to Paracelsus's *De tartaro* and to his own book on astral diseases. This internal reference suggests that Hayne envisioned his three treatises not as three separate texts, but as parts of an integrated, coherent approach to medicine.[69] Interestingly, Hayne noted that venesection is not as dangerous in pestilence as in cases of head and chest diseases, referring the reader to his booklet *De peste*, of which there is today no record.[70]

Hayne concluded his treatise on tartar diseases with a general chapter on "how one should cure all diseases and sicknesses from [that is, caused by] tartar."[71] Here, he began by admonishing the physician to consider first whether he is confronting a case of "astral infection" coming from the upper elements, namely, air and fire. That can be determined by distilling and fractioning the urine to discern the three kinds of poisons, as he described in chapter 16. In such astral diseases, one should *not* begin by purgation and venesection, the most common medieval interventions (after dietary advice), but rather one should

67. Ibid., 161–62.

68. Ibid., 169.

69. "findet solches in *Theophrasti* Schrifften und Lateinischen Büchlein vom *Tartaro*, welches er zu Basel *publice dictirt* hat / darneben / so was mangelt in meinem Buch von den Astralischen Kranckheiten"; ibid., 170.

70. Ibid., 172.

71. "Wie man alle Kranckheiten und Seuchen auss *Tartaro* curiren solle"; ibid., 182.

employ *diaphoretic* medications. Then, when the astral diseases have been transmuted into chronic diseases, purgations and bleeding can be used. He referred the reader to part two, book two of the *Book of Influential Diseases*, that is, diseases caused by astral influences.[72]

The second thing that the physician should determine from inspection of the urine is whether it points to "dissolutische" diseases, presumably those that are characterized by tartar compounds in the body that must be broken up and dissolved by chemical drugs before they can be expelled by the body through one or another *emunctorium*, such as by sweating, urination, or diaphoresis. He gave brief semeiological advice and referred the reader to the third book of part two.[73] Hayne merely specified here that such diseases cannot be cured simply with purgatives, but require specific arcana prepared chemically.

The third and last thing that the physician should consider is whether the disease under consideration is caused by the lower elements, namely, water and earth, which are taken into the body with food and drink. This sort of disease can be diagnosed by uroscopy, which reveals the condition of the three principle digestions. Such tartars can be resolved with subtle, penetrating sulphurous spirits, and the pain can thus be conquered.[74]

Following Paracelsus's lead, Hayne presented recipes for both particular and general medicaments to treat tartar diseases. He then specified two more potions for resolving and removing the prime matter and the *feces* of the tartar from the body,[75] both of which are herbal in composition. He then noted that, with God's guidance, he had discovered a spirit of terebinth that, when added to wine, is healing. Also, a spirit extract of sapphire (only three to five drops are needed!) is useful in all tartar and saline diseases of the back and sides. But the

72. Ibid., 182–83. He did not name the author of this book. Historians have recently drawn attention to the mutability of diseases in the medieval and Renaissance medical conception as a barrier to making modern diagnoses of past descriptions. For example, Stein, "Meaning of Signs," esp. 619, shows that early modern diagnosticians believed that syphilis could become leprosy or vice versa. Hayne's statements here and just after imply that physicians also worked with such disease transmutations in curing diseases. That is, if one could transmute an intractable disease into one that can be easily expelled, using chemical arcana, then one could use known remedies to rid the body of the pathogenic matter through traditional means.

73. Hayne, *Drey unterschiedliche newe Tractätlein*, 183.

74. Ibid., 184.

75. "Ein ander Tranck / so *primam materiam tartari* ausstreibt," and "Noch ein ander *materiam tartari* zu resolviren / und die *feces tartari* ausszutreiben"; ibid., 188. Both recipes feature chicory, *carduus benedictus*, *resta bovis*, polypody root, rhubarb, anise seed and fennel seed, and senna, with several additions that slightly distinguish them.

best remedy for all these tartar diseases is tincture of gold. Lacking these, one can use chicory, *resta bovis*,[76] and *carduus benedictus* as general medicines against blockages. Hayne once more noted that Paracelsus offers further advice in his book on tartar.

A Paracelsian Chemical Uroscopy

The third and by far longest of Hayne's *Three Diverse New Treatises* instructs the physician on how to diagnose diseases by examining the patient's urine.[77] Hayne began this third treatise (in which he referred to it as the second book, suggesting that the manuscript Faber received was not finished for publication) by specifying that it will be divided into three subsections: the first treating urines that indicate "stone" (that is, tartar) diseases; the second on urines showing signs of astral diseases; and a third on urines revealing diseases with "natural" causes. These three subsections appear to correlate respectively with the lower elements earth and water, the upper elements air and fire, and the three "seeds," namely, salt, sulphur, and mercury.[78] A fold-out chart labeled A, which is inserted at the beginning, shows the system of diagnosing diseases of the brain—the highest part of the body—by examining the uppermost portion of the urine: bubbles are indicative of salts, a golden yellow "circle" (the *meniscus*) indicates a sulphur-mucus, and foam reveals the presence of mercury. A second flap inserted here provides a more detailed analytical outline for the organization of the treatise and offers insight into how Hayne conceived pathology. The three subsections mentioned previously correspond now to (1) "body," (2) "spirit," and (3) "seed." "Body" comprises elemental earth and water, which are taken in as food and drink and can result in tartars and excrements in the body. "Spirit" (the life of the body) encompasses elemental fire and air, which are taken into the body as

76. *Aresta bovis*, also called *Arrestia bovis* or Aresteboeuf, is an herb of the family *Ononis* or restharrows, used to treat bladder and kidney stones.

77. "Das Fundament / und den rechten Grund / wie man die Urinen dess Menschen / so wohl Gesunden / als Krancken / nach Spagirischer Art / künstlich judiciren und erkennen möge"; Hayne, *Drey unterschiedliche newe Tractätlein*, title page. Georg Faber's comment (Hayne, *Drey unterschiedliche newe Tractätlein* [1620 ed.], *Dedicatoria*, iv^r) that he was publishing Hayne's treatise on uroscopy—which was based on the Hermetic and Spagyric art and not on the kind practiced by "*pseudouromantes* und Juden Aertzte"—suggests that uroscopy was widely practiced in early seventeenth-century Germany and that there was a need to put it on a sound theoretical basis.

78. Hayne, *Drey unterschiedliche newe Tractätlein*, 195–96.

Paracelsus's theory of tartars as material causes of diseases in the body is fundamental to his conception of medicine and runs through many of his treatises. It is perhaps most clearly explained in his 1531 *Opus paramirum* and *Das Buch von der tartarischen Krankheiten* of 1537 to 1538. But one also finds it expressed earlier, in notes from the lectures he delivered in Basel during that fateful year 1527/28, which marked the beginning and end of his academic career.[81] Owing to his reputation as a healer among leaders of the Swiss Reformation and the patronage of the prominent Basel publisher Froben, Paracelsus was appointed municipal physician in the university town, a post that permitted him to lecture publicly and advise students.

According to the popular telling of the Paracelsus story, Paracelsus set the stage for the autumn term and attained lasting fame as an iconoclastic reformer by throwing, in 1527, a valuable copy of Avicenna's *Canon* into one of the bonfires that are customarily lit on St. John's Eve, which is the celebration of midsummer.[82] The *Canon* was *the* textbook most commonly used to teach medicine in the late medieval and Renaissance university, since it concisely summarized Hippocratic and Galenic medicine in a way that was more accessible to students than were the verbose works of Galen. Paracelsus's act thus symbolized his total rejection of more than thirteen centuries of medical tradition, earning him a reputation among modern scholars as a revolutionary, someone who caused a rupture in medical history, not as a contributor to a continual progressive development.

On 17 August 1527, not quite two months after that signal event, Paracelsus began to lecture. What revolutionary heterodoxy would he present to a mixed audience of Basel's university students, barber-surgeons, and other locals in a city that was in the throes of a religious reformation? He chose as his first topic a treatise that he was in the process of writing, the title of which identified two

81. Weeks, introduction to *Paracelsus . . . Essential Theoretical Writings*, 7: "It is possible that without the Basel quarrel Hohenheim might have pursued less comprehensive objectives and indeed never have become 'Paracelsus.'"

82. This famous episode in Paracelsus's life is the setting for Georg W. Pabst's 1943 film *Paracelsus*, which aptly depicts him ignoring university protocol and lecturing in German—a novelty—on his unorthodox conceptions of disease and healing. What we do not see in the movie, or in the shorter accounts of his life, is the content of these lectures. For example, Goodrick-Clarke wrote in his introduction to *Paracelsus*, 17–18, only that Paracelsus lectured "on such subjects as tumors and wounds, purges, uroscopy, pulse diagnosis, and his work *De Physiognomia*." In *Paracelsus*, Pagel (p. 21) summarized that "[i]n these lectures Paracelsus formed the essential nucleus of his system of medicine." We can form an impression of his teaching from various notes published in the modern Sudhoff edition of Paracelsus's works, pertaining to a series of lectures he delivered in the fall and winter terms.

Another conclusion we can reach from examination of these lecture notes is that Paracelsus regarded the semeiology of pulses and physiognomy to be subordinate parts of uroscopy. This is indicated by the title of the second tractate of book two: "On Mixed Urine and Its Conditions. On Pulse and Physiognomy."[92] The nature of this subalternation is shown in the first outline of this tractate. Mixed urine has four conditions. Two of these, color and substance, are labeled simply as deriving from its condition, but the other two are labeled as coming from an extraneous condition. Of these latter two, one comprises "accidents" that describe the tartar along with three exterior signs, and the other comprises four "portions" of the urine in the *matula,* namely, the urine itself, its foam, its bubbles, and its circle or top surface. The exterior signs also arise in the pulse, physiognomy, and mouth. The pulse is judged at five locations on the body's exterior; physiognomy takes into consideration signs appearing in the eyes, nostrils, external parts of the nose, the mouth, and in coloration. Signs evident from the mouth include air, speech, and breath. Details of the semeiology of the pulses and physiognomy do not directly pertain to chemical uroscopy and need not detain us here, although it is of interest to note that Paracelsus regarded pulse as yielding diagnostic information about the soul, and that it too yielded information about tartarous and mineral diseases.[93] Moreover, *physionomia* included such signs as sunken eyes, which are a sign of death, showing that Paracelsus retained elements of Hippocratic clinical observation in his semeiology, although it was subordinated to uroscopy.[94]

Returning to the outlines of the lectures on internal urine and mixed urine (which, by definition, contains some internal urine), we see that it is here that elements of traditional uroscopy came through most clearly, but were given a Paracelsian twist. He instructed that internal urine or the urine of blood can be analyzed by color, substance, and contents, all traditional diagnostic categories of medieval uroscopy. The colors, however, vary with place and organ, the substance according to the nature of the minerals present, and the contents according to the quality of the disease. The color of the blood itself (and I presume

chapters, with notes, followed by a rudiment of a second tractate, so labeled; ibid., 4:566–74, 591–96. The undeveloped state of the second tractate is supported by the secondary notes, which have only a short annotation pertaining to its topic, "mixed" urine; ibid., 4:573–74, 4:595–96.

92. "Tractatus secundus. De urina permista eiusque conditionibus. De pulsu et physionomia"; ibid., 4:572.

93. "De pulsibus, id est, animæ motu"; ibid., 4:575.

94. Ibid., 4:578.

he meant that this is reflected in the color of the internal urine) has three categories: six colors pertain to particular organs, seven to surrounding parts, and six are characteristic of the excrements. These are all very specific and, although not worked out systematically, were clearly meant to yield distinct diagnostic results.[95] Likewise, annotations to the outline for mixed urine and its conditions reveal a color semeiology that is strikingly traditional in form. Green, translucent urine is a sign of a bad spleen, but if it also contains opacities or clouds, it is a sign of death; reddish urines indicate great hotness and hot diseases; on the contrary, white is a sign of cold disease, but if it is a pale white (*lauter Weiss*), it is a sign of good urine and health; violet or blue-colored urine that is translucent is a sign of death, if the patient is infirm, otherwise it is a sign that he tends toward leprosy; clear, light-brown urine in an acute disease means death, but in a chronic one is a sign of health.[96] And so on. Although the previous parts of Paracelsus's inchoate treatise on urines, pulses, and physiognomy would suggest that all these diagnostic signs are embedded in his characteristic theories of digestions as chemical separations and tartars and minerals as etiological agents of diseases, this list of colors—as it stands in this lecture outline—clearly hearkens back to the traditional medieval uroscopies.

One speculation that we can derive from Paracelsus's incomplete uroscopy treatise is that, at this moment in his life, Paracelsus was engaged in formulating his novel physiology and pathology as a practical, teachable medicine and that he was doing so by applying it to—or perhaps drawing it from—established medical diagnostic and prognostic practices. This aspect of the German reformer's medicine does not present a rupture with traditional medieval medicine so much as a reinterpretation of it, and we find this synthesis of old and new in the formulations of Thurneisser and Hayne as well. Thurneisser's uroscopy reflected the chemical associations that Paracelsus had made between failures of chemical digestions in various parts of the body and their morbid consequences. These should be evident in various parts of the urine, which we can see alluded to in his association between the three chemical principles, salt, sulphur, and mercury; the corresponding *hypostasis*, body, and cloud of the urine in the flask; and corporeal or abdominal diseases, spiritual or thoracic diseases, and psychic diseases of the head. This triplicity of threefold corre-

95. Ibid., 4:566–67, Typus I.
96. Ibid., 4:573.

spondences directly reflects Plato's division of the human soul into three parts, which, in turn, gave structure to medieval Galenic physiology.

Thurneisser's approach transcended Paracelsus's in some ways, by subjecting the urine to distillation, making specific correlations between parts of the distillation flask and the human body, and weighing volumetrically various urine fractions, for which he is taken as a pioneer of modern urology.[97] Johann Hayne's approach to a chemical semeiotics of urines is not quantitative; it is in general much more qualitative, in the same way that Paracelsus had suggested in his lectures, and yet clearly reflects his chemical understanding of urine samples and how uroscopy reveals chemical processes in the body. If we return to the third fold-out chart, which is marked "Tabella C" in Hayne's book and which helps us understand how he organized his ideas, we can see that Thurneisser's Neoplatonic division of corporal, spiritual, and psychic diseases is reflected in the three categories of human nutriments: mercurial at the top, sulphurous in the middle, and a group of various excesses and stellar influences that have a commonality in the saline at the bottom.[98] Although the correlation is not exact, the chart indicates a rough correspondence between the distillations and sublimations of subtle mercury and diseases that we can classify as psychological or neurological, traditionally associated with what Galenists called the animal faculty. Separated sulphur impurities are associated with fevers and inflammations, which were tied to the heart and vital faculty in traditional medicine. Finally, Hayne connected the calcination, resolution, reverberation, and coagulation of the saline spirits with tartar diseases, corrosive fluxes and wounds, putrid skin afflictions, and various fluxes of the digestive and reproductive systems, which are mainly located in the abdomen, the seat of the vegetative faculty according to Galenic medicine and the feminine, appetitive, or concupiscent soul according to Plato's

97. He regarded the visual, qualitative inspection of urine—the primary method of medieval uroscopy—to be of limited usefulness, owing to the weakness and unreliability of sight. One ought rather to base a scientific assessment on quantitative measurement of urine and its fractions. See Bleker, "Chemiatrische Vorstellungen und Analogiendenken," 71. Wellcome in *Evolution of Urine Analysis*, 47; Wershub in *Urology*, 135; and Debus in *Chemical Philosophy*, 328, all attribute to van Helmont the earliest determinations of specific quantities for urine from persons differing in age, sex, and condition of health, but I do not see how his treatment differs in kind from that of Thurneisser, as reported by Reusner, *Vrinarum Probationes*, 286–87. In any case, in *Paracelsus*, 198–99, Pagel noted that Nicholas of Cusa (fifteenth century) had already suggested weighing urine for diagnostic purposes. Newman and Principe have recently argued in *Alchemy Tried in the Fire*, esp. chap. 2, that exact quantitative measures were employed by alchemists and iatrochemists well before the "chemical revolution" of the eighteenth century.

98. Hayne, *Drey unterschiedliche newe Tractätlein*, Tabella C, inserted between pages 308 and 309.

Timaeus. This interleaving of ideas from various physiological systems, philosophical schools, and historical eras renders Hayne's medical theory confusing and opaque to the modern reader, but provided a historical continuity that would have made it familiar in various ways to his contemporaries, for whom the preservation of familiar elements of traditional Galenic and astrological medicine would have smoothed the path for the integration of the new chemical ideas of Paracelsus. Therefore, it is not surprising that Paracelsus's teaching of uroscopy would be refined and understood in the context of an eclectic approach to medicine by his followers, among whom were Thurneisser, Hayne, Nolle, Zobell, and even George de Benneville, who carried it with him to the New World.

Historiographical Consequences

The multiple editions of Johann Hayne's *Tractätlein* in seventeenth-century Germany and George de Benneville's adoption of elements of Hayne's system into his eclectic medical manual of the mid-eighteenth century—in places quite literally—hint at the broader significance of Hayne's work in particular, and Paracelsian therapeutics in general, than is apparent from our post-Enlightenment perspective. That perspective, which in large measure shaped twentieth-century history of science and medicine, took a very dim view of astrological medicine and diagnosis by inspection of the urine, regarding them as vestiges of the superstition and irrationality of the medieval past. At the same time, Enlightenment historians celebrated Paracelsus for breaking with medieval Galenism by challenging scholastic medicine and introducing chemical philosophy. According to this traditional historiography, Paracelsus's chemical philosophy was assimilated into traditional Galenic therapeutics. Both gradually were accommodated to early modern corpuscular ideas about material composition and eventually to a materialist conception of physiology, ultimately contributing to the modernization of medicine.[99] Twentieth-century historiography inherited both the idea that Paracelsus contributed to the downfall of medieval medicine and the idea that medieval medicine was characterized by pseudo-scientific uroscopy, which

99. Paracelsus's place in the history of medicine as a reformer of medicine and intermediary between medieval and modern medicine and the founder of iatrochemical medicine is evident already in Sprengel, *Versuch einer pragmatischen Geschichte der Arzneikunde*.

was duly discarded during the course of modernization.[100] Textual evidence that a Paracelsian chemical uroscopy was practiced in seventeenth- and eighteenth-century Europe and America therefore requires some explanation.

Present historiography holds that the mere empirical diagnosis of disease by inspection of the urine, sometimes called *uromancy* or pisse-prophesy to distinguish it from scientific urology—which is a part of the modern medical curriculum—had become the standard practice of the charlatan and quack by the sixteenth century, when it began to be regulated by law and as learned treatises denouncing uroscopy began to appear in print.[101] Petrus Forestus's 1589 book *On the Uncertainty and Fallacy of the Judgements of Urines* is typically invoked as an indicator of a new critical attitude toward a semeiotics of disease based on the inspection of urine.[102] Shakespeare's references to uroscopy, some of them in a farcical vein, suggest that by the early seventeenth century, the seeing of urines was losing credibility among the public, even as it continued to be widely practiced.[103] Even the recent, more nuanced study of the early modern

100. The occasional late practitioner of uroscopy is noted by Garisson, *Introduction to the History of Medicine*, 388; and Desnos, *History of Urology*, 137 (an eighteenth-century Swiss uroscopist named Schuppach), but the general tenor of twentieth-century scholarship agreed with the view that uroscopy was, by the mid-seventeenth century, something of a relic of a pre-scientific age. This is captured by an anecdote attributed to Dr. Radcliffe, founder of the Radcliffe Library at Oxford. When presented with a urine flask by a woman who said it was from her husband, a boot maker, the doctor emptied it out and filled it with his own and bade the woman take it to her husband and ask that he use it to make a pair of boots for him on the basis of its inspection; reported by Wellcome, *Evolution of Urine Analysis*, 51.

101. Like so many foundational narratives in the history of scientific progress, the dismissal of uroscopy was codified during the Enlightenment. In *A New Medical Dictionary, or General Repository of Physic*, Motherby noted under the entry *urina* that "[m]any are the signs which different authors have pointed out in the *urine*, by which to judge of the different states of the body, but Dr. William Hebberden well observes that this excretion affords the practitioner but little useful information," unpaginated. No doubt many English readers of Benjamin Rush's generation agreed with Hebberden. Kiefer, "Uroscopy," 166; and Desnos, *History of Urology*, 137, note that Thomas Linacre, a founder of the Royal College of Physicians in London, formulated statutes to regulate apothecaries' autonomy in prescribing on the basis of uroscopic diagnosis and also physicians' diagnosis and prognosis solely on the basis of examining the urine, without examination of the patient. This is a bit confusing, since Murphy, "Art of Uroscopy," 884, claims that the Royal College of Physicians passed a statute forbidding physicians to give advice on the basis of urine alone in 1601, indicating that similar prohibitions were issued in other parts of Europe. In *Evolution of Urine Analysis*, Wellcome (p. 42) commented that "[i]t will be seen that the art of uroscopy had made little or no progress for three or four hundred years, and its practice about this time had drifted largely into the hands of the quack doctors and charlatans, who travelled the country and preyed on the credulity of the ignorant."

102. Foreest, *De incerto, fallaci urinarum judicio*. Foreest's work is cited as indicative of late fifteenth-century condemnation of diagnosis based solely on uroscopy. See, for example, Kiefer, "Uroscopy," 166; Desnos, *History of Urology*, 137; and Murphy, "Art of Uroscopy," 884.

103. Shakespeare, *Macbeth*, 5.3; *Two Gentlemen of Verona*, 3.1; *Twelfth Night*, 3.4; *King Henry IV, Pt. 2*, 1.2. See also Marlowe, *Tamburlaine, Pt. 2*, 5.3.

decline of uroscopy among learned practitioners by Michael Stolberg persists in this general view: "From the early sixteenth century, uroscopy lost much of the great appeal it had possessed among medieval physicians. Once valued as an outstanding diagnostic tool which ensured authority and fame, it became an object of massive criticism if not derision," but Stolberg acknowledges that "in the end, in spite of the physicians' massive campaign against it, uroscopy remained very much alive" because of the forces of the marketplace.[104] According to this historiography, the use of uroscopy as a specific semeiological tool for diagnosis and prognosis was a distinguishing mark of unsophisticated medical practice by the eighteenth century, surviving only in conservative folk practices and as quaint vestiges of medieval superstition in the occasional medical treatise. Thomas Brian's 1637 publication, *The Pisse-Prophet, or, Certaine Piss-Pot Lectures. Wherein are newly discovered the old fallacies, deceit, and jugling of the Pisse-pot Science, used by all (whether Quacks and Empiricks, or other methodicall Physicians) who pretend knowledge of Diseases, by the Urine, in giving judgement of the same,* exemplified the new attitude.[105] This narrative persists in recent medical historiography, hailing Paracelsus as a harbinger of the new science, rejecting medieval uroscopy as superstitious quackery, and replacing it with quantitative scientific analysis.[106]

However, this traditional historiography does not explain the detailed exposition of uroscopy that Paracelsus presented in his lectures at Basel, vestiges of which were published by Paracelsus's followers in the late sixteenth and early seventeenth centuries and collected by Karl Sudhoff in the early twentieth. Nor does it prepare us to understand the work of Hayne and Thurneisser in the late

104. Stolberg, "Decline of Uroscopy," 313.

105. Publication information is given as London: Printed by E. P. for R. Thrale, 1637. There was also a 1679 edition.

106. Nance, *Turquet de Mayerne as Baroque Physician,* 75–76: "Uroscopy may have been a common practice of the day, but by the early seventeenth century, learned physicians were rejecting it as fraudulent and invalid." Porter, *Greatest Benefit to Mankind,* 232: "Bedside practice changed little. Physicians would . . . feel the pulse and inspect urine samples—though uroscopy lost credit, becoming the trademark of the quackish 'pisse-prophet' and being attacked with 'cozening quacksalvers, women physitians and the like stuffe' by status-anxious physicians such as the Netherlander Pieter van Foreest (1522–1597)." In *History of Urology,* Desnos credited "the revolutionary ideas of Paracelsus" with reorienting medicine away from dependence on uroscopy and toward increased "attention to the symptoms and examination of the patient," which led to a declining importance of uroscopy in the seventeenth century (137). Desnos acknowledged that uroscopy was not entirely abandoned in the eighteenth century, pointing to the popularity of de la Rivière's *Mirror of Urines,* but noted that "after this period, uroscopists were nothing but charlatans." In *Urology,* Wershub (p. 135) argued that "[u]ntil nearly the end of the seventeenth century, the diagnosis of disease was accomplished by visual examination of the urine (uroscopy)."

sixteenth century and the apparent influence of their kind of medicine in seventeenth- and eighteenth-century therapeutics: namely, an approach to practical medicine that was based squarely on Paracelsian medical and chemical ideas and depended on Paracelsus's uroscopy as a diagnostic system.[107]

Michael Stolberg offers a perspective that begins to contextualize the persistence of uroscopy in Paracelsian praxis. He argues that sixteenth- and seventeenth-century Galenist physicians sought to distance themselves from uroscopy, owing to a growing awareness of its unreliability and its stagnation as a developing scientific approach to diagnosis. To support this conclusion, he points to his failure to identify a single Galenist physician in printed editions after 1500 "who chose to be depicted as holding a urine glass in his hand" and concludes that this "iconographic evidence . . . suggests a drastic revaluation of uroscopy among learned physicians."[108] The qualifiers in his finding deserve emphasis, in part because, as I have shown here, the chemical physicians who incorporated Paracelsus's diagnostic uroscopy into their practices are exceptions that help to validate Stolberg's basic conclusions. Although Johann Hayne was long dead when the third edition of his *Drey underschiedliche newe Tractätlein* was printed in 1683, its publisher saw an advantage to having the author pictured on the frontispiece holding a *matula*. This suggests that while uroscopy may have been seen increasingly as a liability to learned Galenists, as Stolberg claims, eclectic practitioners like Johann Hayne, George de Benneville, and an unknown number of users of the uroscopy manuals of Reusner and Martinius saw uroscopy in a more positive light. The fact that the authors presented here are closely associated with a culturally German medical marketplace, including de Benneville's practice among eighteenth-century Pennsylvanian Germans, provides additional evidence to support Stolberg's conclusion that rejection of uroscopy was most vociferous among German-speaking writers, owing perhaps to the late prevalence of uroscopic practice among lay practitioners in

107. The direct influence of Paracelsus's lecture notes on the work of Hayne, Thurneisser zum Thurn, Nolle, and others is suggested not only by the similarity of ideas and expressions, but (for Hayne) also by Georg Faber's 1620 *Dedicatoria* to Hayne's book, fol. iii^v–iv^r, which identifies Hayne's *Tractätlein* as based on the author's personal experience *and* the writings of Paracelsus: "Wann dann / Gnädiger Fürst und Herr / vor etlichen Jahren der Hochgelarte D. Johann Hayne / weyland Fürstl. Hoff *Medicus* zu Stetin in Pommern / in beyder Medicin ein berümbter / geubter und erfahrner Mann etliche dergleichen seine Chymische Tractätlein so wol auss seiner selbst eygnen langwierigen *praxi* und *observationibus*: als auch auss obermeldten Theophrasti Schrifften sein kurtz und rund zusammen gezogen."

108. Stolberg, "Decline of Uroscopy," 321.

the German cultural area.[109] If this view is correct, then further attention to the manuscript records of lay healers and to neglected printed volumes should bear out the perception that eclectic practitioners—who retained older therapeutic practices that were in popular demand and who perhaps adapted them to different theoretical systems to suit local belief systems—served post-Reformation German communities that sought a measure of isolation from state authority and the academic medicine that went with it.

Works Cited

Bleker, Johanna. "Chemiatrische Vorstellungen und Analogiedenken in der Harndiagnostik Leonhart Thurneissers (1571 und 1576)." *Sudhoffs Archiv* 60 (1976): 66–75.

Debus, Allen G. *The Chemical Philosophy: Paracelsian Science and Medicine in the Sixteenth and Seventeenth Centuries.* 2 vols. New York: Science History Publications, 1977. Reprint, Mineola, NY: Dover, 2002.

———. *The English Paracelsians.* London: Oldbourne, 1965.

———. "The Paracelsian Compromise in Elizabethan England." *Ambix* 8 (1960): 71–97.

de la Rivière, Davach. *The Mirror of Urines.* Paris, 1696.

Desnos, Ernest. *The History of Urology.* Translated and edited by Leonard J. T. Murphy. Springfield, IL: Thomas, 1972. Originally published as *Histoire de l'urologie* (Paris: Doin, 1914).

Foreest, Pieter van. *De incerto, fallaci urinarum judicio, quo uromantes, ad perniciem multorum ægrotantium, utuntur: & qualia illi sint observanda, tum præstanda, qui recte de urinis sit judicaturus, libri tres.* Leiden: Plantin, 1589.

French, Roger. "Astrology in Medical Practice." In *Practical Medicine from Salerno to the Black Death,* edited by Luis Garcia-Ballester, Roger French, Jon Arrizabalaga, and Andrew Cunningham, 30–59. Cambridge: Cambridge University Press, 1994.

———. *Canonical Medicine: Gentile da Foligno and Scholasticism.* Leiden: Brill, 2001.

Garisson, Fielding. *An Introduction to the History of Medicine.* 4th ed. Philadelphia: W. B. Saunders, 1929.

Goodrick-Clarke, Nicholas, ed. and trans. *Paracelsus: Essential Readings.* Wellingborough, UK: Crucible, 1990.

109. Ibid., 336.

Hamand, Henry. *Ourography or Speculations on the excrements of urine: with the distinctions, causes, colours, and contents thereof: and other symptomes observed in nature. Also, a philosophicall discourse of the colours of urine, with the art of mixing them, according to quantity, number, and weight.* London: printed by R. D. for Francis Eglesfield, 1656.

Hansen, Jan-Erik Ebbestad. *Jacob Böhme: Liv, tenkning, idéhistoriske forutsetninger.* Oslo: Solum forlag, 1985.

Hayne, Johann. *Drey unterschiedliche newe Tractätlein, Deren Erstes von astralischen Kranckheiten, . . . Das Andere, von tartarischen Kranckheiten . . . Das Dritte, begreifft in sich das Fundament . . . wie man die Urinen des Menschen . . . künstlich iudiciren und erkennen möge.* Frankfurt am Main: Paul Jacob and Johan Dreutels, 1620.

———. *Trifolium medicum, oder: Drey höchst nützliche Tractätlein, Deren Erstes von astralischen Kranckheiten . . . Das Andere, von tartarischen Kranckheiten . . . Das Dritte von den rechten Fundament und Grund wie man die Urinen . . . erkennen möge.* Frankfurt am Main: George Heinrich Oehrling, 1683 and 1700.

Hohlweg, Armin. "John Actuarius' *De methodo medendi*—on the New Edition." *Dumbarton Oaks Papers* 38 (1984): 121–33.

Jacquart, Danielle. "Theory, Everyday Practice, and Three Fifteenth-Century Physicians." *Osiris* 6 (1990): 140–60.

Jones, Peter Murray. *Medieval Medical Miniatures.* London: British Library, 1984.

Kiefer, Joseph H. "Uroscopy: The Clinical Laboratory of the Past." *Transactions of the American Association of Genito-Urinary Surgeons* 50 (1958): 161–72.

Martinius, Henricus. *In nomine archiatri summi Henrici Martinii Dantiscani Doctoris Medicinæ Anatomia Urinæ Galeno-Spagyrica. Ex Doctrina Hippocratis & Galeni nec non Recentiorum, imprimis Theophrasti Paracelsi, & Leonhardi Thurnheuseri, aliorumque Chymiatrorum Principium Scriptis adornata. Cui accesit Eusd. Ars Pronuntiandi ex Urinis tam rationalis quam mechanica. Et Cæsaris Odoni de urinis libellus posthumus.* Frankfurt: Sumpt. Georgius Fickwirtus, 1658.

Martinón-Torres, Marcos. "The Tools of the Chymist: Archeological and Scientific Analyses of Early Modern Laboratories." In *Chymists and Chymistry: Studies in the History of Alchemy and Early Modern Chemistry*, edited by Lawrence M. Principe, 149–63. Sagamore Beach, MA: Chemical Heritage Foundation and Science History Publications, 2007.

Morys, Peter. *Medizin und Pharmazie in der Kosmologie Leonhard Thurneissers zum Thurn (1531–1596).* Geschichte der Medizin und der Naturwissenschaften 43. Husum: Matthiesen, 1982.

Motherby, George. *A New Medical Dictionary, or General Repository of Physic.* London: Printed for J. Johnson, St. Paul's Church-Yard, 1785.

Murphy, Leonard J. T. "The Art of Uroscopy." *Medical Journal of Australia* 2 (1967): 879–86.

Nance, Brian. *Turquet de Mayerne as Baroque Physician: The Art of Medical Portraiture.* Amsterdam: Rodopi, 2001.

Newman, William R., and Lawrence M. Principe. *Alchemy Tried in the Fire: Starkey, Boyle, and the Fate of Helmontian Chymistry*. Chicago: University of Chicago Press, 2002.

Nolle, Heinrich. *Hermetical Physick: Or, The Right Way to Preserve, and to Restore Health.* Translated by Henry Vaughn. London: Moseley, 1655.

———. *Systema medicinæ hermeticæ generale in quo I. medicinæ veræ fundamentum, II sanitatis conservatio, III morborum cognitio, & curatio methodo explicantur.* Frankfurt, 1613.

Nummedal, Tara. *Alchemy and Authority in the Holy Roman Empire.* Chicago: University of Chicago Press, 2007.

Pagel, Walter. *Das medizinische Weltbild des Paracelsus: Seine Zusammenhänge mit Neuplatonismus und Gnosis.* Wiesbaden: Franz Steiner Verlag, 1962.

———. *Paracelsus: An Introduction to Philosophical Medicine in the Era of the Renaissance.* 2nd rev. ed. Basel: Karger, 1982.

Paracelsus. *Sämtliche Werke I: Medizinische, naturwissenschaftliche und philosophische Schriften.* Edited by Karl Sudhoff. 14 vols. Munich: R. Oldenbourg, 1922–33.

———. *Volumen medicinæ paramirum.* Edited and translated by Kurt F. Leidecker. Supplements to the Bulletin of the History of Medicine 11. Baltimore: Johns Hopkins University Press, 1949.

Porter, Roy. *The Greatest Benefit to Mankind: A Medical History of Humanity*. New York: Norton, 1999.

Pumfrey, Stephen. "The Spagyric Art; Or, The Impossible Work of Separating Pure from Impure Paracelsianism: A Historiographical Analysis." In *Paracelsus: The Man and His Reputation, His Ideas and Their Transformation*, edited by Ole Grell, 21–51. Leiden: Brill, 1998.

Reusner, Hieronymus. *Vrinarum Probationes D. Iodoci Wilichii Reselliani: Illustratæ Scholis Medicis, Hieronymi Reusneri Leorini D. Med. in quibus principia solidæ Vroscopiæ, ad solidæ Philosophiæ fontes reuocantur: multique medicorum errores deteguntur. His accessere variæ matularum delineationes: atque genuini urinarum colores. Remedia item plurima ex urina desumpta: maxima uero ex parte Chemica.* Basel: Sebastian Henricpetri, 1582.

Ross, Malcolm Mackenzie. "George Herbert and the Humanist Tradition." *University of Toronto Quarterly* 16 (1947): 169–82.

Sarton, George. *Introduction to the History of Science*. Vol. 3, pt. 1. Baltimore: Williams and Wilkins, 1947.

Shackelford, Jole. "Paracelsian Uroscopy and German Chemiatric Medicine in the *Medicina Pensylvania* of George de Benneville." In *Medical Theory and Therapeutic Practice in the Eighteenth Century: A Transatlantic Perspective*, edited by Jürgen Helm and Renate Wilson, 13–35. Stuttgart: Franz Steiner Verlag, 2008.

Siraisi, Nancy. *Medieval and Early Renaissance Medicine*. Chicago: University of Chicago Press, 1990.

Smith, Pamela. *The Business of Alchemy: Science and Culture in the Holy Roman Empire*. Princeton, NJ: Princeton University Press, 1994.

Soukup, R. Werner. "Crucibles, Cupels, Cucurbits: Recent Results of Research on Paracelsian Alchemy in Austria around 1600." In *Chymists and Chymistry: Studies in the History of Alchemy and Early Modern Chemistry*, edited by Lawrence M. Principe, 165–72. Sagamore Beach, MA: Chemical Heritage Foundation and Science History Publications, 2007.

Sprengel, Kurt Polycarp Joachim. *Versuch einer pragmatischen Geschichte der Arzneikunde*. Vol. 3. Halle: Johann Jacob Gebauer, 1794.

Stein, Claudia L'Engle. "The Meaning of Signs: Diagnosing the French Pox in Early Modern Augsburg." *Bulletin of the History of Medicine* 80 (2006): 617–48.

Stolberg, Michael. "The Decline of Uroscopy in Early Modern Learned Medicine (1500–1650)." *Early Science and Medicine* 12 (2007): 313–36.

Thurneisser, Leonhard. *Bebaiosis agonismou. Das ist Confirmatio Concertationis oder ein Bestettigung . . . die Neuwe und vor unerhörte erfindung der aller Nützlichesten und Menschlichem geschlecht der Notturftigesten kunst dess Harnnprobirens ein zeitlang gewest ist. Welcher kunst Grundt und Fundament hierin durch den Inuentorem Leonharten Thurneisser zum Thurn*. Berlin: Grauwen Closter, 1576.

———. *Pison*. Frankfurt an der Oder: Johan Eichhorn, 1572.

———. *Προκατάληψις* [*Prokatalepsis*]. *Oder Præoccupatio, Durch zwölff verscheidenlicher Tractaten, gemachter Harm Proben*. Frankfurt an der Oder: Johan Eichorn, 1571.

Wallis, Faith. "Inventing Diagnosis: Theophilus' *De urinis* in the Classroom." *Dynamis* 20 (2000): 31–73.

———. "Signs and Senses: Diagnosis and Prognosis in Early Medieval Pulse and Urine Texts." *Social History of Medicine* 13 (2000): 265–78.

Weeks, Andrew. *Boehme: An Intellectual Biography of the Seventeenth-Century Philosopher and Mystic*. Albany: State University of New York Press, 1991.

———. *Paracelsus: Speculative Theory and the Crisis of the Early Reformation*. Albany: State University of New York Press, 1997.

———, ed. *Paracelsus (Theophrastus Bombastus von Hohenheim 1493–1541): Essential Theoretical Writings*. Leiden: Brill, 2008.

Wellcome, Henry S. *The Evolution of Urine Analysis: An Historical Sketch of the Clinical Examination of Urine*. Lecture Memoranda, American Medical Association Los Angeles, 1911. London: Borroughs Wellcome, 1911.

Wershub, Leonard Paul. *Urology: From Antiquity to the 20th Century*. St. Louis, MO: Warren H. Green, 1970.

Willichius, Jodocus. *Commentarivs Anatomicus, in quo est omnium partium corporis humani diligens enumeration*. Strassburg: Mylius, 1544.

———. *Ivditia Vrinarvm Doctoris Iodoci Vuilichij Medicosi doctissimi & celeberrimi*. Wittenberg: Creutzer, 1560.

———. *Physiognomonica: Add. est eiusdem interpretis oratio in laudem Physiognomoniæ*. Wittenberg: Schirlentz, 1538.

———. *Wie man sich in einer Stadt für der Pestilenz behüten soll und möchte*. Frankfurt: Johannes Eichorn, 1549.

———. *Wie man sich vorhalten und bewaren sol in den Heuseren, in welchen iemandes an der Pestilentz gestorben ist*. Frankfurt: Johannes Eichorn, 1550.

Zobell, Friedrich. *Chymische Medicinische Perle, Darinn herrliche, köstliche Medicamenta zufinden, Nebst deren gründlichem Bericht und ausführl. Ursachen derer Kranckheiten, auch deren Curation nach spagyrischer Arth und Wiese, Durch Communication* Matthæi Pauli, Med D. *zum Druck befördert*. Dresden: Michael Günther, 1686.

Chapter 3

Andreas Libavius and the Art of *Chymia*

Words, Works, Precepts, and Social Practices[*]

Bruce T. Moran

When composing letters intended for publication addressed to Zacharius Brendel (the elder) (1553–1626), the sometimes-cantankerous schoolteacher, alchemical author, and physician Andreas Libavius (d. 1616) adopted an attitude of notable humility and respect. After all, Brendel was at that time a professor of philosophy and medicine at the University of Jena. Libavius, by contrast, was a city physician and school superintendent in the more obscure town of Rothenburg ob der Tauber. Libavius recognized the social distance that separated him from his correspondent, and in his letters to Brendel, as well as to others, he began to focus as much upon the social world as upon the material and textual worlds in thinking about the necessary practices that would give artistic shape to the formless business that many people called *chymia*.

*I wish to express my gratitude to Dr. J. Mark Sugars for help with difficult Libavian passages in Latin and Greek.

Brendel, he acknowledged, was an outstanding philosopher who had expanded medical knowledge and had improved the study of chemical medicines. In letters to the Jena professor, part of a publication project in which the letters may or may not have actually been sent, Libavius was clear about his purpose. He wanted to start a discussion about chemical essences, and, especially, to determine more precisely to what the term referred. Brendel, he thought, could offer insight, but he was adamant that a discussion of essences should not be a conversation about metaphysics. A chemical essence, he insisted, had to be discussed as something "chemical," that is, as something in the physical, elementary world. "I keep myself in the chemical marketplace," and "I do not wander about abstractly, nor am I tricked by a concern for phantasms and visions,"[1] he observed. Libavius intended to keep the focus upon the terrestrial realm, moving nature by means of nature, not by means of what was beyond nature,[2] as did some "chymists," who considered essences to be tied to the heavens. At the same time, however, he rejected the popular, street-vendor opinions of those he called "parachymists," who called anything that was only slightly altered an "essence," including the waters, oils, powders, and salts that they then offered for sale. Clearly, some chymists were different from others, and Libavius looked to Brendel to help sort out not only what a chemical essence was but also what the essence of the art of true *chymia* ought to be.[3]

Multiple Uses, Multiple Meanings

Sometimes we are too quick to impose general meanings upon the terms we use. The early modern era referred to *chymia, alchymia, Hermetica, physica,* and *magia* in various ways dependent upon specific contexts and accepted views of nature. Competing definitions within a mixture of contexts have especially made identifying *chymia* the kind of effort that requires hitting a moving target.[4] Both Libavius and the Paracelsian writer and publisher of Paracelsian

1. "In foro chymico me contineo" and "Abstracte non vagor neque phantasmatum et specierum cura tangor"; Libavius, *Rerum chymicarum . . . liber primus,* 92–93.

2. "Naturam mouet per naturam, sicut agricola et hortulanus ex pomo producit pyra"; ibid., 31.

3. Ibid., "Epist. IX. De Essentiae vocabulo," 92–97. See also Moran, "Essences and Mostly Clean Hands."

4. See Park and Daston, "Introduction: Age of the New." On further historiographic problems, see Newman and Principe, "Alchemy vs. Chemistry"; and Principe and Newman, "Some Problems with the Historiography of Alchemy."

texts Gerhard Dorn (ca. 1530–84), for instance, thought that *chymia* was the means by which hidden powers could be extracted or separated from material substances, and both employed similar procedures, like distillation and sublimation, in their specific operations. Nevertheless, they could not have been further apart in defining what those hidden powers were, where they came from, and what chemical knowledge was actually the knowledge of. And it is not just the word *chymia* that was understood so differently. Alchemy too had its various definitions. Some had assigned the art of *alchymia* solely to the preparation and transmutation of metals. Some thought of it in terms of making medicines. Some limited it further to making a single medicine, an elixir. For some, it was *halchymia,* or the knowledge of preparing (mineral) salts.[5] Some thought of alchemy in purely mystical terms, following Hermes. Some, like Girolamo Cardano (1501–76) and Hieronymus Rubeus (that is, Girolamo Rossi, ca. 1539–1607), wanted to separate the art of distillation from alchemy. Others, like the medieval thinkers Geber and Avicenna, thought that alchemy produced waters and oils by means of distillation and sublimated a mercurial elixir for purposes of transmutation. Interpretive traditions also varied. Some followed Aristotle; some also connected their ideas to medieval authors. Some thought that Paracelsus (1493–1541) was alchemy's true inventor.

As he made clear in one of his most imposing texts, Libavius considered *alchymia* to be a reference to an entire art, part of a category of knowledge called *ergalikos,* which offered explanations or precepts for things taken in hand (*egcheireseon*). Its practices included the extraction of *magisteria* and essences and required proficiency in the use of instruments and procedural know-how.[6] In its entirety, the art of *alchymia* comprised two parts. One part was *manuarius,* related to those things belonging to the hand, which were mechanic. The other part he called *ousiodes,* or essential. The former described specific kinds of physical labors, while the latter was both didactic, explanatory, and methodical, laying open to the eyes, and by means of precepts also to the mind, the rationale for special processes, natural powers, and varieties of practice.[7] *Alchymia*

5. Libavius, *Syntagmatis . . . tomus primus,* 4.

6. Ibid., 2. Müller-Jahncke, "Andreas Libavius im Lichte der Geschichte der Chemie"; Debus, *Chemical Philosophy,* 169ff.; Debus, "Guintherius, Libavius, and Sennert." See also Meitzner, *Die Gerätschaft der chymischen Kunst;* Newman, "Alchemical Symbolism and Concealment"; Newman, *Atoms and Alchemy,* 68ff.; Moran, *Andreas Libavius and the Transformation of Alchemy;* Moran, "Less Well-Known Libavius: Spirits, Powers, and Metaphors in the Practice of Knowing Nature"; and Forshaw, "'Paradoxes, Absurdities, and Madness.'"

7. Libavius, *Syntagmatis . . . tomus primus,* 3.

Figure 3.1: Detail from title page of Andreas Libavius, *Syntagma selectorum undiquaque et perspicue traditorum alchymiae arcanorum [tomus primus]* (Frankfurt, 1611–13). L0045181, Wellcome Library, London.

revealed what was hidden in mixed things. It extracted essences, collected them, and made them pure and stable. Whatever, then, had been drawn out into the open from the hiding places of nature, whether from substances or from qualities, could be attributed to *alchymia,* as long as it took place by means of chymical (and Libavius meant physical as opposed to metaphysical or divine) *artificii.*[8]

Chymia also suffered from multiple uses and meanings. Paracelsians, Libavius believed, had corrupted the term, making it signify secret wisdom and practices seemingly compelling but methodically haphazard. Paracelsus had tried to seize preeminence in the art for himself. But while laying down precepts about synthesizing *magisteria* and quintessences in his book *Archidoxa,* there were no precepts of general operations from which the methods of special operations could be derived. "What good is it if an oil is ordered to be prepared by distillation, if you do not know what the method of distillation is?"[9] How was one able to transfer the procedures of one operation to another without understanding the consistencies of the natural world? "It seems to me that those people are simply teaching disputation who drop into [the discussion] something

8. Ibid., 3nC.

9. "Quid prodest iuberi destillatione oleum conficere. Si ignores quae sit destillandi ratio?"; Libavius, *Rerum chymicarum . . . liber primus,* 114.

that is axiomatically true or false on the basis of their assessment of the nature of things before they have explained the permanent condition of nature herself, since as long as this is hidden, the judgment of what is true is necessarily [also] hidden."[10]

At times, *alchymia* and *chymia* seem so closely related in Libavius's discussions as to appear identical. Writing of the etymology of the word *chymia* ("notatione chymiae"), Libavius noted that the *chymici* were once the most excellent philosophers who mostly sought out what was hidden in nature. Their art was called *alchymia* by the Arabs and Egyptians, "who, as everyone agrees, got that name from the Greeks."[11] At other times, however, *chymia* appears to have a more specific meaning. Although the word is difficult to define, Libavius said that "in my opinion it should be defined as the art [*ars*] or knowledge [*scientia*] of elaborating, through the separation of things inwardly mixed together, the essences and parts of essences of substances concrete in themselves . . . and of bringing them to the purest and most efficacious state, or effecting the absolute best in the same species." Those who thought of *chymia* as knowledge of distilling or of transmuting metals offered, in this regard, only an imperfect definition. These were only particular operations of the art, not the art itself.[12] *Chymia*, as the art of elaboration, brought out what was best in nature. It excelled nature when it rendered things more suitable for use "which the spirit of vitriol bears witness, which everyone knows is a remedy helpful to the stomach, when the stomach is seized by putrid humors, or is burdened by nausea, . . . and yet undiminished vitriol is very harmful to it. Thus, arsenic is a poison, but once elaborated by the art [of *chymia*] it unlearns its harmfulness [*noxam dediscit*] so that it can be safely applied to wounds."[13]

While helpful to medicine, *chymia* was not a category of medicine. Rather, it was part of the proper vocation of natural philosophers. Physicians needed

10. "Ita videntur mihi dialecticam docere, qui instillant axiomatum veritatem et falsitatem ex rerum natura aestimandam, prius quam ipsam naturae affectionem, qua latente iudicium veri latere necesse est, explicarint"; ibid., 115.

11. "Unde Alchymia dicta Arabibus et Ægyptiis, quos a Graecis consentaneum est in nominis accepisse"; ibid., 82–83.

12. Ibid., 86–87. For a discussion of the cultures of *chymia* and *alchemia* as Libavius viewed them, see Moran, *Andreas Libavius and the Transformation of Alchemy*.

13. "veluti spiritus vitrioli contestatur, quem constat ventriculo gratum esse remedium, ubi a putridis humoribus occupatur, aut nausea, atonia et aliis quibusdam gravatur; atqui vitriolum integrum illi est infestissimum. Ita arsenicum venenum est, at arte elaboratum noxam dediscit, ut ad vulnera posit tuto adhiberi"; Libavius, *Rerum chymicarum . . . liber primus*, 88.

to study it, however, as *physica*. At one point, indeed, Libavius said that *chymia* is *physica*. Granted, those physicians who were occupied with the calamities of the sick, and thus unable to spend time and energy contemplating natural philosophy, could be excused from the omission, just as long as they did not forget what they had learned in their years of schooling. They needed to be able to instruct the apothecary and make good judgments concerning the works of *chymia*. However, "those who taught at leisure [*cum otio*] in the universities I would not set free of the charge of neglecting nature. For *physici* [university medical professors] to this extent ought not to be such by contemplation and doctrine only, but also by exercise."[14] *Alchymia* and *chymia* shared the same precepts of practice, but *chymia* was a didactic discipline, something that could be taught as part of natural philosophy within the university.[15]

Chymia also possessed a certain patrimony, a social and intellectual history that connected it to the most ancient knowledge and peoples and separated it from the views of *recentiores* or moderns. With the exception of theology, Libavius observed, no art was older.[16] The biblical Tubal-Cain, as the discoverer of metallurgy, was its true founder. Its first appearance, Libavius believed, was in ancient Babylonia and Chaldea, and its arrival in Egypt after the Mosaic flood made organized society possible. "Then, as it were, human beings began to be born and brought into the more admirable form of a republic and an empire."[17] Things produced by means of *chymia* gave structure to social life, and the correct practice of the art of *chymia*, he would argue, required abiding by specific social norms.

Making Does Not Make an Art

Not all had insisted upon such a patrimony, or even that *chymia*, as an art, needed to possess any sort of existence apart from the manual crafts. Girolamo Cardano, for instance, had very few words to describe *chymia*, but he knew what it was when he saw it. He simply referred to the "chymical art" (chymistica ars) as the

14. "Qui vero cum otio in academiis docet; eos crimine neglectae naturae non liberem. Physici enim illi non contemplatione et doctrina duntaxat esse debent"; ibid., 52.

15. On didacticism and early chemistry, see Hannaway, *Chemists and the Word*.

16. Libavius, *Rerum chymicarum . . . liber primus*, 108.

17. "et tunc quasi homines inciperent nasci, inque formam Reipubl. Et imperii ornatiorem redigi . . ."; ibid., 110; Epist. XII. De constitutione artis chymiae, 107–16, esp. 109–11.

knowledge and skill involved in making things. Some of those things, he noted, were admirable, some worthless, some dubious, some beautiful, some healthful, some efficacious, and some almost divine.[18] His, however, was a purely empirical definition. And Libavius observed that Cardano never reflected upon the nature of the art itself or upon what the art of *chymia* was per se. *Chymia*, in the sense that Cardano had used the term, was an umbrella reference for a variety of artisanal creations and performances. Examples included stretching glass into long strands, interweaving glass with white threads,[19] hardening glass, softening glass without fire, making false gems (from glass), engraving or etching images into glass, making artificial amber, mixing and altering metals, making white gold (*electrum*), refining metals, making vessels from molten metals, purifying camphor, producing waters and oils by means of circulations and extractions involving alcohol, making and using vessels for use in distillation, bleaching silk and whitening flowers by means of sulfuric vapors, making true purple, making hard stones, making waters for dissolving gold and silver, and composing the finest waters for penetrating spaces (*dimensiones*) judged impossible by nature. By means of their art, Cardano's chymists produced saleable goods, softening horns and bone to make handles and sword hilts, making inks, cosmetics, combs, cases, and other containers.[20]

Libavius did not doubt that these things were valuable, often beautiful, and required artisanal skill to make, but, he asked, by what instruction and by what method were the processes of making things to be communicated? In Cardano's view, the "art" of *chymia*, since it was the art of making things, extended even to those who knew how to burn charcoal. This was a definition by way of attractive artifact, manual dexterity, and popular opinion. But such, Libavius lamented in another place, was the depraved way of the world. For ages, the artless had passed judgment on the arts.[21] The *ars chymistica* needed principles to follow,

18. Cardano, *Hieronymi Cardani mediolanensis medici*, fols. 269v–271r.

19. Cardano referred here to the production of filigrano or filigree glass, in which white canes of lattimo (or milk) glass, made by adding tin oxide to glass, were pressed into hot glass to make various patterns, usually stripes, spirals, or twists. See McCray, *Glassmaking in Renaissance Venice*, 124–25. See also Baretta, *Alchemy of Glass*.

20. Cardano, *Hieronymi Cardani mediolanensis medici*, fols. 269v–271r. For alchemy in commercial and industrial contexts, see Nummedal, "Practical Alchemy and Commercial Exchange." See also Smith, *Business of Alchemy*. For alchemical procedures in the context of an "artisanal epistemology," see Smith, *Body of the Artisan*.

21. "Est ille quidem seculorum mos deprauatus, ut de artibus non iudicent nisi inertes"; Libavius, *Rerum chymicarum . . . liber primus*, 51.

principles that allowed for the extension of general knowledge to a variety of specific undertakings. For principles, he explained, were "like a mother to other arts, since from her breast come forth things almost innumerable which then withdraw into unique companies," so that while the bronze worker pursued his own study, he had originally received his tenets from *chymia*. This is the same, Libavius continued, in regard to other things as well, so that if the teacher of *chymia* would wish to change professions and work as an artificer and metallurgist, he could, just as a physician could, following the principles of medicine, practice surgery.[22]

That was not how *chymia* was encountered on the street by people who made things to sell in the marketplace, however. There, the term gained meaning by means of utility and as a result of what people bought and sold. Market girls and unguent sellers flaunted their waters and extracts, and the popular view, Libavius complained, attributed more of the art to little old women than to any philosopher or doctor. And well it might, he added, since in comparison to the genuine products one found in the marketplace, the medicine boxes of physicians contained products that were decidedly inferior (*posteriores*).[23] For some, then, *chymia* was a kind of local, even domestic, talent, but for Libavius, there was a difference between this type of homespun, trial-and-error-driven aptitude and the practices of those who possessed expertise. A real art had nothing to do with luck, or accident (*fortuna*), or an aptitude for technique. Someone who stitched together a shoe, he argued, could not be considered an *artifex* unless he understood the principles and causes for doing what he did.[24] It was through a knowledge of precepts and axioms that *chymia* became worthy of a liberal person, and this included, he proclaimed, many expert artisans (*solertes*) who had gained the support of flourishing republics, as anyone could see, he added intriguingly, who lived near England or had Denmark in view.[25]

Clearly, artisans and philosophers had a great deal to learn from one another. But there were limitations in the methods of each. On the one hand, Libavius admitted, the precepts of *chymia* had to be discovered "by practice and in imitation

22. "Vi principiorum multis aliis artib. Quasi mater est, cum ex eius sinu fere innumerabiles prodierint, quarum quaelibet in singulares factiones secessit"; Libavius, *Syntagmatis . . . tomus primus*, 4.

23. Libavius, *Rerum chymicarum . . . liber primus*, 51.

24. Libavius, *Rerum chymicarum . . . liber secundus* (1595), 524.

25. Libavius, *Rerum chymicarum . . . liber primus*, 51–52. For a recent discussion concerning the idea of expertise, see Ash, "Introduction: Expertise and the Early Modern State." See also Ash, *Power, Knowledge and Expertise in Elizabethan England*, esp. 19–54 (chapter on "German Miners").

of craftsmen." But not everything could be communicated through the eyes, just as not everything could be explained in words. One needed both, words and eyes, in order to uncover what was eternal and generic in any art. Words always needed to serve things, for words, Libavius was fond of saying, "are later than things" (rebus enim posteriora sunt verba).[26] Nevertheless, once created, words circumscribed the arts by means of axioms.[27] Those who followed the chymical art on the basis of axioms knew its limits. *Chymia* was bounded by both precepts and nature. By controlling the fire, the chymist removed the chains that had barred entrance to the innermost substances of things. When this was done, the art ceased.[28]

Natural vs. Supernatural *Chymia*

Among the storehouse of problems that affected defining the art of *chymia* was the problem of how the subject should be learned. Some thought that *chymia* was a category of the mind, knowledge of which rested essentially upon revelation rather than strict rational inquiry. This was the view of those who followed Paracelsus, and, in this regard, Libavius pointed especially to what Alexander von Suchten (1520–75) had to say in his *De secretis antimonii liber unus*. There, von Suchten referred to alchemy as "a pure and eternal virgin" who would not let a "rational man" (hominem rationale) approach her. She would embrace only a "man of mind" (hominem mentalem). For, von Suchten said, "reason seduces and cannot exist without being led astray since it depends a great deal upon a specter." Only in mind was there "true intelligence" (intelligentia vera). "We should follow this, not that [called] rational."[29] If one accepted von Suchten's view, the *chymicus*, Libavius commented, would simply proclaim oracles. Was *chymia* to be a category of the divine or supernatural? Some thought so, and Libavius referred to Gerhard Dorn as one who speculated about the practice of the supernatural.[30] Indeed, Dorn wrote extensively concerning supernatural practices (*de artificio supernaturali*), which he identified both as a kind of

26. Libavius, *Syntagmatis . . . tomus primus*, 3.

27. Libavius, *Rerum chymicarum . . . liber secundus* (1595), 4.

28. Libavius, *Rerum chymicarum . . . liber primus*, 30–31.

29. "Die Alchimia ist ein reine und ewige jungfraw / lest keinen hominem rationale zu ir / sie will hominem mentalem haben. . . . Niemand lass sich sein rationem verfüren / sie kan nicht sein jrzung / dan ir hangt ser vil vom Idolo an: in mente ist intelligentis vera, der sollen wir folden / nit rationi"; Suchten et al., *De secretis antimonii liber unus*, 62.

30. Libavius, *Syntagmatis . . . tomus primus*, 7.

chemical separation of the elements and as a metaphysical operation.[31] Common separation, he argued, consisted of separating qualities. The *artificii supernaturalis* surpassed this art, separating a "heaven" from the impure elements. It thus fabricated the "Chaos" out of which physical substances emerged. In doing this, the artist not only broke the bounds of nature, but surpassed them. Supernatural artisanry began where the artistry of nature left off. By means of supernatural craftsmanship, the practitioner recreated a chaos identical with the initial Chaos out of which God separated the heavens and the elements. Dorn thought of the operation as within nature, but called it *metaphysica*. It was precisely that, Libavius complained, metaphysics, and belonged not just above nature but outside nature, a category of the mind maybe, but in practice reserved as an act of God.[32]

"Now you may understand," Dorn announced, "why we say that our metaphysic is superior to and surpasses the work of nature. For the separation of heaven from the elements, or the first separation of the general chaos, was not a natural, but rather a divine, work of art." This was the only way, Dorn proclaimed, to make the universal and supernatural medicine that could heal every illness without distinction. To make the *Quinta Essentia* meant engaging with a metaphysical chaos.[33] The true chymist, Libavius averred, knowing the limits of his art, "moves nature by means of nature."[34] Dorn and others, on the other hand, moved nature by means of what was above nature. Dorn had created fantasies, Libavius commented, that he had dreamed either at the summit of heaven or had grasped in a Platonic cave. Many more such illusions had been proposed by Paracelsus, "an inventor of marvelous dreams."[35] Michael Toxites (1514–81), like Dorn an editor and interpreter of Paracelsus's writings, also turned the *chymicus* into a prophet. Paracelsus's writings, Toxites announced, were not tossed together from other books but were based in nature and understood "by means of true, natural magic which is a mother of all hidden things . . . given by god to

31. Dorn, "De artificio supernaturali."

32. On distinctions between what counts as supernatural and preternatural, compare Daston, "Preternatural Philosophy."

33. "Jetzund wirstu meines erachtens verstehen mögen / warumb wir sagen das unser Metaphysic der Naturwerck superier und ubertreffe. Dann die separation und scheidung des Himmels von den Elementen / oder die erste absönderung des allgemeinen Chaos, ist kein Natürlich / sonder Göttlich Kunstwerck gewesen"; Dorn, "De artificio supernaturali," 271.

34. "Naturam mouet per naturam"; Libavius, *Rerum chymicarum . . . liber primus*, 31.

35. "mirificum somniorum inventor"; Libavius, *Syntagmatis . . . tomus primus*, 7.

mankind by means of revelation. . . . Thus God provides the revelation of secret understanding from his [own] school. From there every man of sound judgment will easily take away [a knowledge of] what true *alchimia* is and how this is necessary for the physician."[36]

To Libavius, this sort of view posed a double threat. It violated trusted logic, and it promised to demolish the institutions and disciplines that depended upon it.[37] Roger Bacon had defined alchemy as a knowledge that taught how to make a medicine called elixir, and some thought of *chymia* in the same way, not as a part of medicine, but as a medicine, that is, as a medicament. To make matters worse, this was not the kind of medicine that rested upon the art that the Greeks called *iatrikon*. It was a universal medicine that had led to the fashioning of what Libavius described as a "parabolic chymiatria" taught by Paracelsus, Dorn, Severinus, and Quercetanus. These had repudiated Galenic and Hippocratic medicine. But there was no reason, Libavius objected, for thus throwing disciplines into confusion.[38] True *chymiatria* (chemical medicine) was based in medical axioms that had been discovered and proven through use and experience. It had been declared through the ages by practitioners, who followed ancient precepts and who had long made chemical medicaments and put them to use.

With such sweeping variations in the use of the term *chymia*, it might have been expected that the reference would, at a certain point, be made into a term of abuse, and Libavius was ever sensitive to the fact that for some, the term, without artistic guidance, had become equated with fraud. When someone in his own day, he explained, wanted to call a man a useless liar, he called him a *calendariographus* or common astrologer. When someone wanted to call a person a wicked impostor, he called him a *chymicus*. Where theory and didactic method could correct the deficiencies of the artisanal chymist, and a philosophy built upon reason and Aristotle rather than Plato and revelation might bring the supernatural artificer back down to earth, there was nothing to be done for frauds except rid the world of them. If there were no such thing as a *chymicus* of

36. "Alle seine schreiben aber warden nit auss andern büchern zusamen geflickt / Sunder auss der natur gegründt warden durch die warhafftige / und natürliche Magicam / welche ein muter ist alles verborgnen ding so dar natur zuwissen sein / unnd von Gott von wegen der offerbarung dem menschen gegeben ist. . . . Also gibt Gott die offenbarung heimlichs verstands auss seiner schul. Hieraus würd ja ein jed verstendiger leichtlich abnemen / wz die warhafftig Alchimia / und wie notwendig dieses dem medico sey"; Toxites, *Vorrede* to Suchten et al., *De secretis antimonii liber unus*, 10, 21.

37. On epistemological concerns, see Hannaway, *Chemists and the Word*.

38. Libavius, *Syntagmatis . . . tomus primus*, 5, 7nA.

this sort, Libavius proclaimed, the world would be free of an exquisite infamy and a most filthy villainy. These were the sorts of "chymists" who were often guilty of making poisons, adulterating coins, and making false gems. This *chymicus* was a plague upon society, and, Libavius remarked, plagues ought to be removed from the memory of mankind.[39]

The frauds of Paracelsian chymists were among the worst sorts of chemical deceit. Their medicines were poisons, and their practices did not assist the health of their patients but ravaged it. They used antimony and mercury in their medicines in such a way that they lacerated the brain and every part of the body. They administered cinnabar, both natural and artificial, made from mercury and sulfur. They ordered the sick to drink the water of calcanthum (copper sulfate) and caused such great agony that some people had been torn apart, their throats and stomachs eroded. The cruelest lie was that they denied that their medicines were at all corrosive. And yet, those same medicines were prepared by the most savage fires, needed to be kept in the thickest containers, and, when put to the test, scarred not only wood and glass but also metal. If this was what it meant to be a "chymist," no wonder the skin of all good men grew stiff at the mere mention of the name.[40]

Requirements and Obstacles of Good Practice

Libavius admitted that while the art of *chymia* was one according to its essence, it varied in its practice and appearance (*habitus colorque*). Other arts—including theology, astronomy, physics, jurisprudence, and medicine—varied in this way as well. However, *chymia* had been especially prone to having been misused by bad people. Some of them had called themselves philosophers. These were impostors, social renegades who possessed no patrimony except what their own friends supplied to them. Libavius knew of one solution. His intention, he said, was to make *chymia* carry out its business in public.[41] What was false could then be exposed, and those who mutilated and distorted chymical knowledge could be sent away in disgrace. In public, one would recognize good habits, the good practices of true *chymia*.

39. Libavius, *Rerum chymicarum . . . liber primus*, 9.
40. Ibid., 11–13.
41. "Hanc in public prostituere"; ibid., 19.

What would that look like? How should one recognize the social practices of true *chymia*? Over several epistolary essays in his collection of chemical letters, Libavius tells us just what he thinks those practices are. The genuine chymist, he informed his readers, operated in his own laboratory, one well stocked with supplies of natural substances. When a work was completed, he did not hide it, but brought it out for public use. He expounded the theory behind his miracles and set things up so that they might be found satisfactory in an accustomed manner. This is what good people did. They required faithful witnesses of their works. The professor of *chymia* shared his products and submitted to expert judges. He entered the laboratory not in a way that would prove useless to other aspects of life, but he cultivated the art in such a way that he did not neglect what pertained to the divine, did not reject the good of the republic, and did not desert his home. The true chymist did not turn away those who were wise from his laboratory, but encouraged them to enter. He did not make use of obscure words, which was a practice far from the *consuetudine* (habit) of true *chymia*. Genuine *chymia* did nothing with the pretext of a feigned religion. It cherished humanity, composing, by means of the manual arts, goods for the convenience and health of all. The genuine chymist did not think what was produced by way of the elements was drawn from the heavens, but approved oil as oil, water as water, granting to natural things their own perfection.[42]

Chemical artificers knew that solitary speculation did not profit human society if replicable practices were not combined with it, and that no art was divine and excellent in which theory was not followed by action.[43] Thus, the conscientious chymist did not withdraw from people, except to separate himself from the crowds of the unskilled. He kept the safety of the republic in view and did not run risks in the laboratory that led to harm or loss of health.[44] He preferred the crowns of eminent philosophers to the magnificence of palaces and was content that his own fame should flourish after his death, for the fame of the true chymist was conferred by posterity.[45] If the chymist were also a doctor, his adornment would be all the more illustrious, since to his wonderful

42. Ibid., 23–28.

43. "Sciunt enim artifices chymici non tam prodesse societati humanae solitariam speculationem, quam cum ea coniunctam praxin usumque salutarem, et nullam artem diuinam excellentemque esse, quae non actionem theoriae habeat consequentem"; ibid., 41–42.

44. Ibid., 34.

45. Ibid., 36–37.

inventions he would have joined methods of using them in the best possible way.[46] However, a chymist per se did not function as a physician. If he healed, he was a physician at the same time that he was a chymist and acted in accordance with principles established over time by means of experience, rational judgment, and the axioms of Hippocrates and Galen.[47] He offered service to other arts, however his first and proper goal was not servitude, but a certain blessed self-sufficiency (*autarkeia*). Knowing the limits of his art, the chymist contemplated nature's innermost recesses so that he might know what phenomena and powers were in things themselves. "I believe," Libavius exclaimed, "that there is in no part of *physica* as much delight as there is in *chymia* alone." The chymist explored sympathies and antipathies, causes, effects, and other things in nature one at a time, and thus did not know them indefinitely and vaguely, but definitely and with certainty.[48] Thus the practice of *chymia* entailed, on the one hand, teaching in schools and interpreting the decrees of teachers. Just as important, however, it also required attending with the hand to "the anatomies of nature."[49]

True *chymia*, as *physica*, a term that for Libavius was synonymous with the proper *magia* of the ancients, opened up the nature of the universe in theory and in practice. Yet, in the texts of those ancient authors who had expounded the art, he admitted that there were rough spots and obscurities (*salebra et obscuritates*). This, however, was not due to the art itself. Those who practiced *chymia* wrote with sincerity and integrity but expressed themselves with the peculiarities of their own language. To shed light upon what was obscure in ancient practices, what one had to do was to understand the discourse of artificers. "If you have not paid attention to this, you have not yet completed your apprenticeship," Libavius admonished. "If you do understand this, half your work is done."[50]

In establishing the art of *chymia*, linguistic preparation was both an intellectual and a social practice. Nowhere else did language have a more private meaning than in the arts, because, Libavius explained, language in the arts came about from experience and from the specific conditions of a particular artifice.

46. Ibid., 38.

47. Ibid., 40.

48. "In nulla physicae parte tantum credo oblectamenti esse quantum in sola chymia"; ibid., 39.

49. "Alia hora docebimus in schola et magistrorum placita interpretabimur. Alia manu administrabimus naturarum anatomias"; ibid., 33.

50. "Si sermonem artificum non intelligis, nondum fecist ityrocinum . . . Haec si intelligas, dimidium operae est"; ibid., 48.

Novices sometimes complained that they could not understand what chymists said. But if those same people, he continued, went into a cobbler's or metalworker's shop, they would not understand what they heard concerning practices by those craftsmen, either. The fundamental practice of *chymia*, in other words, was to learn the meaning of words and the use of terms of both ancient authors and contemporary craftsmen. Doing the latter became a social practice when the novice crossed the threshold of the artisanal workshop. Words and works enlightened one another. "For custom has made law," Libavius said, "so that to the extent one is skilled in the work of chymists one knows easily in turn what an author is getting at."[51] Moreover, many things had not been written down. In this case, "autopsy and the application of the hands must assist in this failing." For, "no one comes out an artificer without practice in the art," and "practice brings many things to mind that are [otherwise] concealed in precepts."[52]

The practice of combining words and works was not something that one could find within the academy, and Libavius hoped that Brendel would help change that. The academies, Libavius complained, had neglected the art, teaching by contemplation not by exercise. As a result, those who wanted to learn *chymia* had sought instruction elsewhere, turning to craftsmen, who supplied no precepts for what they did but whose reliability and experience had been publicly established.[53] Princes had proven to be of very little help. Although they had, in Libavius's view, not only a responsibility for nourishing and supporting the academies, they also bore an obligation, for the general good, to promote *chymia*. The situation was desperate. Everything seemed to be in decline. Schools were deserted more and more every day, he lamented, and if the academies had not already been founded by the ancients, he wondered whether anyone in his own day would have thought to create them. In terms of princely support within the university, no *stipendium chymicum* was to be expected.[54] Brendel, Libavius thought, could take matters into his own hands. True, progress in the art required expense, but while chymical practice might seem to require much in the way of instruments and materials, the expense was only prohibitive if the goal were

51. "Consuetudo enim legem fecit, adeo ut peritus operum chymicorum, et ordinis, facile intelligat, quorsum tendat autor"; ibid., 49.

52. "Huic defectui succurrere debet αὐτοψια, et manuum admotio. Artifex nemo euadit sine artis usu . . . Multa admonet usus, quae in praeceptis occultabantur"; ibid., 50.

53. Ibid., 52.

54. Ibid., 53.

riches. If one made experiments on a small scale, the expense to students could be controlled. It is not the art that is expensive, he enjoined, but the temerity of the one practicing it.[55]

Ancient Texts, Linguistic Skill, and Artisanal Expertise

He who wanted to practice true *chymia* needed to have both intellect and an abundance of experience gained through the senses (*oportet sensibus et experientia pollere*). In her study of *chymia* within the Academia Naturae Curiosorum, Margaret Garber notes ways in which *chymia* came to possess a new epistemic footing.[56] In Libavius's view, *chymia* as *physica* required not just a university foundation but a place within the faculty of arts. Brendel, he thought, possessed everything that was necessary to establish the art of *chymia* within that setting. "Now you are thinking about the subject of the art of *chymia,* in what way it should be defined . . . [and] . . . you have set out that which the practice of the artificer has observed."[57] Brendel had fastened together logical inferences (*consectaria*) and brought forth as a result splendid knowledge. He had instructed in necessary procedures and had set forth each one on the basis of infallible premises and principles. Libavius called upon him to go public: "How much good will be bestowed at once upon humankind when, by means of clear knowledge, the false is refuted by what is genuine? Students [of the art] will owe you everything."[58]

With practice and precepts, students would no longer think that shadows were real. Frauds could be exposed and what was obscure made clear. In the type of instruction Libavius hoped to see, one would need to study ancient opinion and develop the necessary linguistic ability to do so, but one would also need to engage artisanal practice in its own vernacular spaces and terms in order to separate out what was true and false among ancient authors. Precisely this, I think, is one of the messages embedded in part of a small poem, addressed to students of

55. Ibid., 50.

56. See the chapter by Garber that follows in this volume.

57. "Iam cogitas de subiecto artis chymiae, quodnam id sit ponendum. Ponis autem hoc quod usus artificum notauit"; Libavius, *Rerum chymicarum . . . liber primus,* 123.

58. "Quanto afficietur humanum genus bono, cum illustri noticia adulterinum a genuino statim redarguetur? Tyrones tibi Omnia debebunt"; ibid., 124.

chymia (*studiosis chymiae*), that Libavius attached like a signpost in the opening pages of his book of chemical letters.

> Learn, Tyro, what the Stagyreian
> Hero [Aristotle] makes known, [and] what rings forth
> From Galen and the divine Hippocrates.
> But join [to this] the wonderful mysteries of the smith of Lemnos [that is, Hephaestus]
> And like an artifex dissolve [them].[59]

Whether or not the elder Brendel felt inspired by Libavius, Brendel's son, also called Zacharius (1592–1638), sometime later also turned his attention to defining the art of *chymia* and published at Jena in 1630 his *Chimia in artis formam redacta,* a book edited by Werner Rolfinck (1599–1673). At the outset of the text, the younger Brendel noted what Theobald von Hogelande (ca. 1560–ca. 1608) had said in the preface of his *De Alchemiae difficultatibus Liber,* namely, that among the arts that divine counsel had bestowed upon human kind for its common welfare, *chymia* shone forth with rays of nobility and the splendor of truth. He added reference also to the infallible truth claimed for *chymia* by Johann Wolfgang Dienheim (fl. 1610), an apparent witness to transmutation, in his *Medicina universalis* (1610). As with the later history of *chymia* that brought transmutation into the category of curiosities,[60] *chrysopoeia* and *argentopoeia* remained tied, in Brendel's view, to the art of *chymia,* while *chymia* itself remained tied to the practices of medicine. None of the arts, he noted, had been so sacred as not to have been profaned by somebody, and some, he observed, when they had heard the word *chymia* had immediately wanted to vomit. Brendel had Galenist physicians especially in mind, and he reiterated what Thomas Moffett (1553–1604) had said about them:[61] that they should be seasoned with salt like herring and afterwards left to hang in the smoke.[62] Those who wanted to remove *chymia* from dogmatic medicine were like those who wanted the sun to be removed from the world. The term *chymia,* Brendel admitted, was ambiguous, and he made a point of declaring that he did not mean by the term what "philosopher chemists" had claimed for it, namely, that it was the *scientia scientiarum.* He wanted to refer to

59. "Cognosce Tyro, quid Stagyreius; Argutetheros, quid Galenus; Personet, Hippocratesque dius. Coniunge sed mysteria Lemnii; Miranda fabri, solvereque artifex . . ."; ibid.

60. See the chapter by Garber that follows in this volume.

61. Moffett, *De iure et praestantia chymicorum medicamentorum,* 13–14.

62. Brendel and Rolfinck, *Zachariæ Brendelii Chimia in artis formam redacta,* 5.

it as a *habitus*, as the precepts for a set of practices that, among other things, united it with medicine. While medicine did not rule over the chymical art, it was inseparable from it. "Let us unite medicine and *chymia*," he proclaimed, "each one as master and slave to the other."[63] Through personal experience and by understanding the precepts of practice, the chymist could advance quickly. Without such instruction and experience, he would be like a man wandering in an unknown forest without a guide.[64]

Libavius understood the need for a guide, one who understood both practices and precepts. The combination of the two made, in his view, *chymia* an art. Philosophers needed to understand the language of artisans, and artisans needed to learn the theory that turned what they did into what not only they, but anyone, could know. But, as Libavius was fond of admonishing, nothing should be received into *chymia* that was not of *chymia*.[65] The forest for which one needed a guide was an earthbound terrain, difficult enough to find one's way around in without the seductions of paths leading to realms of the metaphysical, celestial, or supernatural.

WORKS CITED

Ash, Eric. "Introduction: Expertise and the Early Modern State." In *Expertise: Practical Knowledge and the Early Modern State*, edited by Eric Ash, 1–24. Osiris, ser. 2, 25. Chicago: University of Chicago Press, 2010.

———. *Power, Knowledge and Expertise in Elizabethan England*. Baltimore: Johns Hopkins University Press, 2004.

Baretta, Marco. *The Alchemy of Glass: Counterfeit, Imitation, Transmutation in Ancient Glassmaking*. Sagamore Beach, MA: Science History Publications, 2009.

Brendel, Zacharias, and Werner Rolfinck. *Zachariæ Brendelii Chimia in artis formam redacta* . . . . Jenae: typis Blasii Lobensteins, sumibus Johannis Reiffenbergeri, 1641.

Cardano, Girolamo. *Hieronymi Cardani mediolanensis medici De rerum varietate libri XVII adiectus est capitum, rerum & sententiarum notatu dignissimarum index*. Basileæ: Per Henrichum Petri, 1557.

63. "utramque, medicinam et chimicam, ceu dominam et servam"; ibid., 7.

64. Ibid., 9–10.

65. "Iuxta hanc doctrinam nihil in chymiam recipietur quod non sit chymicum"; Libavius, *Rerum chymicarum . . . liber primus*, 119.

Daston, Lorraine. "Preternatural Philosophy." In *Biographies of Scientific Objects*, edited by Lorraine Daston, 15–41. Chicago: University of Chicago Press, 2000.

Debus, Allen G. *The Chemical Philosophy: Paracelsian Science and Medicine in the Sixteenth and Seventeenth Centuries.* 2 vols. New York: Science History Publications, 1977.

———. "Guintherius, Libavius, and Sennert: The Chemical Compromise in Early Modern Medicine." In *The Chemical Promise: Experiment and Mysticism in the Chemical Philosophy, 1550–1800: Selected Essays of Allen G. Debus*, edited by Allen G. Debus, 129–44. Sagamore Beach, MA: Science History Publications, 2006.

Dorn, Gerhard. "De artificio supernaturali." In *Schlüssel der Chimistischen Philosophy: mit welchem die heimliche und verborgene Dicta und Sprüch der Philosophen*, by Gerhard Dorn. Strassburg: Verlegung Lazari Zetzneri, 1602.

Forshaw, Peter. "'Paradoxes, Absurdities, and Madness': Conflict over Alchemy, Magic, and Medicine in the Works of Andreas Libavius and Heinrich Khunrath." *Early Science and Medicine* 13 (2008): 53–81.

Hannaway, Owen. *The Chemists and the Word: The Didactic Origins of Chemistry.* Baltimore: Johns Hopkins University Press, 1975.

Libavius, Andreas. *Rerum chymicarum epistolica forma ad philosophos et medicos . . . liber primus, secundus.* Francofurti: excudebat Ioannes Saurius, impensis Petri Kopffij, 1595.

———. *Syntagmatis selectorum undiquaque et perspicue traditorum alchymiae arcanorum, tomus primus.* Francofurti: excudebat Nicolaus Hoffmannus, impensis Petri Kopffij, 1611–13.

McCray, Patrick. *Glassmaking in Renaissance Venice: The Fragile Craft.* Aldershot, UK: Ashgate, 1999.

Meitzner, Bettina, trans. and ed. *Die Gerätschaft der chymischen Kunst: der Traktat "De sceuastica artis" des Andreas Libavius von 1606: Uebersetzung, Kommentierung und Wiederabdruck.* Stuttgart: Steiner, 1995.

Moffett, Thomas. *De iure et praestantia chymicorum medicamentorum dialogus apologeticus.* Francofurti: Apud haeredes Andreae Wecheli, 1584.

Moran, Bruce T. *Andreas Libavius and the Transformation of Alchemy: Separating Chemical Cultures with Alchemical Fire.* Sagamore Beach, MA: Science History Publications, 2007.

———. "Essences and Mostly Clean Hands: Preparing to Teach Chemistry with Libavius and Aristotle." *Science and Education* 15 (2006): 173–87.

———. "The Less Well-Known Libavius: Spirits, Powers, and Metaphors in the Practice of Knowing Nature." In *Chymists and Chymistry: Studies in the History of Alchemy and Early Modern Chemistry*, edited by Lawrence Principe, 13–24. Sagamore Beach, MA: Science History Publications, 2007.

Müller-Jahncke, Wolf-Dieter. "Andreas Libavius im Lichte der Geschichte der Chemie." *Jahrbuch der Coburger Landesstiftung* 17 (1972): 205–30.

Newman, William R. "Alchemical Symbolism and Concealment: The Chemical House of Libavius." In *The Architecture of Science*, edited by Peter Galison and Emily Thompson, 59–77. Cambridge, MA: MIT Press, 1999.

———. *Atoms and Alchemy: Chymistry and the Experimental Origins of the Scientific Revolution*. Chicago: University of Chicago Press, 2006.

Newman, William R., and Lawrence Principe. "Alchemy vs. Chemistry: The Etymological Roots of a Historiographic Mistake." *Early Science and Medicine* 3 (1998): 32–65.

Nummedal, Tara. "Practical Alchemy and Commercial Exchange in the Holy Roman Empire." In *Merchants and Marvels: Commerce, Science, and Art in Early Modern Europe*, edited by Pamela H. Smith and Paula Findlen , 202–22. New York: Routledge, 2002.

Park, Katherine, and Lorraine Daston. "Introduction: The Age of the New." In *The Cambridge History of Science*, vol. 3, *Early Modern Science*, edited by Katherine Park and Lorraine Daston, 1–18. Cambridge: Cambridge University Press, 2006.

Principe, Lawrence, and William R. Newman. "Some Problems with the Historiography of Alchemy." In *Secrets of Nature: Astrology and Alchemy in Early Modern Europe*, edited by William Newman and Anthony Grafton, 385–431. Cambridge MA: MIT Press, 2001.

Smith, Pamela H. *The Body of the Artisan: Art and Experience in the Scientific Revolution*. Chicago: University of Chicago Press, 2004.

———. *The Business of Alchemy: Science and Culture in the Holy Roman Empire*. Princeton, NJ: Princeton University Press, 1994.

Suchten, Alexander von, Georg Fedro von Rodach, Georg Forberger, and Paracelsus. *De secretis antimonii liber unus Editus Germanice quidem anno 1570*. Basileae: Per Petrum Pernam, 1575.

Chapter 4

Chymical Curiosities and Trusted Testimonials in the Journal of the Leopoldina Academy of Curiosi

Margaret D. Garber*

Alchemy was once an integral part of medicine. Central European medical practitioners had been infusing bodies with mineral, metal, and curious chemical medicaments since the Middle Ages.[1] However, as the standard account goes, historical actors unfastened alchemy from its medical moorings in the eighteenth century, and chemistry set sail on its own as a discipline based upon manual procedures, standardized language, and useful symbols. The history of this conjunction and decoupling gave rise to debates that motivated much of Allen

*I thank Bruce Moran and Karen Parshall for their helpful comments and suggestions. I am also grateful to Mary Fuller, Marcy Norton, Tara Nummedal, and the volume's editors and anonymous readers for their excellent recommendations. Any remaining errors are my own.

1. Chemical and medical interests crosscut both medical theory and productions of medicaments, and were called either *chymiatria*—pioneered through the writings of sixteenth-century authors as dissimilar as Paracelsus and Libavius—or iatrochemistry through the seventeenth-century works of authors such as van Helmont and de la Boe. See Moran, "Survey of Chemical Medicine in the 17th Century."

Debus's scholarship during his long and productive career.[2] At odds with Butterfield's persistent "delayed chemical revolution," which claimed that Enlightenment rationalists created chemistry by shearing away its medico-alchemical past, Debus insisted that chemistry's academic acceptance depended upon its pharmaceutical value to schools of medicine—a move that established chemistry's university footing once physicians rejected chemical philosophy as a viable explanatory source for physiology. Pushing Debus's claim further, I argue that physicians and chemists received far more from alchemy than philosophy and pharmaceuticals. The rise of the "expert chemist" was linked to an earlier transformation of alchemical adept to physician adept, one whose chemical experience was measured by chemical productions. It was the gold-making transmutation believers themselves who initiated these efforts to standardize language and agree upon procedures and symbols.

In late seventeenth-century central Europe, physicians in the Leopoldina Academy of Curiosi illustrated their prominence in more than medical theory or pharmaceuticals, portraying additionally their dominance in skills involving ovens, fires, acids, minerals, and metallic salts, evidence of proficiencies that etched a design of legitimacy onto the pages of their embryonic medical journal, the *Miscellanea curiosi*.[3] This demonstration of proficiency issued in part from some members' apparent competence with transmutation, the transformation of baser metals such as lead into more noble metals such as gold or silver, a process known to historical actors as *chrysopoeia* or *argyropoeia*, respectively. The Curiosi, as a first society of physicians in the German territories, were not shy in seeking the status of *adepti*, those who excelled in the making of the most sought after of arcana, by actively engaging in chemical productions and appropriating alchemical recipes. Social legitimacy for physicians of the Academy of Curiosi rested heavily on their self-portrait as purveyors of curiosities and conveyors of medical authority, an identity that included practice in alchemical activities in their journal of curiosities.

2. Indicative of Debus's thesis, throughout a massive corpus, is notably *Chemical Philosophy*, 2 vols.; *Chemistry and Medical Debate*; and "Iatrochemistry and the Chemical Revolution."

3. Initially known from 1652 as the Academia Naturae Curiosorum, after 1687 upon imperial patronage, they renamed themselves Leopoldina Academia Naturae Curiosorum. Their journal was initially entitled *Miscellanea curiosa sive ephemeridum medico-physicarum germanicarum curiosarum*. With the exception of Büchner and Barnett (noted below), references to this society are contained within broader topics (listed chronologically): Büchner, *Academia Naturae Curiosorum Historia*; Ornstein, *Rôle of Scientific Societies in the Seventeenth Century*, 169–75; Thorndike, *History of Magic and Experimental Science*; Evans, *Making of the Habsburg Monarchy, 1550–1700*; and Barnett, "Medical Authority and Princely Patronage."

To amplify the Curiosi's practice with these activities, the term "chymica" figured prominently on the title page of the periodical's first thirty annual ephemerides (each of which contained some two hundred observations of singular, bizarre, or novel recipes, procedures, or events), even as it comprised only a small fraction of the journal's multiple curiosities. *Chymica*'s prominence in the Curiosi's journal begs the question of exactly what its members meant by the term. Did it signify chemical medicine (*chymiatria*), alchemy, early chemistry, or all three? Recent scholarship has demonstrated the ahistoricity of defining alchemy and chemistry as related, but oppositional binaries, in such examples as vitalist/mechanist, irrational/rational, obscure/clear, occult/open (where alchemy occupies the former of the dualisms and chemistry the latter), leading to a positivistic *terminus ad quem*, pseudo-science/science.[4] Deploying such dichotomies retroactively onto the respective Latin derivations (namely, *alchymia/chemia)* distorts seventeenth-century authors' uses of these terms. As Lawrence Principe and William Newman have demonstrated, such divisions emerged partly as the result of an error that unreflectively continued to be cited and partly as useful metaphors for Enlightenment authors who portrayed chemistry as the illuminated path out of the darkness of alchemy.[5] Since late sixteenth- and seventeenth-century historical actors used the terms "alchymia" and "chymia" interchangeably, historians employ the term "chymistry" to prevent importing false binaries between alchemy and chemistry that have been enshrined in the historical literature of this science, especially that of the so-called "delayed scientific revolution."[6] Despite the laudability of the etymological correction, such a move should not obstruct historians from locating contexts in which historical actors *did* make other distinctions between these Latin terms. While "alchymia" and "chymia" were used interchangeably in many contexts, the Curiosi distinguished *chymia* or *chymica* (hereinafter *chymia*) by using this as a subset of *alchymia* or *alchimia* (hereinafter alchemy), without entertaining the aforementioned oppositional dichotomies.

4. Newman and Principe, "Alchemy vs. Chemistry"; and Principe and Newman, "Some Problems with the Historiography of Alchemy."

5. Principe and Newman describe Enlightenment writers' use of light/dark metaphors to distinguish chemistry/alchemy in "Some Problems with the Historiography," 386.

6. Newman and Principe note that the distinction between *alchymia* and *chymia* resides in the definitive Arabic article "al." To avoid presentist sensibilities when using the terms, they advise employing the archaic "chymistry." See "Alchemy vs. Chemistry."

This chapter examines observations in the *Miscellanea curiosa* during the last quarter of the seventeenth century and asks how the Curiosi's knowledge about *chymia*, a subject often infused with secrecy, was displayed in the very public, literary, and social space of a journal of curiosities. What precisely the Curiosi meant by "chymia" is worthy of consideration, since it appears as more of a category than a term. In the *Miscellanea curiosa*, the category *chymia* referred to three separate, yet interrelated approaches: *chymia* as witnessed histories that served to defend metallic transmutation; *chymia* as firsthand material practice, that is, the circulations of recipes, procedures, and warnings; and *chymia* as social practice, that is, the normative rules, standards, or principles that applied to metals and minerals.

Examining the Curiosi's journal exemplifies the fact that clarifying *chymia* and standardizing its use emerged from seventeenth-century *chrysopoeians* themselves and not from Enlightenment figures. By aiming to uncover what it meant collectively to do *chymia* and what the practical understandings and display of the term "chymia" signaled for historical actors who dispatched observations to this physicians' journal, this chapter proposes that efforts to standardize alchemy as an experimental practice shaped both the reform of chemistry and the format of science journals. Here, a brief history of the Academy of Curiosi precedes a tour of the linguistic landscape of *chymia* within journal observations (*observationes*), a format that confined and shaped *chymia*. The chapter closes with an explanation of how these varying uses of *chymia* were deployed to create a vibrant social network in the service of epistemic authority.

THE CURIOSI

The Curiosi's journal was a place for both knowledge making and community building due to the early modern appetite for news about natural rarities within the cultural space of curiosities. Within societies of *virtuosi* and other purveyors of curiosities, with whom the Academy members identified, the journal became a social space for creating a community devoted to locating nature's rare and strange particulars.[7] The establishment of their Academy was itself curious in the

7. On *virtuosi* and particulars of nature, see especially Daston and Park, *Wonders and the Order of Nature*, and Eamon, *Science and the Secrets of Nature*. There is a vast literature on curiosities, which I address in my book project, "Exotic Medicine: Chymistry and Medical Curiosities in the Leopoldina Academy of Curiosi

sense that it originally consisted of just four physicians in remote Schweinfurt (a city in Lower Franconia of the German territories). Their ambition was to create encyclopedic knowledge on curiosities from *A* to *Z*, an endeavor they hoped to realize by organizing a social body of physicians throughout the Holy Roman Empire. The four founding members—Johann Bausch (1605–65), Johann Michael Fehr (1610–88), Georg Balthasar Metzger (1623–87), and Georg Balthasar Wohlfarth (1607–74)—began this task based upon their own knowledge of medicine, botany, and *chymia*, with their stated goal of imitating Francis Bacon's ideals of collecting and sharing new *observationes*.[8] Despite devastating interruptions during the Thirty Years' War, each managed to acquire a cosmopolitan medical education by matriculating through a number of universities in Germany, Switzerland, and Italy, including the University of Jena, at a time when Zacharius Brendel the Elder (1553–1626) and the Younger (1592–1673) taught preparations of chymical medicaments, and at the University of Marburg in Hesse-Kassel, when the influences of Johannes Hartmann (1568–1631), the first chair of *chymiatria* (the study and preparation of chymical medicines), were still palpable.[9]

On New Year's Day, 1652, the Curiosi established their Academy and elected its first president, Johann Bausch, *Stadtphysicus* (city physician) of Schweinfurt, a man notable for his cabinet of curiosities. As academy membership expanded spatially across the vast German and Bohemian territories, the Curiosi made a uniquely strategic move to require *no* face-to-face public meetings. Instead, the dispersed Curiosi retained collective coherence by transforming their intimate Academy into a literary medical society that moved from communicating monographs to circulating observations. It was mostly through the social networking and editorial efforts of Philipp J. Sachs von Lewenheimb (1627–72), a town physician in Silesia and master of epistolary exchange, that the Curiosi established the first medical journal in the Holy Roman Empire, one that drew upon the burgeoning medical genre of observations.[10]

(1670–1740)."

8. On the history of these early members, see Barnett, "Medical Authority and Princely Patronage," 53–56; and Büchner, *Academia Naturae Curiosorum Historia*, 19–21.

9. On Bausch's medical instruction for two years at the University of Jena, including his work at the time of Brendel, and at Hesse-Kassel, see Barnett, "Medical Authority and Princely Patronage," 40–44. On Brendel and Libavius, see Moran's chapter in the present volume. On Hesse-Kassel's Hartmann, see Moran, *Alchemical World of the German Court*.

10. Membership was open to all licensed physicians, including town, court, and imperial physicians, as

By its third year, the Curiosi's title page announced its contents as "observationes medicas, physicas chymicas," which signaled to readers that entries would be short (with longer treatises confined to appendices) and would be partial to practice over theory. While theoretical concepts understandably undergirded practical concerns, exclusively theoretical contexts were circumscribed to optionally include learned commentary in the form of *scholia* submitted by a member or an editor. By moving from the genre of monograph to what Gianna Pomata terms the "epistemic genre" of *observationes*, the Curiosi continued the founders' ambition for collective, encyclopedic knowledge.[11] According to Pomata, the term *observationes* was a late sixteenth-century neologism emerging from the discipline of medicine that indicated short firsthand reports, the separation of case study from commentary (*scholium*), and the hierarchical privileging of practice over theory (with theory relegated to *scholia*). Such a format permitted the town or court physician (*practicus*) to distinguish his interests from the university physician's focus on genres, such as *quaestiones disputatae*, that were ensconced within university traditions.[12] The turn towards a genre of short, collective, empirical observations focused primarily on practice would have consequences for the presentation of *chymia* and its particular secrets, especially by curtailing its hitherto lengthy narratives.

To highlight the Curiosi's chymical interests, the term "chymicas" appeared frequently in the *Miscellanea curiosa*. It adorns the title page, at first as part of a longer list of observations that were anatomical, surgical, pathological, botanical, and therapeutic, under the category "medico-physico." Then, by 1673, *chymicas* was added to create the *observationes* triad: "medicas, physicas, chymicas."[13] *Chymia* embellishes early frontispieces, figuratively announced from dual horns of plenty: one side boasting *anatomia* and the other *botanica*, trumpeting the kingdoms *mineralia*, *animalia*, and *vegetalia*, publicizing the chymical physicians' province to know not only about anatomy, botany, metals, and minerals of a region but also medicinal uses, natures, and histories of these kingdoms from

---

well as university professors of medicine at both Catholic and Protestant institutions. Publications were to be only in Latin. See "Leges," *Miscellanea curiosa* 2 (1671).

11. Pomata, "Observation Rising." Pomata distinguishes the use of *observationes* from the verb "observatio," since the latter signified "observance" until the late Middle Ages and later denoted "diligence." Significantly telling for the Curiosi, whom she invokes as an exemplar of the genre, "the social profile of the authors of *observationes* is that of the *practicus*, often with a leaning to medical heterodoxy or Paracelsianism"; ibid., 59–62.

12. Ibid., 48.

13. Title page, *Miscellanea curiosa* 3 (1673–74).

ancient to contemporary times. Medicine and natural history were intertwined disciplines in the seventeenth century; scholars who studied medicine studied natural history.

The publishing impulse rewarded the Curiosi with a substantially increased membership. Between the Academy's founding in 1652, and prior to its journal's publication, there were just thirty-five members. After its journal began in 1670, membership increased strikingly. By 1695, the Curiosi had grown to over two hundred members.[14] Although the journal was primarily for members, its goal was to extend membership and circulation throughout Europe, an ambition that was partially realized by its translation into English in the *Philosophical Transactions of the Royal Society of London*.[15]

Chymia in Chrysopoetic Contexts

Of all of the potential curiosities that could be published within the journal, *chymia* offered anticipated chymiatric and enticing chrysopoetic possibilities for readership. Practical information about the efficacy and production of metallic and mineral medicaments became an increasingly significant component of pharmacy and medical training at some universities, so the exchange of useful news about *chymiatria* (chemical medicine) would be expected in a journal for physicians. The society announced this publicly. Not only did *chymicas* appear prominently on its title page, but its contributors also enacted it privately; chymical medicaments were to be included within the journal according to bylaw XI.[16] The Academy's membership demonstrated interest in acquiring experienced, hands-on chymical knowledge by extending membership—usually limited to licensed physicians—notably to chymists enlisted to resolve contentious chymical disputes.[17]

14. As membership grew, the society and its library moved wherever the newly elected president resided, then settled into its present headquarters in 1878 at Halle. See Barnett, "Medical Authority and Princely Patronage"; and the website of the Nationale Akademie der Wissenschaften: http://www.leopoldina.org/en/academy/history.html.

15. The journal was reviewed in the Royal Society's *Transactions* for the first five years and periodically until 1931. Over the next century, the *Transactions* cited several of the observations under numerous changing titles for the journal and the society.

16. "Medicamenta tam vulgaria quàm Chymica"; see "Leges," *Miscellanea curiosa* 2 (1671). All translations are my own.

17. The Academy admitted chymist and professor of rhetoric J. Kirchmaier, the lawyer and chymist

The secretive history of *chymia* made it enticing as well. A journal that could offer rare and strange secrets, especially those of metallic transmutation, could satiate the eager physician's appetite for curiosities if it served up something substantively informative. Publications of entries devoted to chrysopoetic (gold-making) and argyropoetic (silver-making) transmutations attest to the perceived appetite for *chymia* on the part of its readership and suggest that the society's own interest in *chrysopoeia* may actually have shaped its early identity. Initially, the Curiosi associated their Academy with Jason and the Argonauts. Members selected as pseudonyms the names of those heroic hunters in search of the Golden Fleece, a choice suggestive of self-conscious adornment with a chrysopoetic image. Jason's fleece was called "golden," since fleece was often used as a sieve to trap gold particles when gold-bearing water was poured over it. Afterwards the fleece would be burned, allowing any gold to be recovered from the ashes.[18] As Antoine Faivre posits, this was sufficiently well known that a number of early modern authors, such as Pico della Mirandola (1463–94), Michael Maier (1568–1622), and Johann Valentin Andreae (1586–1651), used the expedition of Jason and the Argonauts as an allegorical image for their own chrysopoetic writings.[19]

The editorial flair of the *Miscellanea*'s first secretary, Sachs von Lewenheimb, imbued the journal with a transmutational flavor when he introduced observations of metallurgical matters to an international audience by publishing a treatise defending *chrysopoeia*.[20] His "Aurum chymicum" (chymical gold) highlighted the gold-making arts of adepts, whose artifacts rivaled those creations nature grew in the most intimate viscera of the earth. Following a lengthy rebuttal to Athanasius Kircher's widely known dismissal of *chrysopoeia* (published in his 1664 *Mundus subterraneus*), Sachs weighed the testimony of valid chymical arts against the acknowledged trickery of "imposters," who sullied the reputation of true "adepti."[21] Insisting that Curiosi did not rely simply on the bygone world of

---

Christian Adolph Balduin (joined 1673), and Johann Kunckel (joined 1691), who, although without formal training, was judged sufficiently proficient in chymical matters to become director of the alchemical laboratory of the Great Elector of Berlin and minister of mines to Sweden's King Charles X. Barnett relates their admission to authors' interest in phosphorous; Barnett, "Medical Authority and Princely Patronage," 228–29.

18. Tylecote, *Early History of Metallurgy in Europe*, 47.

19. Faivre, "Approach to the Theme of the Golden Fleece in Alchemy."

20. Sachs von Lewenheimb, "Aurum chymicum," 68–70.

21. Ibid., 69. On Kircher's alchemical interests, see Hirai, "Kircher's Chymical Interpretation of the Creation and Spontaneous Generation."

past authors, such as Ramón Llull (ca. 1232–ca. 1315), Arnold of Villanova (ca. 1238–ca. 1310), Paracelsus (1493–1541), and others, Sachs characterized his academy as cosmopolitan and contemporary, epitomized through a familiarity with sixteen well-documented cases of recently witnessed transmutations over a twenty-year period.[22] While Sachs did not personally witness any of the listed transmutations, the histories were vindicated by authorial excellence (such as Jean Baptiste van Helmont [1579–1644]), famous events (such as the 1666 transmutation at The Hague by Helvetius [1625–1709]), or testimonial status of unimpeachable witnesses (such as Emperor Ferdinand III), as this paraphrase exemplifies:

> II. Cornelius Martin of Antwerp, a lawyer judge, Physician and Philosopher not only performed a transmutation but made other metallic tinctures.
>
> III. Johannes Baptista van Helmont that most noble Philosopher produced a gold-making and a silver-making stone [lapidem aurificum & argentisicum]...
>
> V. Baron Moncony, accepted a mysterious powder from another secret adept and was able to produce greater than 24-carat gold [plusquam 24 caratto-rum].[23]

Sachs drew attention to traditional tropes of alchemical literature, in noting such things as a "mysterious powder" and a "secret adept," and to the almost-stock character of Baron von Chaos, whose transmutation of mercury into gold was "witnessed in 1648 by the Holy Roman Emperor Ferdinand III in Prague."[24] Yet he relied equally on the powerfully normative social operations of trust and authority to counter gold-making deniers, who claimed that only frauds practiced these arts.[25] In pre-Enlightenment Europe, fraud and alchemy were not exclusively correlated. Tara Nummedal demonstrates this point in her study of sixteenth-century criminal trials surrounding alchemical fraud, which reveals that issues of intent to deceive via fraudulent transmutation were the focal point

22. On transmutation histories, see Smith, *Business of Alchemy*, 173–227; Newman, *Gehennical Fire*, 1–11; and Principe, *Aspiring Adept*, 93–98.

23. Sachs von Lewenheimb, "Aurum chymicum," 65–75. See his transmutation history on 71–73. I have described this in Garber, "Con or Craft."

24. Baron von Chaos, also known as Conrad von Richthausen (d. 1663), was an alchemist who also served in the mines as treasurer. See Smith, *Business of Alchemy*, 181.

25. On the issue of trust in the seventeenth century, see Shapin, *Social History of Truth*.

of trials and weighed far more heavily than simply failed alchemical projects.[26] "Fraudulent" and "true" alchemy were obverse sides of the same coin. As Nummedal explains, "Stories of true and false alchemy . . . emerged in tandem."[27]

Discussions of transmutations often took the form of a defense in the *Miscellanea curiosa*, and such defenses often made the same moves. They appealed to firsthand experience by trusted witnesses and positioned their evidence as contemporary, while they characterized antagonists as shortsighted novices who relied upon outdated book knowledge. With Gabriel Clauder (1633–91), court physician to the Duke of Saxony, we see a move from authoritative knowledge of transmutation histories to authoritative experience with transmutation, especially in his 1686 transmutation defense entitled "Possibilities of Metallic Transmutation." Like Sachs, Clauder took aim at Kircher, but in this case as a denier of modern learning.[28] Scarcely referencing Kircher's massive tome, Clauder dismissively quipped, "there is as much old as there is new in the *Mundus subterraneus*" and, further, he denounced Kircher as a *Coryphaeus* (head of a chorus) of deniers, who rejected both transmutation and modern reasoning.[29] His criticism of Kircher rested upon a seeming contradiction. How could Kircher, who otherwise accepted the metamorphoses of plants and animals by means of plastic *semina* (seeds that served as sources of generation), reject metallic transmutation as impossible?[30] Kircher's rejection of the seminal origin of minerals and metals was based upon his belief that these were inanimate, a view Clauder judged as inconsistent and old-fashioned:[31]

> Many metamorphoses and transmutations are mediated by virtue of plastic *semina* in the animal and plant kingdoms; yet reasonably, I do not see why

26. On alchemical fraud, see Nummedal, *Alchemy and Authority in the Holy Roman Empire*.

27. Ibid., 171.

28. For a discussion on Clauder's book that discusses Kircher and transmutation, see Priesner, "*Defensor Alchymiae*."

29. "Negantium modernorum . . . Coryphaeus extra dubium est immortalis Athanasius Kircherus . . . In Mundo suo subterraneo, tam veterum, quam neotericorum"; Clauder, "Transmutationis Metallorum Possibilitas," 372–74. Clauder became a member in 1687 and published in *Miscellanea curiosa* on topics such as mirrors, storage of acids, mercury, vitriols, and sal ammoniac. Later, as adjunct, he was responsible for extending membership and for correspondence with other societies.

30. On *semina* as the source of disease external to the patient's body, see Shackelford, *Philosophical Path for Paracelsian Medicine*. On its medieval roots, see Emerton, *Scientific Reinterpretation of Forms*.

31. Clauder, "Transmutationis Metallorum Possibilitas," 372–74. On Kircher's notion of generation, see Hirai, "Kircher's Chymical Interpretation of the Creation and Spontaneous Generation."

> minerals are denied the same. For however much metals and minerals are not apprehended in the act of living or having a perfected life; nevertheless, they live a life analogous to those of plants and animals because they grow and are strengthened.[32]

Clauder's discourse on theories about metamorphoses of plants, animals, and insects from seeds or *semina* drew a consistent analogy with the belief of *semina* as the source for the growth for minerals and metals, a belief that served for him as the basis for the possibility of transmutation. To conduct a transmutation meant to hasten, by artificial means, the naturally slow growth of metals from imperfect and immature states (such as lead) to perfected maturation (such as gold). Therefore, metallic growth was a central tenet of transmutation. Clauder praised the blending of older and more recent views: "the best people stand firm with the supporters of alchemy on the common features of metallurgy. More and more is becoming known in this our century on account of the many favorable and fertile causes, methods, and ways by which the earlier supporters changed imperfect metals into more noble ones and also divided them into salts."[33] By suggesting that alchemy was progressive, intellectually consistent, and based upon past and present practical knowledge, Clauder magnified the Curiosi's access as contemporary and valuable, while he distanced Kircher's criticism as outdated.

Clauder's transmutation defense not only posed the question of who could assume the role of powerful detractor, it also reformulated the answer to who could assume the role of trusted eyewitness. Whereas in the journal's first volume Sachs was anxious to rest credibility on trusted eyewitnesses such as an emperor (Rudolf II and Ferdinand III), archduke (Leopold Wilhelm), or baron (Moncony), whose unimpeachable status brought an air of indisputability to the cases, by 1686 Clauder placed trust in his own ocular experiences and those of fellow physicians.[34] Relying upon the private wisdom of those, such as himself, who saw operations with their own eyes ("occulatus possum esse testis"),

32. "Plures metamorphoses & transmutations mediante virtute seminali plasticá in regnô animale & vegetabile contingent; cure eaedem minerali sint denegandae, sanè non video. Quamvis enim metalla & mineralia perfectam vitam vivere non deprehendantur; vivunt tamen illis analogicam, cùm crescent & augeantur"; Clauder, "Transmutationis Metallorum Possibilitas," 372.

33. "Optimè, insuper constat alchymiae cultoribus imò communibus metallurgis, hôc nostro seculô prioribus multas ob causas feliciore ac fertiliore, magis magisque innotescere vias ac modos quibus metalla imperfectiora in nobiliora"; ibid., 374.

34. On not daring to gainsay men of high status, see Shapin, *Social History of Truth.*

he celebrated those who manually brought forward the techniques of the previous generation's alchemical cultivators, and those contemporary physicians who knew firsthand the characteristics of metals and fires. *They* were the witnesses who turned that wisdom into fruitful methods for transmuting ignoble metals not only into nobler metals, but also into useful salts such as *sale armoniaco, sale tartari, nitre, mercurio sublimato* (sublimated mercury), and *arsenico*. By recasting the role of the witness in accounts of *chymia* and *chrysopoeia*, to those who performed chymical procedures, Clauder attempted to transfer chymical credibility to the Curiosi.

Chymia as Firsthand (*Autopsia*) Material Practice

Fruitful productions of useful salts depended upon mastering methods, causes, and practical knowledge of chemicals and powerful foundry fires through the arts of *chymia* that, according to Clauder, consisted of both chemical knowledge and *enchirises*.[35] Clauder's term *enchirises* is interesting on two levels. On the one hand, it indicated a procedure done manually, using apparati, and one that included understanding operations of fire. On the other, it gestured to a division within alchemy, traceable to Libavius, of which *encheria* and *chemia* (the making of chemicals) were subsets.[36] True knowledge, for Clauder, depended upon firsthand manual experience. In his conclusion to "Possibilities of Transmutation," Clauder ranked the knowledge that resulted from hands-on practical experience with chymical arts and those tests witnessed with one's own eyes as epistemically higher than philosophical book knowledge: "Whether it is the Philosopher's [province] to dispute against *autopsia*, others may judge."[37] Armchair philosophy could not match true hands-on experience.

Clauder's selection of the term "autopsia" (in "contra-autopsiam") is significantly reflexive, since it places emphasis upon knowledge gained by one's *own* eyewitnessing of the object, event, or medical case; *autopsia* is an observation

35. "Allisque, v.g. Sale Armoniaco, Sale Tartari, Nitrô, Borrace, Arsenico, Mercuriô sublimatô, Lapide calaminari &c. ope ignis fortioris fusoria per arcanas artis chymicae enchirises exaltentur"; Clauder, "Transmutationis Metallorum Possibilitas," 372.

36. On both points, see Newman and Principe, "Alchemy vs. Chemistry," 44. On Libavius's emphasis on knowledge of the hands, *enchiremata*, see Moran, *Andreas Libavius and the Transformation of Alchemy*, 300.

37. "An igitur sit Philosophi, contra autopsiam disputare, judicent alii"; Clauder, "Transmutationis Metallorum Possibilitas," 374.

witnessed firsthand.[38] Taken together with *observationes,* mentioned above, Gianna Pomata argues that these two terms ushered in a type of experience that was markedly different from Aristotelian *experientia,* since the latter concerned the grounds of certain knowledge arrived at by the senses certainly, but manifested through the cognitive extension of multiple memories and sharpened philosophical training. In contrast, *observationes* produced a "cognitive activity with a distinct literary format" that preferentially favored practice over theory (albeit practice informed by theory).[39] Employing such terms as "enchirises," "autopsiam," and "observationes" emphasized the firsthand witnessing of transmutations of metals and the production of metallic salts through practical training and hands-on experience, a kind of knowing-by-doing.[40] By positioning himself with this firsthand knowledge in opposition to the book-trained philosopher (Kircher), in effect Clauder rhetorically asked the reader to judge who had the better understanding of transmutation.

Clauder's defense of transmutation—which in the journal format depended on a critique of others' shortsighted theory, methods, and inexperience—contrasts strikingly with his earlier 1678 book on transmutation, which according to Claus Priesner described a pilgrimage to success with the Philosophers' Stone, the epistemic virtues of the alchemist, and supernatural powers involved in transmutation.[41] In this latter presentation, Clauder described his own alchemical pilgrimage, in which he met a stranger who helped him prepare an elixir that turned three hundred pounds of mercury into gold. In order to produce this Philosophers' Stone, Clauder required God's help for his own mystical transmutation into adept.[42] The significance of the two very different texts points to the fact that Clauder *could* have provided the latter, more traditional narrative for the *Miscellanea curiosa.* Yet he did not. The format of observations, which by

38. Pomata, "Observation Rising," 66; and Daston, "Empire of Observation (1600–1800)."

39. Pomata, "Observation Rising," 69. See her discussion about *observationes,* esp. 65–69.

40. Today, this term might be called "tacit knowledge," which is a topic of significance to history of science and science and technology studies. Originally coined by the philosopher and chemist Michael Polanyi in *Tacit Dimension,* tacit knowledge is that which cannot be made explicit. The meaning of "tacit knowledge" has been debated and sharpened, especially in the work of Harry Collins, who proposes that tacit knowledge *requires* explicit knowledge. Collins's burden is to disentangle confounded notions of tacit knowledge into three types of phenomena: "contingencies of human life (relational tacit knowledge), the nature of the human body and brain (somatic tacit knowledge), and the nature of human society (collective tacit knowledge)." See Collins, *Tacit and Explicit Knowledge,* x.

41. Priesner, "*Defensor Alchymiae.*"

42. Ibid.

1686, had taken on a standard, shortened, one-to-four-page length, required the stripping away of extraneous and cryptic language. Rather than the more flowery narrative, Clauder's journal entries provided abbreviated observations with practical outcomes: *chymia* was conducive to purer salts or medicaments and to competitive advantage. Proof of the adept was in the product. In Clauder's book, by contrast, proof of the adept was in the apprenticeship, namely, direct contact with a mysterious, vanishing stranger distributing secret knowledge. Clearly such beliefs were still in vogue, as evidenced by one of Sachs's early transmutation histories that featured a "mysterious powder from a secret adept."[43] Yet, the public face of *chymia*, as recorded in the new journal format of the Curiosi's first three decades, discouraged fancifully secretive narratives (even in the reporting of transmutations) and favored new types of abbreviated personal experiences of short, firsthand observations. These format decisions shaped the newly emerging face of both journals of science and chymists.

One further point is noteworthy. In the several observations Clauder penned, he used the terms "chymica" and "chymia" interchangeably. He described the efficient production of mineral salts from metals as part of the arts of *chymica*; he identified chemical physicians as *chimicos*, and what they do as *chymicae scientiae*, defined as the art of making tinctures, spirits, and salts, whether fixed or volatile; and, finally, he attributed to *chymiatrica* the arts of making specific medicaments, for instance those made with mercury or antimony. Thus *chymica* and *chymia* related to the making of chemicals including mineral salts. When speaking of knowledge of manual operations such as foundry fires, he used the term "encheria," and for transmutation, he used "transmutatio." In only one instance, however, did he invoke the term "alchymia": when welcoming the great fortune of his time to have learned about fruitful productions, he expressed gratitude to the supporters of *alchymia*. It was their experience and knowledge of the characteristics of metals that led to the contemporary success in both the transmutation of metals and fruitful metal and mineral salts. "Alchymia," in Clauder's use, gestured to a *broader* category of practice and knowledge of which *chymia* and *chymiatrica* were subsets.

Clauder's use of "chymia" as a subset of alchemy, and that by other Curiosi, may hark back to Libavius's understanding. This transition to using "chymia" as a subset of alchemy, at least for this group, may suggest an attempt to

43. "Ille homo acceperat pulveram ab alio, mysterium"; Sachs von Lewenheimb, "Aurum chymicum," 70.

appropriate—for the province of physicians—*chymia* beyond the making of medicaments (*chymiatrica*) to include the making of mineral and metal salts and to transmutations of gold and/or silver. At the same time, it suggests that while the Curiosi felt indebted to alchemical traditions, their use of "chymia" stopped short of the practices of alchemy broadly construed: artisanal productions of things such as paints, gems, or glass.

The Curiosi exchanged firsthand observations of recipes and procedures under the title of *chymia* (or *chymica*) that constituted a dual material and social practice; engagement with chymical materials nurtured the social transformation of physician into artifex, an experienced creator of chymical things. Crucial to this transformation were observations that fostered curiosity about specific agents, such as recipes for the universal solvent, an elixir of longevity known as the alkahest, and procedures for metallic transmutation. Practical advice, helpful hints, and warnings may have enabled a reader familiar with chymical practice to have some reproducible success when guided by the physician-artifex. By inferring through such observations that they were especially skilled to perform difficult procedures, and to create end-products of desirable rare recipes, the Curiosi cultivated an image of themselves as chymical adepts. Chymical observations offered readers proof that *adepti*-physicians had hands-on experience in the practice and understanding of chymical curiosities. As such, the material and social practices were intertwined; success in the material practice of *chymia* improved the social standing of the physician and authority of the Academy of Curiosi.

Chymia as Social Practice

Of all of the characteristics that kept *chymia* secretive, and thereby all the more desirable, its obscure language ranks as one of the most prominent. Several secretive texts on alchemy and *chymia* were published over the centuries, frequently in perplexing, puzzle-ridden, and metaphorical language that rewarded *adepti* and their students while it obstructed novices or outsiders. In contrast, the journal offered to present this rare knowledge on its pages in fairly straightforward language, and it began to refer to metals, salts, and chymical operations in consciously standardized ways.

Many of these increasingly standardized terms were taught in universities, such as the University of Jena, where the Academy member and physician Georg Wolfgang Wedel (1645–1721) taught. In 1673, a year after joining the Curiosi,

Wedel replaced his teacher Werner Rolfinck (1599–1673) as "director of chymical exercises" and became the first chair in practical medicine and chymistry in 1719.[44] A frequent contributor to the *Miscellanea curiosa*, a true believer in transmutation of metals, and known to later generations for his *Opiologia* (the most definitive text on opium until the nineteenth century), Wedel often treated first principles or standards in his entries. In an early observation of 1675 entitled "De Principiis chimicis," Wedel explained that he was driven to write in order to standardize some of the basic rules of *chimica*. So many opinions had emerged that conceal nature's arcana, he complained, that he needed to explain the chymical principles so that "unanimity in theorematic presuppositions could be established."[45]

Wedel best exemplifies the attempt to clarify the language of *chymia* and also the presuppositions that supported the understanding of the chymical principles of salt, sulphur, and mercury, especially their essences, affections, and effects. In his "Chimical Principles," he laid the groundwork for just such an effort. The clarification of symbols, already apparent in Wedel's 1675 observation, was further promoted in his 1705 textbook entitled *Introductio alchimiam*.[46] Like many professors of his time who wrote and published their own textbooks, Wedel focused his propaedeutic interests on simplification. He provided "correct" chymical symbols for students to memorize: salt, sulphur, and mercury as well as the elements earth, air, water, fire, and minerals and metals. Alchemy, Wedel warned, was a peculiar art that attracted people who liked obscure meanings conveyed by means of the hieroglyphics it shared with astrology. Wedel proposed that uniformity and consensus of symbol-use ought to replace the idiosyncrasies that obscure language promotes. Symbols, he suggested, were convenient notations for basic components, yet sometimes symbols were wisely ambiguous; therefore, it was incumbent upon students' judgment to distinguish and recognize the diversity of mercuries, sulphurs, and salts, such as incombustible sulphurs and corrosive salts.[47]

44. On Rolfinck at University of Jena, see Debus, *Chemistry and Medical Debate*, 24–25; Debus, "Chemistry and the Universities in the Seventeenth Century"; Moran, *Distilling Knowledge*, 106; and Newman and Principe, "Alchemy vs. Chemistry," 52.

45. "Ut concordia in praesuppositis theorematibus firmetur"; Wedel, "De Principiis chimicis," *Miscellanea curiosa* 6–7 (1675–76): 334–36.

46. Wedel, *Introductio in alchimiam*.

47. "Et in hac tamen aequivocatione latet sapientia, cum diversi sint mercurii . . . diversa sulphura, diversa salia: & in ipso mercurio omnino fit acrimonia salina & sulphur volatile incombustibile. Satis est, in sua arte id

Significantly, the case of Wedel demonstrates that the urge to clarify language and standardize knowledge about *chymia* emerged *not* from later eighteenth-century chemists but rather from seventeenth-century chrysopoeians (transmutation believers). This focus on clarity of symbols followed a long tradition, already prevalent when Andreas Libavius (d. 1616) asked that words be used clearly (rather than obscurely) and proposed that great improvements to the arts of *chymia* would stem from a more public, learned, and witnessed presentation of it, as Bruce Moran discusses in the preceding chapter.[48] Interestingly, it was to the University of Jena, and specifically to a much earlier professor of philosophy and medicine, Zacharius Brendel (the Elder), who predated both Wedel and Clauder (and their teacher Rolfinck), that Libavius hoped to express his anticipated reforms to the arts. What this movement toward a standardizing strategy reveals was that rather than a move away from the exchange of rare and secret curiosities, there was instead the move toward shared communication enhanced by a language common to physicians and erudite chymists, if still perplexing to those unfamiliar with (or uninterested in) an understanding of *chymia*.

In addition to the focus on standardization, Wedel's *Introductio in alchimiam* offered a course on alchemy (broadly understood), which is instructive with regard to his distinction of the terms "chimia" and "alchimia." While he debunked frauds and fallacies of *alchimists* and the abuses of *alchimia* that pertained to the fraudulent productions of the Philosophers' Stone, he supported "true" uses of the transmutational arts. He explained that *alchimia* was an ancient art derived from the Hermetic tradition, whose etymology stemmed from the addition of the definitive article "al" to the Greek word "chimia."[49] Further, his text offered recent experimental histories, including famous transmutation events, considerations of vegetables, animals, and minerals, a section on apparati and furnaces, and separate chapters on metals like antimony, silver, and mercury. All of these subjects fell under the category of *alchimia*, including the uses, symbols, and history of the arts. Yet when writing observations for the *Miscellanea curiosa* that involved principles of salt, sulphur, and mercury or that concerned the medical uses, medicaments, and even metallic transmutations, he used the term *chimica*. Perhaps like Clauder, Wedel used "chymiatria" for the making of medicaments

sibi licere, arbitrantes nosse ac distinguere"; ibid., 24.

48. On Libavius's concerns with regard to *chymia*, see Moran's chapter in this volume and his *Andreas Libavius and the Transformation of Alchemy*.

49. Wedel, *Introductio in alchimiam*, 15.

(what Rolfinck called "chymia medica") and used "chymia" for both the transmutation of metals and the production of metallic salts. They reserved "alchimia" for the broader knowledge to which all chymists were indebted.

The Curiosi's exploration of *chymia* within the journal's observations, however, did not entertain all possibilities. Among the many observations of *chymia,* none is devoted to the broader set of practices, such as glassmaking, the making of dyes and paints, or the making of gems or etchings. Such practices were part of a broader and older history that was referred to as "alchimia." In contrast, throughout their journal, the Curiosi employed the term "chymia," and as indicated above, used it in three interrelated ways: when defending transmutation (including histories of transmutational events), when referring to firsthand material practice (including descriptions and recipes of transmutation), and when engaging in social practices aimed at producing normative rules and standardized language. These distinctions between terms "alchimiam" and "chymia" differ greatly from later eighteenth-century binaries; they also differ from the earlier interchangeability of alchemy and *chymia,* since, at least on the pages of the *Miscellanea curiosa, chymia* was a subset of *alchimia.*

CONCLUSION

The exchange of chymical curiosities on the pages of the *Miscellanea curiosa* sanctioned its members to be collectors of the rare, strange, and useful, that is, to be Curiosi. This society of physicians' interest in *chymia* was not confined to chemical medicaments (*chymiatria*) but expanded into discussions surrounding transmutation, powers of mineral and metallic salts, and the production of reagents used to make all of the aforementioned. Moreover, the journal provided a public place to do work that bettered the social standing of the Curiosi. By categorizing certain groups as outsiders—fraudulent alchemists or outdated, bookish philosophers—the Curiosi exercised their own social flexibility, propelling themselves into positions that demarcated legitimate from illegitimate, at least virtually, at a time when this was not otherwise the case. The traffic in chymical matters privileged Curiosi as those who controlled the language, symbols, and principles of the subject, as reliable witnesses of true transmutations and as adepts of rare procedures and recipes. In their observations of *chymia,* the Curiosi raised their status to *adepti,* practitioners who shared a usable language and superior chymical knowledge.

The presentation of *chymia* in many observations of the *Miscellanea curiosi*, such as those of Sachs and Clauder, attempted a more contemporary spin. While Sachs relegated such masters as Ramón Llull, Arnold of Villanova, and Paracelsus to texts of the past, he distinguished his own Academy as not being as much intrigued by them, as it was informed by witnessed accounts of more recent, contemporaneous transmutation events. Likewise, Clauder defended *chymia* against Kircher's critique by arguing that combining past knowledge of alchemy with contemporary skills in metals and ovens made his own century especially productive. This shift in emphasis from the past to more contemporary knowledge and practice is strikingly unusual. And this shift dovetails with the self-portrait of Curiosi as being as experienced in *chymical* matters as they were confident in their powers to produce rare medicaments. In contradistinction to the past century's devotion to narratives featuring authors who received special powders from monks or mysterious strangers, the Curiosi valued practical manual knowledge. In short, they became *adepti*-physicians.

Just as the transmission of *chymical* curiosities helped shape the journal into a mouthpiece for authoritative physicians in the Holy Roman Empire, so too the journal's format of observations helped shape *chymia*. By privileging short, firsthand observations, editors sheared away the lengthier narratives characteristic of secretive adepts. In the final quarter of the seventeenth century, the format of the *Miscellanea curiosa* shaped *chymia* by favoring short observations, creating standardized terms and symbols, and focusing more on practice than theory, highlighted by firsthand manual skill. The very public face of the journal created conditions for making statements of fact, those that required witnessed observations and evidence of having useful knowledge of the arts of *chymia*. As if by a transmutation in authority, witnesses of chymical events changed from trusted public figures to newly trustable physician-adepts of the Curiosi. Physician-adepts through their journal observations contributed much more to *chymia* than merely philosophy and pharmaceuticals. Despite the changes that the new public format introduced, the desire to investigate bizarre, novel, and seemingly inexplicable events continued to drive publications, and it was through the publication of rarities—and each author's distributed performance—that what it meant to be a Curiosi physician, at least for some, materialized into what it meant to do *chymia*.

WORKS CITED

Barnett, Frances Mason. "Medical Authority and Princely Patronage: The Academia Naturae Curiosorum, 1652–1693." PhD diss., University of North Carolina, 1995.

Büchner, Andreas E. *Academia Naturae Curiosorum Historia*. Halle: J. J. Gebauer, 1755.

Clauder, Gabriel. "Transmutationis Metallorum Possibilitas." *Miscellanea curiosa . . .*, ser. 2, vol. 5 (1686): 372–74.

Collins, Harry M. *Tacit and Explicit Knowledge*. Chicago: University of Chicago Press, 2013.

Daston, Lorraine. "The Empire of Observation (1600–1800)." In *Histories of Scientific Observation*, edited by Lorraine Daston and Elizabeth Lunbeck, 81–116. Chicago: University of Chicago Press, 2011.

Daston, Lorraine, and Katharine Park. *Wonders and the Order of Nature*. New York: Zone Books, 1998.

Debus, Allen G. "Chemistry and the Universities in the Seventeenth Century." *Academia Analecta: Klasse der Wetenschappen* 48 (1986): 13–33.

———. *The Chemical Philosophy: Paracelsian Science and Medicine in the Sixteenth and Seventeenth Centuries*. 2 vols. New York: Science History Publications, 1977.

———. *Chemistry and Medical Debate: van Helmont to Boerhaave*. Canton, MA: Science History Publications, 2001.

———. "Iatrochemistry and the Chemical Revolution." In *Alchemy Revisited: Proceedings of the International Conference on the History of Alchemy at the University of Groningen, 17–19 April 1989*, edited by Zweder R. W. M. von Martels, 51–66. Leiden: E. J. Brill, 1990.

Eamon, William. *Science and the Secrets of Nature*. Princeton, NJ: Princeton University Press, 1994.

Emerton, Norma. *The Scientific Reinterpretation of Forms*. Ithaca, NY: Cornell University Press, 1984.

Evans, Robert J. W. *The Making of the Habsburg Monarchy, 1550–1700*. Oxford: Clarendon Press, 1979.

Faivre, Antoine. "An Approach to the Theme of the Golden Fleece in Alchemy." In *Alchemy Revisited: Proceedings of the International Conference on the History of Alchemy at the University of Groningen, 17–19 April 1989*, edited by Zweder R. W. M. von Martels, 250–55. Leiden: E. J. Brill, 1990.

Garber, Margaret. "Con or Craft: Defending Chrysopoeia in a Later Seventeenth-Century Medical Journal." *Cultural and Social History* 3 (2005): 264–72.

Hirai, Hiro. "Kircher's Chymical Interpretation of the Creation and Spontaneous Generation." In *Chymists and Chymistry: Studies in the History of Alchemy and Early Modern Chemistry*, edited by Lawrence M. Principe, 77–87. Sagamore Beach, MA: Science History Publications, 2007.

"Leges." *Miscellanea curiosa . . .* 2 (1671).

Moran, Bruce T. *The Alchemical World of the German Court: Occult Philosophy and Chemical Medicine in the Circle of Moritz of Hessen (1572–1632)*. Stuttgart: F. Steiner Verlag, 1991.

———. *Andreas Libavius and the Transformation of Alchemy: Separating Chemical Cultures with Polemical Fire*. Sagamore Beach, MA: Science History Publications, 2007.

———. *Distilling Knowledge: Alchemy, Chemistry and the Scientific Revolution*. Cambridge, MA: Harvard University Press, 2005.

———. "A Survey of Chemical Medicine in the 17th Century: Spanning Court, Classroom and Cultures." *Pharmacy in History* 38, no. 3 (1996): 121–33.

Newman, William R. *Gehennical Fire: The Lives of George Starkey, an American Alchemist in the Scientific Revolution*. Cambridge, MA: Harvard University Press, 1994.

Newman, William R., and Lawrence M. Principe. "Alchemy vs. Chemistry: The Etymological Roots of a Historiographic Mistake." *Early Science and Medicine* 3 (1998): 32–65.

Nummedal, Tara. *Alchemy and Authority in the Holy Roman Empire*. Chicago: University of Chicago Press, 2007.

Ornstein, Martha. *The Rôle of Scientific Societies in the Seventeenth Century*. New York Academy of Medicine Library, History of Medicine Series 6. Chicago: University of Chicago Press, 1938.

Polanyi, Michael. *The Tacit Dimension*. Garden City, NY: Doubleday Books, 1967.

Pomata, Gianna. "Observation Rising: Birth of an Epistemic Genre." In *Histories of Scientific Observation*, edited by Lorraine Daston and Elizabeth Lunbeck, 45–80. Chicago: University of Chicago Press, 2011.

Priesner, Claus. "*Defensor Alchymiae*: Gabriel Clauder versus Athanasius Kircher: Defense Strategies of Alchemists in the Seventeenth and Eighteenth Century." In *Alchemy Revisited: Proceedings of the International Conference on the History of Alchemy at the University of Groningen, 17–19 April 1989*, edited by Zweder R. W. M. von Martels, 229–38. Leiden: E. J. Brill, 1990.

Principe, Lawrence M. *The Aspiring Adept: Robert Boyle and His Alchemical Quest*. Princeton, NJ: Princeton University Press, 1998.

Principe, Lawrence M., and William R. Newman. "Some Problems with the Historiography of Alchemy." In *Secrets of Nature: Astrology and Alchemy in Early Modern Europe*, edited by William R. Newman and Anthony Grafton, 385–431. Cambridge, MA: MIT Press, 2001.

Sachs von Lewenheimb, Philipp J. "Aurum chymicum." *Miscellanea curiosa* . . . ser. 1. vol. 1 (1670): 68–70.

Shackelford, Jole. *A Philosophical Path for Paracelsian Medicine*. Copenhagen: Museum Tusculanum Press, 2004.

Shapin, Steven. *A Social History of Truth: Civility and Science in Seventeenth-Century England*. Chicago: University of Chicago Press, 1994.

Smith, Pamela. *The Business of Alchemy: Science and Culture in the Holy Roman Empire*. Princeton, NJ: Princeton University Press, 1994.

Thorndike, Lynn. *History of Magic and Experimental Science*. Vols. 7–8. New York: Columbia University Press, 1958.

Tylecote, Ronald F. *The Early History of Metallurgy in Europe*. London: Longman Group, 1987.

Wedel, Georg Wolfgang. "De Principiis chimicis." *Miscellanea curiosa* . . . 6–7 (1675–76): 334–36.

———. *Introductio in alchimiam*. Jemae: Sumptibus Johannis Bielkii Christophori Krebsii, 1705.

Chapter 5

Phlogiston and Chemical Principles

The Development and Formulation of Georg Ernst Stahl's Principle of Inflammability

Ku-ming (Kevin) Chang*

The foundation of modern chemistry has traditionally been viewed as arising out of the debunking, first, of alchemy and, then, of the phlogistic chemistry of Georg Ernst Stahl (1659–1734). That historical narrative considers alchemical practitioners as credulous and unscientific, and interprets Stahl's chemistry as erroneous and detrimental to the development of what ultimately became "modern" chemistry. Trained to resist the received view of their subject, historians of alchemy, among them the students of Allen Debus, look past the branding as an obstruction of the work of Stahl by mainstream history of science.[1] Indeed,

* I wish to thank Karen Parshall for her careful reading of several versions of this article and for her advice for revision.

1. There is another reason why students of Allen Debus would be interested in Stahl. In the many books he published and in the courses he taught on the history of Paracelsianism, Debus almost always closed briefly with two early eighteenth-century chemists, Hermann Boerhaave (1668–1738) and Stahl. While Boerhaave is, relatively speaking, well studied, many questions about Stahl remain. As Allen Debus's student at the University of Chicago, I had the privilege to ask him about Stahl. He always very modestly replied that he did not know enough to say more than he had written. It was clear to me that this was his way of telling me to begin on

this chapter aims to elucidate why historians of science *should* move beyond this branding of Stahl's work and take a serious look at, for example, his place in the history of chemical principles.

If Stahl has not been viewed as one of the major contributors to the history of modern chemistry, neither has his work been totally dismissed nor ignored by historians. Hélène Metzger, for instance, credited Stahl with two contributions. First, she judged that Stahl "transformed and broadened [the] concept of combustion by ranging calcination amongst the combustion phenomena: the calcination of metals was the same as the combustion of organic bodies and sulfur." Second, she held that Stahl was able "to prove his contention experimentally by causing 'combustibility' to pass from charcoal and pitch into both the metallic limes (reviving the metals) and vitriolic acid (restoring sulfur), as it were reversing the process of combustion."[2] This nicely highlights Stahl's two main contributions at the same time that it omits his by-now notorious notion of phlogiston. Many eighteenth-century chemists and savants, however, were happy to acknowledge both of Stahl's contributions by connecting them to phlogiston.

Thus, instead of avoiding phlogiston, this chapter analyzes Stahl's formulation of it in the context of the history of chemical principles. Unlike previous studies on the history of phlogiston,[3] however, it does not focus solely on historical discussions of combustion, a phenomenon with which phlogiston has long been associated. Rather it examines the development of the sulfurous principle in terms of what the chemical community expected of the chemical principle. Such a study requires an examination of several phases in the development of chemical principles from the late sixteenth to the eighteenth centuries. That development began with the three principles, or *tria prima*, that Paracelsus and his followers championed. It continued with the early acceptance of the principles in the first half of the seventeenth century, was met by criticisms of the *tria prima* at mid-century, and was followed by chemists who, unwilling to give up the three principles completely, adjusted their formulations of chemical principles in response to such criticisms. These formed the background for Stahl's knowledge of the sulfurous principle.

---

a research agenda of my own.

2. Metzger, *Newton, Stahl, Boerhaave, and Chemical Doctrine,* 213. The metallic limes in this sentence would be better translated as "metallic calces."

3. See, for example, White, *History of the Phlogiston Theory*; and Coleby, "Studies in the Chemical Work of Stahl."

The chapter next explores the interpretations of chemical principles at different points of Stahl's career, analyzes his formulation of phlogiston, and addresses his importance in carrying the principle of sulfur into the eighteenth century. The analysis of Stahl's work focuses on his definitive formulation of the sulfurous principle or phlogiston in *Zufällige Gedancken und nützliche Bedencken über den Streit, von dem so genannten Sulphure* of 1718 (hereinafter *Treatise on Sulfur*). It also addresses his theoretical generalization and experimental observations of that principle and highlights his effort to assess the adequacy and material consistency of chemical principles. This chapter thus elucidates the historical significance of Stahl's formulation of phlogiston by contextualizing his work within both his own intellectual development and the broader development of chemical thinking in Europe.

The Paracelsian *Tria Prima*: Early Debates

The *tria prima* is often traced to the work of Paracelsus (1493–1541). He built his doctrine, on the one hand, on the Geberian theory of sulfur and mercury and, on the other, on Aristotle's elemental theory of earth, air, fire, and water. The Geberian corpus had taught that mercury and sulfur were the constituent materials of metals, although they were each a product of the four elements.[4] Paracelsus added salt to mercury and sulfur, made them the *tria prima,* and viewed them as present not only in metals but also in all natural substances. As in Geberian chemistry, sulfur denoted the substance that lent gold its distinctive yellow color and accounted for the inflammability of substances, like common sulfur, that caught fire easily. Mercury continued to signify the lustrous and malleable material nature that all metals shared, while it also became known as the principle of spirit, representing volatile "spirits" (vapors) that easily evaporated upon heating and condensed as distillates when cooled. Salt signified the quality of natural material that resisted fire or that coagulated as a solid. The *tria prima* can be said to have reflected observations of chemical operations with fire such as distillation and calcination. In a very general sense, these principles, like the Aristotelian elements, accounted for material qualities by their presence.

Paracelsus and his early followers presented a somewhat complicated framework in which the four elements and the three principles coexisted. For Paracelsus,

4. For Geberian matter theory, consult Newman, *Summa Perfectionis of Pseudo-Geber.*

the Aristotelian elements served not as elementary constituents of objects but rather as matrices in which objects were generated. Not actual chemical substances, the *tria prima* worked to inform substances with their material qualities and were compared to the soul working in and on the body or to seeds that embodied the character of each individual and species.[5] Although producing physical qualities, the principles were spiritual forces or form-conferring agents rather than material constituents. They differed qualitatively in all substances. "There were," Paracelsus asserted, "as many sulfurs, salts, and mercuries as there are objects."[6] Each object, then, had its own sulfur, salt, and mercury. Peter Severinus (1542–1602), credited for bringing Paracelsus's ideas to the attention of the scholarly world, continued the scheme that accepted both the three principles and the four elements.[7] Joseph Duchesne (1544–1609), the physician to King Henry IV of France, who led the Paracelsian physicians in a battle against the recalcitrant medical faculty at Paris, taught that the three principles—sulfur, mercury, and salt—must be present in all things.[8] They worked as active principles, while the traditional elements were passive.[9]

Pro-Paracelsian figures were not alone in their acceptance of the three principles; avowed foes of Paracelsian alchemy and eclectic chemical authors accepted them as well. Andreas Libavius (1555–1616), a fierce critic of Paracelsian chemistry, embraced the *tria prima* in spite of the fact that he accused the Paracelsians of taking the notion from Aristotle.[10] Compared with steadfast proponents of Paracelsianism like Severinus and Duchesne, Daniel Sennert (1572–1637) was at best an eclectic. For him, the three principles were formed of the four elements, which were the ultimate constituents of bodies. They, however, served to explain phenomena—odors, tastes, colors, solidity, inflammability, etc.—that the four elements could not.[11]

Practical chemists also accepted the *tria prima*. Jean Béguin (1550–1620), for example, opened the second chapter of his textbook with this definition:

5. Pagel, *Paracelsus,* 103, 129.

6. Cited in Debus, *Chemical Philosophy,* 79.

7. Shackelford, *Philosophical Path for Paracelsian Medicine,* 106, 166, 182.

8. Debus, *French Paracelsians,* 55.

9. Fire, however, was seen as heaven itself, and thus was not an element. In this, Duchesne followed Paracelsus and Girolamo Cardano; Debus, *Chemical Philosophy,* 161.

10. Ibid.

11. Clericuzio, "'Sooty Empiricks' and Natural Philosophers," 333; Clericuzio, *Elements, Principles and Corpuscles,* 26–27.

"Chemical solution is an operation wherein the natural mixt by separation of heterogeneal parts, is deduced to its own three principles, viz., Mercury, Sulfur, and Salt, of which by Nature it was first made."[12] The three principles in his teaching then became concrete materials that could be extracted from natural substances. This reflected a trend toward the materialization of chemical principles that only increased over the course of the seventeenth century.[13]

Despite different interpretations, the doctrine of the three principles was widely accepted by chemists in the seventeenth century. A variant on it added earth and water to the *tria prima* to make five chemical principles, as chemists tried to account both for the dregs or solid residues that remained after heating or dissolution and for the enclosed fluid or inherent fluidity of natural substances. Whether three or five, the chemical principles were so successful that they replaced the Aristotelian elements as the fundamentals of nature in the Paris Academy of Sciences of the 1660s,[14] not to mention in chemical communities elsewhere in Europe. In fact, before then, the *tria prima* had become such an entrenched part of general chemical teaching during the lifetime of Jean Baptiste van Helmont (1579–1644) that he felt the need to pen his disagreements.

Van Helmont took issue with the Paracelsian thesis that every natural body was formed of all three principles and doubted that natural bodies could be reduced to the three principles by fire (which implied direct heating and distillation). He argued that what chemists collected at the end of their fire analyses were new substances that heating produced. It was thus hopeless to reduce natural bodies to their original constituents, let alone to try to prove that all three of the principles were present.[15]

When the posthumous publication of this Belgian chemist's *Ortus medicinae* (1648) did not stop interest in the chemical principles, Robert Boyle (1627–91) continued van Helmont's critique in his well-known *Sceptical Chymist* (1661). Although he reiterated much of van Helmont's critique, Boyle also added new points. He noted that some "mixts," that is, chemical compounds, were divisible into more than five distinct substances, thus suggesting that the number

12. Béguin, *Tyrocinium Chymicum,* 19.

13. See also Kim, *Affinity, That Elusive Dream,* 27.

14. Jacob, *Strangers Nowhere in the World,* 54. Jacob's remark is based on her work in the archives of the French Academy.

15. See van Helmont, *Ortus medicinae,* 399–405; and Debus, *Chemical Philosophy,* 320–24.

of chemical principles might not be limited to five.[16] On the other hand, as he pointed out, there were substances—gold and silver, for example—that could not be proven to be composed of these principles.[17] He especially questioned the nature and consistency of the products obtained in fire analyses or the principles in general. These products were not always pure and simple. A substance was at some times considered the sulfur of a body and at other times its mercury; and a sulfur was said to be inflammable in some instances and incombustible in others.[18] For Boyle, to be the constituent substance of all things, the principles would have had to exhibit homogeneity and consistency when extracted from different objects. Like van Helmont, Boyle challenged the validity of fire analysis as well as the universality and consistency of the chemical principles.

Criticisms regarding consistency targeted not the early Paracelsian conception of the principles as spiritual or form-conferring agents but rather their identity as purely material constituents. In early Paracelsians' conceptions, as seen above, the chemical principle as a spiritual agent was meant to vary from one body to another. Such a principle diverged greatly from the targets of van Helmont's—and especially Boyle's—outspoken criticisms. These two critics—and, just as important, those who accepted their criticisms—internalized the materializing trend of the chemical principles. They wanted the principles, if any, to be pure, homogeneous, and consistent, and ultimately to constitute the matter of physical bodies.

MODIFIED CHEMICAL PRINCIPLES IN THE SECOND HALF OF THE SEVENTEENTH CENTURY

While the teaching of the chemical principles continued after the criticisms of van Helmont and Boyle, its remaining proponents—and there were many in the chemical community—were compelled to make adjustments. Thomas Willis (1621–75), Nicolas Lémery (1645–1715), and Johann Joachim Becher (1635–82) serve as three examples. They, like many other contemporaries, opened their chemical works with a section that presented the chemical principles.

16. Boyle, *Sceptical Chymist*, 191.

17. Ibid., 213.

18. Ibid., 201, 209. Also see the analysis in Principe, *Aspiring Adept*, 44–46.

Willis taught five principles in his well-received *De fermentatione* (1659), a text, first appeared between the publications of van Helmont's and Boyle's criticisms, that deliberately introduced corpuscular thinking into the interpretation of the chemical principles. Although the Aristotelian forms were helpful at times, Willis suggested that the "Epicurean Philosophy" explained better the inner working of material change with the model of corpuscularian union. The corpuscular model, for example, helped explain the degrees of the principles' activities. The principles of spirit and sulfur were particularly active because they easily "unlocked" the particulate union of the mixed body. Salt resisted inflammation, for it "detains the sulfurous particles in its bosom."[19] Thus it played only a supportive role to the other two active principles. Altogether, the particles in motion and their proportions in compounded or "mixt" bodies caused the "beginnings and ends of things."[20] Aware of the uncertainties of the distinction of the principles, Willis acknowledged that spirit and sulfur were much alike, as they were both inflammable. Yet he insisted that they were different, as sulfur existed in bodies that were almost devoid of spirits.[21]

Willis proposed a pragmatic definition of the principle, suggesting that the chemist accept the smallest sensible products of analysis as principles. Though he assumed that all the principles consisted of particles, he hardly speculated on the shapes and sizes of the principles' particles. They were, Willis supposed, "not simple and wholly uncompounded entities, but such kind of substances only, into which physical things are resolved . . . last sensible."[22] He acknowledged the limits both of the human senses and of chemical analysis in his day. As both prevented the chemist from knowing what the particles of the ultimate principles were, he had to be content with the smallest sensible products of analysis.

Lémery presented another response to the critique of the chemical principles. His *Cours de chymie* (first published in 1675) went through at least eleven editions in French and many translations into English, German, Dutch, and Italian, some of which were repeatedly published until the mid-eighteenth century. In his text, he taught the existence and qualities of five principles. In his view, fire

19. Willis, *Dr. Willis's Practice of Physick*, 5.
20. Ibid., 2.
21. Ibid., 6.
22. Ibid., 2.

altered substances and produced new ones, as the "Scepticks" indicated.[23] Yet it could not be denied that the oils, that is, "sulfurs" in olives, almonds, and many other nuts and fruits, could be drawn out by simply pressing and heating. Chemical analysis was therefore not entirely unreliable for extracting the constituent principles of material substances.[24] However, in remarks that immediately followed the section on chemical principles and that were apparently inserted in later editions, Lémery disputed the truth of the principle of spirit (also known as mercury). For him, three kinds of substances were best qualified as spirits: the "spirits of animals" (such as the spirit of hart's horn), the "burning spirit of vegetables" (such as the spirit of wine and of juniper), and what was known as the "acid spirit." The first, however, was nothing but a salt dissolved by a little phlegm (which is the water principle), the spirit of wine was only an "oil exalted" (oil is a sulfur), and the acid spirit was a salt turned fluid. As a result, "it must be concluded, that the Spirit or Mercury which Chymists have talk'd of, is a mere Chimaera, that serves only to confound mens [*sic*] minds, and render Chymistry unintelligible."[25] Although at first proposing the chemical principles, Lémery openly questioned the distinct nature of the mercury principle and even disputed its adequacy as a principle.

The discussions of the mercurial principle, also known as spirit, in the works of Willis and Lémery reveal the inherent problem about its character. In the Geberian and Paracelsian models, the mercurial principle imparted malleability and meltability to metals. As the material essence of metals, it also shared their qualities of heaviness and denseness. Increasingly, however, chemists also took the mercurial principle to represent the fluidity and volatility of spirits such as alcohol (the spirit of wine). Responsible for the weight of metals as well as the volatility of spirits, the principle of mercury/spirit had a problematic double identity.

Becher proposed the version of the chemical principles that Stahl followed early in his career. In the *Physica subterannea* (1669), he echoed Boyle's critique of the Paracelsian principles.[26] Unlike Boyle, however, Becher maintained that

23. Lémery, *Course of Chymistry*, 4.

24. Ibid., 4–5.

25. Ibid., 5–6. This exclusion of spirit or mercury from the principles did not appear in the 1677 edition. Having first appeared in the 1686 edition, it then remained in the book until the last edition, which appeared in 1746.

26. Becher, *Physica subterranea*, 53–61; and Clericuzio, *Elements, Principles and Corpuscles*, 195. Becher's *Physica subterranea* was first published in 1669 as *Actorum laboratorii chymici Monacensis, seu Physicae subterra-*

there were only two principles: water and earth. Three kinds of earth existed in what he called the subterranean substances, meaning minerals and metals. They were fusible or vitrifiable earth, the principle of solidity and fixity; oily or fatty earth, the principle of inflammability; and fluid earth, the principle of malleability. These principles were explicitly compared to salt, sulfur, and mercury,[27] although they differed from their Paracelsian predecessors in at least two ways: they were applied only to minerals (which included metals), and they were species of the earthy principle. As in Willis's teaching, these earthy principles were mixed substances rather than simple uncompounded material. This needs to be understood in terms of Becher's scale of corpuscular composition, which was characterized by an increasing order of combinatory complexity. The simple, uncompounded particles of matter were *simplicia*. The different orders of compounded particles were *mixta*. Particles of *simplicia* united with one another to form *composita*, the first-order compounds. Compounds of the next order were *decomposita*, and those of still higher order were *superdecomposita*.[28] For Becher, the principles were not simple substances, but *mixta* or compounds of simple particles.

The works of Willis, Lémery, and Becher encompassed several features of the chemical teachings prevalent in Stahl's formative years. Most chemical authors taught or accepted the chemical principles, even though the number often expanded from three to five. The critics and the proponents of the principles almost all advocated corpuscularian interpretations of matter. They all acknowledged the difficulty or impossibility of proving that the principles were simple, uncompounded substances. They thus adjusted the chemical principles in such a way that they became either compounds of simple substances or materials last reducible by available chemical analysis (thus not necessarily simple), or both. Critics and proponents alike all pressed for the principles to be materially distinct and consistent. If not adequately meeting the distinction or consistency requirement, individual principles such as mercury could be eliminated. All these features ultimately played a part in Stahl's formulation of the principle of inflammability in his *Treatise on Sulfur*.

---

neae libri duo. It was republished by Stahl in 1703 as *Physica subterranea*; I use the 1738 republication of Stahl's 1703 edition. For the broader significance of Becher's chymical work, see Smith, *Business of Alchemy*.

27. Becher, *Physica subterranea*, 61–84.

28. Clericuzio, *Elements, Principles and Corpuscles*, 196; and Becher, *Physica subterranea*, 273.

The Development and Formulation of Stahl's Inflammable or Sulfurous Principle

At the beginning of his career, Stahl taught the Becherian system of chemical principles; he gave chemical lectures at Jena from 1684 to 1686, immediately after receiving his medical degree there. Some three decades later, a compilation of lecture notes taken by a certain student was published as *Chymia rationalis et experimentalis* (1720) and used by Peter Shaw as the basis of his translation, *Philosophical Principles of Universal Chemistry* (1730), the work by Stahl best known to the English audience.[29] The principles Stahl taught then included water and three species of earth. Following Becher, he listed among the earths the *terra pinguis*, or fatty earth, which corresponded to the traditional sulfurous principle. He made no mention of phlogiston in the lectures.

Thereafter, in his earliest published monograph, *Zymotechnia fundamentalis* (1697), Stahl considered the nature of sulfur. Although nowhere in the book did he discuss the number and nature of the chemical principles, he did refer to sulfurous substances as "sulfurs." He had found problematic some previous characterizations of sulfur, such as "sulfur is the mixing cause, induces new forms, mixes itself with other things, is the most active principle, author or *anima* of the whole fermentative motion, and carries through the entire commotion by its impetus [*nisus*]."[30] He rejected all of these characterizations, ultimately understanding sulfur as but an agile particle that was moved by the ultimate instrument of material change, that is, heat. Consisting of inter-corpuscular motions, heat was transmitted primarily by a special medium, æther, an entity purported by a considerable number of mechanical philosophers at the time. For Stahl, æther pervaded sublunar space, surrounded all bodies, and was the easiest particle to move thanks to its extremely small size.[31] All bodies, including sulfur, were passive, for they were just recipients of motion.[32] Sulfur therefore was neither a soul-like agent that induced the form nor an active principle with its own momentum.

29. For the history of the publication of Stahl's chemical lectures, see Chang, "Georg Ernst Stahl's Alchemical Publications," esp. 26–27.

30. "Sulphur esse causam miscentem, formas novas inducere, sese aliis rebus immiscere, principium activissimum, totius motus fermentativi, authorem et veluti animam, peragere totam illam commotionem, nisu suo . . . ."; Stahl, *Opusculum*, 171.

31. Chang, "Fermentation, Phlogiston and Matter Theory," 53–57.

32. Ibid., 57–59.

Stahl first presented phlogiston within a formulation of chemical principles in his *Specimen beccherrianum* (1703), a commentary on Becher's *Physica subterranea* on the occasion of its republication (which Stahl arranged).[33] In that work Stahl asserted that earth, water, and fire were "material principles of mixtion" (principia mixtionis materialium).[34] Shortly afterwards in the book, however, he identified phlogiston as the material of fire, and then reaffirmed the Becherian three species of earth, equating phlogiston as the second earth.[35] He repeated much of what he had written about heat as the principal instrument of natural and chemical reactions and as the effect of inter-corpuscular movements. He, however, replaced the medium of æther essentially with that of air, vaguely suggesting that the existence of æther could not be demonstrated.[36]

The *Treatise on Sulfur* (1718), Stahl's only systematic account of the principle of inflammability, represented his mature thoughts on the subject. He began the *Treatise* by laying out his motivations for its publication. In his view, chemical authors often subsumed under sulfur, mercury, and salt things that they did not understand.[37] He therefore took it upon himself to elucidate the real principle. He recounted that he had studied sulfur in previous works: *Zymotechnia fundamentalis, Specimen beccharianum,* and a 1712 dissertation on the anatomy of artificial sulfur.[38] As these works were published in Latin, they were not available to all those who may have been interested in their content. Such limitation on the readership demanded a publication in the vernacular. The *Treatise on Sulfur,* in particular, promised to give the most exact details of the chemical principle he referred to as the inflammable principle (*brennliches Wesen*) or the principle of inflammability (*Principium der Verbrennlichkeit*).[39] In one place, he distinctly named it "phlogiston."[40]

33. Stahl, *Specimen beccherianum.*

34. Ibid., 16.

35. Ibid., 40, 44.

36. Ibid., 35–36, 43.

37. Stahl, *Zufällige Gedancken und nützliche Bedencken,* 31. I have also consulted the French translation, for this reference; see Stahl, *Traité du soufre, ou remarques sur la dispute qui s'est élevée entre les chymistes,* 18. I will cite both the German original and the French edition (in brackets).

38. Stahl cited this work as *De anatomia artificialis sulfuris.* It must be *Dissertatio medico-chymica inauguralis, qua solutio martis in puro alcali et anatomia sulphuris communis, sistuntur . . .* (Halae Magdeb[urgicae]: Henckel, 1712).

39. Stahl, *Zufällige Gedancken und nützliche Bedencken,* 76–77 [*Traité du soufre,* 54–55].

40. Ibid., 80 [57].

In the *Treatise,* Becher stands out as a trustworthy authority on many points. It is, Stahl suggested, difficult to ascertain when the chemical principles were first invented. At least from the time of Paracelsus, chemists had considered salt, sulfur, and mercury as the principles of all natural bodies. While chemists had previously spoken of them as analogous to material qualities, more subtle chemists had begun to see them as pure and unmixed substances of identical qualities, the combination of which gave rise to all natural things.[41] Becher, on the other hand, had limited his principles to the mineral kingdom and regarded them not as "simple" (meaning uncompounded) substances, but as compounded bodies, or *composita.*[42] In general, Becher provided clear, simple, and evident demonstrations of his subjects and showed the difference between knowing things through the senses and gaining knowledge of them through meditation and reflection.[43]

By comparison, Johannes Kunckel (1630–1703) was Stahl's antagonist in the *Treatise on Sulfur.* The posthumous publication of Kunckel's *Laboratorium chymica* in 1715 apparently motivated Stahl to publish his treatise on sulfur. For him, Kunckel was a "well-experienced and downright thoughtful artisan," who "through precisely observed empirical knowledge and reflections drawn therefrom" dared to understand and conceptualize "total chemical reactions and, especially, the metallic and mineral properties and occurrences."[44] In short, Kunckel was a chemist accomplished in both practice and knowledge. Yet in Stahl's view, this chemist lacked theory and was unable to present his observations and experiments in a clear and detailed way. For example, Stahl found fault with Kunckel's conflicting suppositions that were associated with his proposition that all metals were composed of "a mercuric, a saline, and an earthy essence [or principle]."[45] He sometimes supposed that a viscous material existed between and bound the mercurial and saline principles of the metal, and sometimes that the viscous material existed between the metal's saline and earthy principles. For Stahl, Kunckel failed to determine if this material was itself a principle, or if it was a part or a property of the mercurial

41. Ibid., 69–70 [49].

42. Ibid., 71–72 [50–51].

43. Ibid., 74–75 [53–54].

44. "wohlgeübter, und mit allem Ernst nachdencklicher Kunst-Arbeiter . . . , sich unterfangen, sowohl durch viele genauer beobachtete Erfahrungen als derauß gezogene Betrachtungen, sowohl den gesamten Chymischen Vorfällen, als insbesonderere denen metallisch und mineralischen Eigenschaften und Ereignungen näher zu kommen und einen vernehmlicheren Begriff davon darzustellen"; ibid., 46 [30].

45. "ein Mercuialisches, ein saltziges und ein erdiges Wesen"; ibid., 55 [38].

principle. Worst of all, he never showed that this viscous material was a corporeal substance.[46]

In contrast, Stahl insisted on the existence and consistency of sulfur or the inflammable principle in metals. He paid close attention to the production of metals from ores or *Asche* (rusts or calces).[47] In his view, since foundry workers had never questioned why they covered the ores of lead or other metals under layers of coal in the casting furnace, he was the first to realize that coal played the role of restoring metals. This restoration or "reduction" returned metals to their brightness, ductility, density, and consistency, owing to the sulfurous or inflammable principle the coal imparted. Indeed, metals could also be reproduced, or reduced to their natural states, by mixing their calx with coal.[48] Conversely, metals gave off their sulfur or phlogiston when heated in the open air or combined with materials, such as acids, that were deficient in the inflammable principle. Once deprived of phlogiston, metals became brittle and blackened calces. In this state, they could no longer be affected by acids.[49] For Stahl, the reduction of different metals from their ores or calces by charcoal shows the consistency of sulfur or phlogiston as the essential component of metals.[50]

Stahl showed that the sulfurous or inflammable principle existed consistently not only in metals but in all inflammable substances. For him, this was demonstrated by its ready passage from one substance to another across the kingdoms of minerals, plants, and animals. It was absorbed by plants via the air and the earth and resided as their resinous and oily parts. That this was the case was reflected in the fact that niter, known for a richness in the sulfurous principle greater than that in any other salt, facilitated and accelerated the growth of plants.[51] Since plants served as food for herbivorous animals, and carnivorous animals ate herbivorous ones, the inflammable principle was converted to animal fats and greases, which were inflammable as well.[52] For Stahl, the conversion or circulation of oily and inflammable parts from one vegetable, animal, or mineral substance to another confirmed that these substances all contained one

46. Ibid., 64–68 [43–48].

47. Stahl referred to the rust or calx of the metal (copper, iron, tin, or lead, for example) as "Asche" (ash). It was translated as "cendre" (ash) or more often "chaux" (usually meaning lime) in the French edition.

48. Ibid., 131–35 [106–11].

49. Ibid., 195–96 [163–64].

50. Ibid., 106–32 [82–107].

51. Ibid., 103 [79].

52. Ibid., 84–86 [61–63].

underlying principle of inflammability. This was corroborated by the decomposition of these substances into oily and inflammable parts.[53] The coal-like quality of their ashes or solid residues after burning also helped confirm that they were of the same nature as coal.[54] As Mikulas Teich noted, Stahl gave phlogiston a greater role in unifying the three kingdoms than Becher had done.[55]

Stahl did experiments to show that sulfur, like metals, could be reduced by combining phlogiston with the product of its burning. His experimentation on the composition of sulfur, which has been studied by Jon B. Eklund,[56] would become exemplary in eighteenth-century Stahlian chemistry. His experimentation consisted of both analysis and synthesis. He analyzed sulfur by burning it in the open air. The burning of sulfur gave off a fume, which then turned into vitriolic acid in the presence of moisture.[57] For Stahl, this indicated that vitriolic acid came out of sulfur and was thus a component of the latter. Since vitriolic acid was not inflammable, sulfur must have lost its inflammability when losing its phlogiston into the air. Sulfur therefore must have been a compound of these two components, that is, vitriolic acid and the inflammable principle. This analytic method was known as deflagration.[58] Next, Stahl synthesized sulfur by combining the inflammation principle and vitriolic acid. This had been shown by Boyle, who produced sulfur by mixing vitriolic acid with turpentine oil (a phlogiston-rich substance). Stahl provided another more sophisticated synthetic method that he had proudly presented in prior works.[59] He produced so-called vitriolated tartar (K_2SO_4) by heating fixed alkali (K_2CO_3 or Na_2CO_3) with vitriolic acid. He then placed the product, vitriolated tartar, together with ground charcoal in a white-hot crucible. Stahl reasoned that the vitriolic acid in the compound seized phlogiston from charcoal, resulting in common sulfur and fixed alkali.[60] This can be represented as,

53. Ibid., 106 [82].

54. Ibid., 105 [121].

55. Teich also pointed out that Becher never taught that the *terra pinguis* or phlogiston was released from the burning of inflammable materials; Teich, "Interdisciplinarity in J. J. Becher's Thought," 31.

56. Eklund, "Chemical Analysis and the Phlogiston Theory."

57. As will be seen below, the burning produced sulfurous oxide.

58. Eklund, "Chemical Analysis and the Phlogiston Theory," 24.

59. Stahl, *Zymotechnia fundamentalis*; Stahl, "Experimentum novum verum sulphur arte producendi (1697)"; and Stahl, "De arcani duplicati et tartari vitriolati genealogia (1701)." The second title was published in a monthly serial by Stahl titled *Observationum chymico-physico-medicarum curiosarum*, while the third was published in the third volume of *Observationes selectee*. All three were included in Stahl, *Opusculum*. I will cite the pagination in the *Opusculum*.

60. Stahl, *Zymotechnia fundamentalis*, in *Opusculum*, 142–43.

vitriolated tartar (= vitriolic acid + fixed alkali) + charcoal (phlogiston) —> sulfur + fixed alkali

When the fixed alkali on both sides of the formula cancel each other out, the remaining part of the formula shows that vitriolic acid and phlogiston, when combined, form sulfur. Stahl also showed the reverse of the formula to be true. That is, fixed alkali and sulfur, when combined, could produce vitriolated tartar. This confirmed the composition of vitriolated tartar by synthesis. There was, Stahl pointed out, a medium product of this reaction, namely, *hepar sulphuris* (liver of sulfur), red in color and very bad in smell.[61] Liver of sulfur, which was unstable, could also release sulfur when a weak acid-like vinegar was added. In the *Treatise on Sulfur*, Stahl instead mixed vitriolic acid and charcoal (representing phlogiston) to form liver of sulfur. Since liver of sulfur could release sulfur, this confirmed that the synthesis of vitriolic acid and phlogiston produced sulfur.[62] The composition of sulfur was thus demonstrated by solid experimentation.

Niter was comparable to sulfur in composition, made up, as it was, of the inflammable principle and an acid salt.[63] Citing Becher on the composition of niter, Stahl reasoned that acid salt of niter, which was nitric acid, was formed by a particle of earth and a particle of "very delicate" water. Likewise the saline part of sulfur, that is, vitriolic acid, was a compound of a subtle earth particle and a water particle.[64] In burning both niter and sulfur, the particle of water was expanded rapidly through heating and thereby reduced to a vapor similar to air. Niter burned violently or even exploded, while sulfur burned more gently. This was because the earthy constituent of sulfur was grosser and thus decelerated the penetration and action of the inflammable substance.[65]

This understanding of sulfur and its acid also resulted in a reinterpretation of the actions of acids. Acids, such as those of sulfur and niter, were phlogiston-hungry substances. Their corrosive effects were dictated by their strong tendency to wrest the inflammable principle from substances such as metals. Alkalis, which were not considered in the *Treatise on Sulfur*, became, by implication, substances

61. *Hepar sulfuris* is a mixture not very well defined, consisting of potassium sulfide (K_2S), potassium polysulfide (K_2S_x), potassium thiosulfate ($K_2S_2O_3$), and probably potassium bisulfide (HKS). See Eklund, "Chemical Analysis and the Phlogiston Theory," 29.

62. Eklund, "Chemical Analysis and the Phlogiston Theory," 27–31; Stahl, *Zufällige Gedancken und nütliche Bedencken*, 108–110 [*Traité du soufre*, 83–85].

63. Ibid., 185 [155].

64. Ibid., 113 [87–88].

65. Ibid., 186, 189–90 [156, 159].

rich in the inflammable principle. The interaction of acids and alkalis, which had attracted much attention in the second half of the seventeenth century, thus received a new theoretical underpinning.

Moreover, Stahl's interpretation of the composition of sulfur would eventually replace the traditional sulfurous principle with phlogiston. For Stahl, phlogiston was the inflammable principle in sulfur, and indeed in all inflammable bodies. Sulfur, no longer seen as a simple substance, did not enter into the composition of all inflammable bodies. If sulfur was not in the composition of all inflammable substances, it would be awkward to continue to call the inflammable principle "sulfur" or the sulfurous principle. To avoid confusion, it seems, later chemists simply preferred phlogiston to these two names.

Like many who thought of the chemical principles in corpuscularian terms, Stahl held that the inflammable principle owed its activity to its divisibility and small size. In German, he wrote not directly of particles or corpuscles but of *Stäubchen* (literally, little dust) and *Theilgen* (small parts).[66] The forceful action that the inflammable principle demonstrated in combustion and acid-alkali reactions derived from its subtility (*Zartheit*) and divisibility (*Vertheilung*).[67] The strength of the combination of this principle with other substances could increase as the number of its molecules entering into chemical combination grew. Two molecules of the principle in a combination would occur with greater intensity than one.[68] Although, in general, Stahl endorsed Becher's scale of composite substances, he was silent on whether the inflammable principle was a simple or a composite substance and thus avoided the issue of whether the particle of this principle was made of inert matter, as postulated in rigorously applied mechanical philosophy. It is nevertheless clear that, for Stahl, this principle was of a corpuscularian nature and that all chemical processes, including combustion, resulted from corpuscular combinations or compounding.

Stahl related phlogiston to fire. It was "the corporeal fire that represents the most real matter of fire, the true principle of the movement in all combustive combinations." Moreover, "it makes up and forms a much-divided invisible fire,

66. His French translator, Baron d'Holbach (1723–89), to whom we shall return, consistently rendered these terms in more modern corpuscularian language as "molécule" or "particule." Stahl used "corpuscula" in his Latin works, for example, *Specimen beccherianum*. See Stahl, *Specimen beccherianum*, 22.

67. Stahl, *Zufällige Gedancken und nützliche Bedencken*, 152, 154 [*Traité du soufre*, 123, 125].

68. Ibid., 178–79 [148].

namely, heat."[69] Remarkably, this definition differed greatly from Stahl's earlier conception of the sulfurous principle. As shown above, Stahl had thought that all principles, including the most active sulfurous principle, were passive and that heat was the movement of corpuscles mediated ultimately by æther. Now, he proposed that the sulfurous principle was a source of movement and was heat itself.

Stahl suggested that the principle of inflammability was also the principle of color. Kunckel had noted that "ardent spirits" (inflammable fluids), vinegar, volatile alkalis, fixed alkalis, and niters caused changes in color. He attributed the cause of color changes to salts. Stahl found this attribution of cause problematic for two reasons. First, Kunckel had failed to differentiate the inherent color of a body from its apparent color. Both, Stahl noted, derived from the texture or solid arrangement of parts and owed to the reflection of rays of light as their "formal cause." He seemed to suggest that Kunckel had paid too much attention to the apparent color in his works on the colors of glass, which for Stahl were caused by mixing rather than by chemical combination. Second, he believed that the changes in color resulted from the transfer of the inflammable principle rather than from salts.[70] Burning, calcination and reduction of metals, the reactions of metals and acids, along with glass production by plant ashes were all color-changing processes. That led Stahl to assume that all processes involving transfers of the inflammable principle produced changes in color. The principle therefore also worked as the principle of color. It is important to note that Stahl did not suggest that it imparted its own color to its new host structure. Rather, it was the change in corpuscular structure that produced the new color. The thinking behind these chemical accounts of color was therefore mechanistic.

Stahl also considered what was known as metallic or fixed sulfur, a concept that, while often employed by alchemists, he found obscure. Alchemists claimed that metallic sulfur held the key to the transmutation of all metals into gold. It was, in fact, often identified as the "tincture" of gold.[71] For Stahl, however, the alchemists' metallic sulfur was, at best, the principle and corporeal cause of a metallic consistency. It was thus the inflammable principle that performed

69. "das cörperliche Feuer, die eigentlichste Feuer-Materie, das wahre Grund-Wesen der Feuer-Bewegung in allen brennlichen Vermischungen darstelle. . . . oder doch nur ein weitzertheiltes unsichtliches Feuer, nemlich die Wärme außmache und formiere"; ibid., 78 [56].

70. Ibid., 97–102 [73–78].

71. Ibid., 139, 158 [113, 129].

the function that Kunckel assigned to his salt of metals. Stahl's interpretation essentially deprived metallic sulfur of the aurific virtue that the alchemists had ascribed to it.

In addition to the traditional alchemical interpretation of metallic sulfur, Stahl also confronted the chrysopoetic belief in mercury as the principle of metals. For a variety of reasons, he found implausible the opinion that mercury was the material substratum, or the origin, of all metals. First, he held, if mercury were the material basis of all metals, it would be abundant, yet mercury mines were relatively rare. Second, under the same assumption, mercury would have been found mixed in many of the metallic mines, while it occurred only in mines of mercury or cinnabar. Third, no one had really been able to change mercury into a metal. And, fourth, since such a transformation had proven so difficult, it was hard to believe that mercury, in the womb of the earth, could produce a great number of metals. In fact, no substance had ever been found that could act on mercury to achieve transmutation.[72] This quadripartite argument represented a remarkable rejection of one of the fundamental assumptions of chrysopoetic alchemy, namely, that mercury was a key principle of metals.

The rejection of mercury as a metallic principle—made possible by Stahl's interpretation of the inflammable principle—amounted to the elimination of it as a principle of all matter. Prior to Stahl, metallicness was thought to come from a constitutive mercurial principle. Although there had long been skepticism as to the interpretation of metals as one species of body that shared a common principle, the skeptics had not offered an alternative explanation for the familiar material similarities between metals. That missing explanation was provided by Stahl's interpretation of phlogiston as the principle that gave metals their luster, ductility, and density. Stahl's phlogiston, in fact, assumed many of the functions previously assigned to the mercuric principle and, in so doing, rendered the latter superfluous.

That Stahl self-consciously sought the material consistency of phlogiston in all inflammable substances and metals is clear. As he put it, this principle "is of the same kind and nature as is found in the inflammable body as well as in the ordinary sulfur, in so much as it burns, is colored, or as it colors in a far-reaching way; as the middle point between the metals and the animal and vegetable gen-

72. Ibid., 297–98 [252–53].

eras, it has received the name of the sulfurous principle."[73] In other words, the sulfurous principle was the material that underlay inflammability or combustion in substances whether animal, vegetable, or mineral, and it was materially consistent in all of them.

Important for the later success of phlogiston was the explanatory scheme that phlogiston provided. As seen above, sulfur produced a fume when it burned in the air, a fume that Stahl called "volatile spirit of vitriol [or sulfur]" and known as sulfur dioxide (SO_2) today. He recognized this as a substance different from vitriolic acid (H_2SO_4, the aqueous solution of sulfur trioxide, SO_3) and sulfur, and named it volatile spirit to identify its vapor-like or gaseous status at a time when the chemical community was not yet aware of the existence of different gases apart from the air. The volatile spirit of sulfur was an intermediate substance in the chemical process of producing vitriolic acid and sulfur. It was also intermediate, Stahl figured, in terms of its composition. It contained less phlogiston than sulfur, as the latter gave off some of its phlogiston in burning. But it had more phlogiston than vitriolic acid. Thanks to the phlogiston in its composition, it was more volatile than vitriolic acid, which was phlogiston-free. Also thanks to its phlogiston (the principle of odor), it had a stronger odor than vitriolic acid. In brief, volatile spirit of vitriol had less phlogiston than sulfur but more than vitriolic acid. Here, Stahl set an example to differentiate chemical substances, or account for the change of qualities in chemical reactions (like the formation of acids, acid-alkali reactions, reduction of metals, and almost all that we would today consider oxidations or reductions) by exchange of one or more phlogiston particles. This would become the source of the phlogistic scheme of chemical explanations that was influential in the second half of the eighteenth century.[74]

As seen above, Stahl essentially stripped mercury of its status as the principle of metals. His position on the principle of salt, which was elaborated in his *Ausführliche Betrachtung und zulänglicher Beweiss von den Saltzen*, is also remarkable. For him, there were four species of salts, that is, vitriolic, nitrous, marine—corresponding to the three best-known mineral acids at the time (the

73. "Und zwar scheinbarlich von derselben Art und Geschlecht, welche sowohl sonst in verbrennlichen Dingen, als selbsten in dem gemeinen Schwefel, so weit er brennet, gefärbet ist, oder weitstreckend färbet, das Grund-Wesen abgiebt: und von ihme, also dem Mittel-Ding zwischen denen Metallen, und vegetalisch- und animalischen Geschlechten, die Benennung des schwefelhaften Wesens bekommen hat"; ibid., 152 [123].

74. Eklund, "Chemical Analysis and the Phlogiston Theory."

acids of vitriol, niter, and marine salt)—and borax.[75] Without considering any underlying substance of these species, Stahl seemed to assume that they simply were different substances that accounted for the qualities of these salts and their various combinations. In fact, he was skeptical about Kunckel's postulation that all acids were variants of vitriolic acid.[76] In addition, Stahl made no suggestion that any one of the saline species was the necessary component of matter. That amounted to a denial of a uniform saline principle that was in the composition of all salts, let alone of all matter. Stahl's scheme of chemical principles thus no longer preserved the parity of sulfur and mercury, as in the Geberian presentation, nor the trinity of the Paracelsian *tria prima*.

Stahl transformed the inflammable principle by meeting the contemporary demands for qualities associated with chemical principles. What he received as a student were the widespread teachings of chemical principles, a sense of a growing trend to interpret them in terms of corpuscles, a tendency to question the adequacy of all or individual principles, and a concern about the consistency of the remaining principles. In the *Treatise on Sulfur*, he further elaborated the inflammable principle in corpuscularian terms, eliminated the mercurial principle, and presented to the eighteenth century a formulation of phlogiston that he believed was materially consistent. Stahl developed his notion of phlogiston to the logical conclusion of the chemical principles. This theory of phlogiston, however, arrived at the expense of the principles of mercury and salt.

The Continuation of the Sulfurous Principle into the Eighteenth Century

Although it is well known that Stahl's principle of phlogiston—as a version of the sulfurous principle—was very influential until the so-called Chemical Revolution or even later, it has hardly been noted that sulfur was the only one of the *tria prima* that survived in the century of Enlightenment, thanks to the favorable reception of Stahl's phlogiston principle.

In general, the *tria prima* did not fare as well in the eighteenth century as in the seventeenth. Mi Gyung Kim has argued that Louis Lémery dealt a fatal blow

75. Stahl, *Ausführliche Betrachtung*, 21–31.

76. Ibid., 142–44. Curiously, Pierre-Joseph Macquer attributed this postulation of universal acid to Stahl. See, for example, Macquer, *Dictionary of Chemistry*, 2:11, 444.

to the theory of chemical principles when he concluded that distillation—or fire analysis, in general—failed to resolve substances into ultimate principles.[77] As noted above, however, signs of this shift had begun in the criticisms of van Helmont and Boyle and in the adjustments of Willis, Lémery's father, Nicolas, and Becher. What makes Louis Lémery different was probably the hope that he placed in solution as the method for chemical analysis.

It may be an exaggeration to speak of the "death" of the chemical principles. In place of chemical principles, there was a "return to the four elements" started by Stahl's contemporary, Hermann Boerhaave (1668–1738).[78] Boerhaave advocated a system in terms of fire, water, air, earth, and what he called *menstrua*. Instead of viewing these as material constituents, he thought of them as instruments or agents that effected chemical combination and decomposition.[79] For authors who came after Boerhaave, however, the return to the elements did not rule out the chemical principles, especially not the inflammable principle.

As I have pointed out elsewhere, the chemists at the Royal Academy of Sciences in Paris accepted Stahl's phlogistic scheme of chemical explanation and his experimentation, approvingly or critically, since the 1710s. Etienne-François Geoffroy (1672–1731) accepted Stahl's description of volatile spirit of vitriol in 1713. He also integrated, though implicitly, Stahl's formulation of the combination of vitriolic acid and phlogiston in his famous Affinity Table, first published in 1718. He then discussed it explicitly, positively, and at length in his German colleague's work on vitriolated tartar and liver of sulfur in 1720. Moreover, besides Geoffroy, the great majority of the chemists working in the Paris Academy from 1720 to 1750 adopted Stahl's phlogistic scheme of chemical explanations in the formation of acids, the analyses of acids and alkalis, and the production of vitriolated tartar and liver of sulfur, for example. They also often cited Stahl's experiments or chemical methods as their models, such as those for the reduction of antimony. They referred to almost all of Stahl's major works as well as some of his minor works, whereas many of the subjects they referred to were retold in the *Treatise on Sulfur*. Equally important, the Academy chemists also referred to other phlogistic accounts, such as that of the properties of diethyl ether, that were published in London and provinces of France in

77. Kim, *Affinity, That Elusive Dream*, 129–32. Kim adds that Lémery, in fact, believed that there existed proper, unalterable principles.

78. Siegfried, "From Elements to Atoms."

79. Powers, "Chemistry without Principles."

the 1730s and 1740s.[80] Therefore, Stahl's phlogistic theory and related chemical work had already received some, if not overwhelming, recognition well before the mid-eighteenth century.

Guillaume François Rouelle (1703–70), who was credited for popularizing Stahlian chemistry in mid-eighteenth-century France, introduced four elements as material constituents, thus making them principles.[81] Rouelle, a very popular teacher of chemistry at the Jardin du Roi in Paris from 1742 to 1768, claimed as his students the *philosophes* Denis Diderot (1713–84), Jean-Jacques Rousseau (1712–78), and the person who launched the Chemical Revolution, Antoine-Laurent Lavoisier (1743–94). Rouelle taught a modified version of Stahlian chemistry. For example, in addition to the character of principles, he, like Boerhaave before him, gave the elements the role of instrument that made chemical combinations and recombinations possible. He also assigned a much larger role to affinity in his chemical explanations.[82] Despite some differences, Rouelle propagated important elements of Stahl's teaching on the chemical principles that were presented in the *Treatise on Sulfur*: the corpuscular character of chemical principles, skepticism about the existence of the mercurial principle, belief in the material consistency of phlogiston, and the notion that phlogiston was the principle of combustion and could be transferred between substances to play a role both in restoring metals from calces and in color changes.[83]

Rouelle's teaching was an important source for his two influential students, Pierre Macquer (1718–84) and Gabriel François Venel (1723–75), who have been described as doing missionary work to popularize Stahlian chemistry.[84] Macquer, an influential textbook author, reserved a place for phlogiston in addition to the four elements in his popular *Elements of the Theory and Practice of Chemistry*.[85] Venel named fire, air, water, and earth in his article on "Principes" in the French *Encyclopédie*, and identified fire with Stahl's phlogiston.[86] For the fire element, both Macquer and Venel preserved most of the features of phlogiston

80. For the discussion in this paragraph, see my "Communications of Chemical Knowledge," 135–37.

81. As Kim points out, Etienne-François Geoffroy (1672–1731) had discussed Stahl's work in the 1720s and 1730s; Kim, *Affinity, That Elusive Dream*, 146–51.

82. Rappaport, "Rouelle and Stahl."

83. Ibid., 74–76.

84. Ibid., 95.

85. Macquer, *Elements of the Theory and Practice of Chemistry*, 7–11.

86. Venel, "Principes," in *Encyclopédie ou Dictionnaire*, 13:375.

that Stahl had formulated. It is therefore not the case that fire absorbed phlogiston. Rather, phlogiston replaced the content of the fire element.

While phlogiston was integrated or associated with elementary fire, mercury and salt were left out of eighteenth-century element theory. As noted above, one reason was that the *tria prima* lost its appeal and explanatory power when distillation (or fire analysis, in general) was no longer the preferred method of chemical analysis. Another was that mercury and salt were no longer theoretically necessary in the new element theory. Many chemical and physical features that mercury and salt accounted for could readily be turned over to elementary water and earth. Perhaps equally important, these two principles gave up their functions as the sources of fluidity and solidity. In the formulations of Rouelle and his contemporaries, fire contained in the body was the cause of fluidity. Water was fluid because it contained fire or phlogiston. When water lost its fire, it became solid ice. In fact, such a formulation found a parallel to Lavoisier's notion of caloric.[87] Thus, fire (that is, phlogiston) could account for both fluidity and solidity. In addition, the role of mercury as the principle of metals, as seen above, was instead accounted for by the function of phlogiston. While the mercurial and saline principles lost their value, the importance of the sulfurous principle, phlogiston, was greater than ever.

Stahl's name was always attached to phlogiston in the eighteenth century, since he enjoyed the reputation—or suffered the notoriety—as the person who presented the presence and properties of phlogiston through both experimentation and theoretical generalization. Hailing Stahl's work on the regeneration of sulfur from vitriolated tartar as a great accomplishment, Macquer wrote, "the process in which Sulfur is regenerated, by recombining together the principles of which it was originally composed, is one of the most beautiful experiments that modern Chymistry hath produced. We are indebted for it to M. Stahl."[88] Paul-Henri Thiry, Baron d'Holbach (1723–89), a well-connected Enlightenment savant and the translator of Stahl's *Treatise on Sulfur*, provided the French public with direct access to the German chemist's work. In his preface to the *Treatise*, d'Holbach marveled at the German chemist's ability to grasp and prove the material consistency of the inflammable principle in such a vast array of substances and phenomena. He wrote, "It is M. Stahl who has made us aware of the

87. Rappaport, "Rouelle and Stahl," 79–80, 111.

88. Macquer, *Elements of the Theory and Practice of Chemistry*, 264.

different states in which this principle is found in all three kingdoms of nature, its passage from each of these kingdoms into the others; it is he who has demonstrated to us that it was the principle of colors and odors, that gave metals their malleability."[89] In an age when mercury's validity as a principle was in serious doubt and salt was at best limited to plain mineral substances, Stahl showed the presence and material consistency of the sulfurous principle in substances of different generas that exhibited otherwise very different chemical and physical qualities.

The Place of Phlogiston in the History of Chemical Principles

D'Holbach's admiration for Stahl's work on the sulfurous principle was shared by Immanuel Kant (1724–1804). In the preface to his *Critique of Pure Reason*, the Könisberg philosopher named three scientific exemplars who discovered scientific laws in nature by pursuing the principles of reason; he compared Stahl's conversion of metallic calces to metals via his "principle" to Galileo's work on gravitational acceleration and to Torricelli's on the mercury barometer.[90]

Stahl's *Treatise on Sulfur* was based on notions of chemical principles that had developed in the previous centuries. Paracelsus and his early followers proposed the *tria prima*, three principles that accounted for qualities often observed in distillation or fire analysis. For the early Paracelsians, the principles worked as spiritual agents that induced qualities rather than as real substances that made up material bodies. These principles were accepted by enemies of the Paracelsians, by eclectic authors, and by practical chemists in the first half of the seventeenth century. Gradually, chemists considered the principles as material constituents. Van Helmont and Boyle, critical of the chemical principles, argued that fire analysis was never able to resolve matter into its constituent principles. Thereafter, chemists who continued to accept the principles—like Willis, Lémery, and Becher—modified their theories. Their differences aside, they joined a trend to

89. "[C]'est M. Stahl qui nous a fait connoître les différens états où ce principe se trouvoit dans les trois regnes de la nature, son passage de chacun de ces regnes dans les autres; c'est lui qui nous a démontré qu'il étoit le principe des couleurs & des odeurs, qu'il donnoit aux métaux leur malléabilité"; Baron d'Holbach, "Avertissement du traducteur," in Stahl, *Traité du soufre*, unpaginated.

90. Kant, *Critique of Pure Reason*, xxvii.

interpret the principles in corpuscularian terms, to view the principles as compounds rather than as simple substances, to question the adequacy of individual principles, and to ensure the consistency of the remaining principles.

Stahl's *Treatise on Sulfur* aimed to affirm the adequacy and consistency of the sulfurous principle in response to Kunckel's proposition that metals contained a mercurial substance but nothing sulfurous. Stahl was optimistic that he carried out his objective through theoretical generalization as well as experimentation. He supplied, as further supporting evidence, accounts of practitioners such as gardeners, foundry workers, and miners. "Adequate evidence," he concluded, "is hopefully given that the sulfurous principle must be recognized as a true and real physical being in all senses in which it can be rationally and clearly alleged."[91] While it had been contended that all of the chemical principles existed in all substances, Stahl asserted—as the basis of his empirical observation—that his inflammable principle was present only in inflammable and calcinable substances. This principle existed and transferred in the form of particles. Its transfer explained combustion, calcination and reduction, color changes, and the corrosive effects of acids. At the end, his interpretation of the composition of sulfur made "phlogiston" a much more preferable name for the inflammable principle than "sulfur" or the "sulfurous principle."

Neither Stahl's eighteenth-century admirers nor historians have paid much attention to the transformation of his thinking on chemical principles. As seen above, in the 1680s, he first accepted Becher's system of chemical principles, which consisted of water and three kinds of salts. His notion of phlogiston was first presented in the late 1690s, in his *Zymotechnia fundamentalis* and some articles. Then, in the 1700s, in the *Specimen beccherianum,* he listed phlogiston as one of three earthy species in his elaboration of chemical principles. In the *Treatise on Sulfur,* he formally eliminated the mercurial principle (or mercurial earth) and, along with it, the belief in chrysopoeia.[92] Mercury, which Stahl had accepted in his lectures in the 1680s, became dispensable, for sulfur served almost all the functions that had previously been assigned to it. Without mercury (and salt, as seen above), the parity of the Geberian theory of metallic constitution or the trinity of the Paracelsian *tria prima* no longer held.

91. "Womit dann also hoffentlich hinlänglicht Zeugnisse gegeben seyn wedern, daß das sulfurische Wesen, in allerley Verstand, darinnen es jemals vernünftig und vernehmlich allegiret wird, für ein cörperliches wahres und würkliches Wesen zu erkennen"; Stahl, *Zufällige Gedancken und nütliche Bedencken,* 174–75 [*Traité du soufre,* 144].

92. For Stahl's early belief in alchemy, see Chang, "Georg Ernst Stahl's Alchemical Publications."

Stahl's elimination of the mercurial principle is remarkable, especially considering that Boyle and Boerhaave, both thought to be forward-looking, never gave up working on mercury as the key to transmutation.[93] In fact, from the Middle Ages, if not earlier, chemists had thought of mercury and sulfur as two major constituents of metals. For those who believed it, mercury supplied the material base and sulfur contributed the yellow tinge, or tincture, to metals. Combined in the right proportion, they gave rise to gold. Their combination in different ratios, along with mixed "impurities," produced base metals. This line of thinking constantly associated base metals with inferiority and impurities. When metals no longer shared a common material base such as mercury, they were allowed to retain distinct qualities that could not be summarized as "impurities." Each of them could then be investigated for its own sake and, for example, take its own place in the affinity or element tables.

Another notable development in Stahl's chemical works was his elimination of the æther that had been postulated in the *Zymotechnia fundamentalis* (1697). Æther as the medium of heat explained the passive nature of all substances, including chemical principles. In the *Specimen beccherianum* (1703), Stahl on the one hand reserved the role of heat as the principal instrument of chemical reactions and on the other eliminated æther in his chemical accounts and equated phlogiston with the matter of fire. In the *Treatise on Sulfur*, phlogiston became the principle of movement in combustion and was even identified as heat. Once phlogiston could serve as heat and the source of the movement that caused chemical decomposition and combinations, a hypothetical medium of heat that was supported by little empirical evidence was no longer necessary. Reflecting that change of thought, the German translation of the *Zymotechnia* (first published in 1734, the year of Stahl's death) consistently replaced the occurrences of "æther" in Latin with "die subtilesten Lufft" (the subtlest air).

Sulfur thrived in the form of phlogiston in a century that saw a move toward the elements. Mercury was gone, and the validity of salt was limited. By contrast, phlogiston essentially replaced the fire element in terms of features and functions. It was thanks to Stahl, especially in his *Treatise on Sulfur*, that the sulfurous principle endured into the eighteenth century. On the eve of the Chemical Revolution, Stahl's sulfurous principle was the only one of the *tria prima* to survive.

93. For Boyle's interest in aurific mercury, see Principe, *Aspiring Adept*, 54–62, 148–78. For Boerhaave's work on mercury, see Debus, *Chemistry and Medical Debate*, 202–5.

The responses of d'Holbach and Kant illustrate that, even though scientists today may find Stahl's "proofs" shoddy, they were not perceived that way in the eighteenth century. Given the technical constraints of the late seventeenth and eighteenth centuries, the equation of calcination with combustion and the experimental demonstration of the transferability of phlogiston were celebrated as great discoveries by mid-eighteenth-century chemists and savants. As Bernadette Bensaude-Vincent and Isabelle Stengers suggest, the inability to isolate phlogiston was no problem to the community of chemists who operated at the levels of compounds, levels that concerned unions of the heterogeneous rather than the extraction of the homogeneous.[94] Although the weight gain of metal after calcination did appear as a problem to Lavoisier a few years into his chemical career, it was not until pneumatic chemistry and the precision balance were in place that Lavoisier was able to address that problem.[95]

It is important to note that Stahl's phlogiston theory applied to more than the phenomena of combustion; as seen above, the exchange of phlogiston was also applied to mid-eighteenth-century discussions of the formations and differentiations of acids, the reactions of acids and alkalis, the reduction of metals and salts, and indeed to many reactions that today's chemists consider oxidations and reductions. Its wide applicability contributed to its overwhelming success. Although Stahl did not leave behind a flawless theory for explaining all of these phenomena, he nonetheless imparted to his followers a theoretical framework and experimental observations for understanding important chemical phenomena. His followers like Macquer continued to expand this framework to include more chemical phenomena. They, for example, endeavored to apply Stahl's work on the sequential scale of volatility and odor of sulfur and its acids (volatile acid of vitriol and vitriolic acid) to niter and its acids (and to common salt and its acid).[96] When faced with difficulties, many of them worked to resolve them with new interpretations, ad hoc or general, and experimentation, instead of abandoning the framework.

In the 1760s, the 1770s, and especially the 1780s, phlogiston was the theoretical focus of the international competition to discover and understand new gases. Lavoisier's identification and interpretation of oxygen eventually debunked the belief in phlogiston. That the so-called Chemical Revolution is

94. Bensaude-Vincent and Stengers, *History of Chemistry*, 62–63.

95. Ibid., 82–91.

96. Eklund, "Chemical Analysis and the Phlogiston Theory," 151–87.

synonymous with the overthrow of the phlogiston theory, however, confirms phlogiston's prominence. By extending the life of the sulfurous principle into the eighteenth century, Stahl's formulation of phlogiston connected Paracelsian chemistry—a subject Allen Debus helped bring to prominence—with the Enlightenment.

WORKS CITED

Becher, Johann Joachim. *Physica subterranea profundam subterraneorum genesin, e principiis hucusque ignotis, ostendens*. Lipsiae: Ex officina Weidmanniana, 1738.

Béguin, Jean. *Tyrocinium Chymicum, or, Chymical Essays Acquired from the Fountain of Nature and Manual Experience*. London: Printed for Thomas Passenger, 1669.

Bensaude-Vincent, Bernadette, and Isabelle Stengers. *A History of Chemistry*. Translated by Deborah van Dam. Cambridge, MA: Harvard University Press, 1996.

Boyle, Robert. *The Sceptical Chymist or Chymico-Physical Doubts & Paradoxes, Touching the Spagyrist's Principles Commonly Call'd Hypostatical* . . . . London: Printed by J. Cadwell for J. Crooke, 1661.

Chang, Ku-ming (Kevin). "Communications of Chemical Knowledge: Georg Ernst Stahl and the Chemists at the French Academy in the First Half of the Eighteenth Century." In *Chemical Knowledge in the Early Modern World*, edited by Seymour H. Mauskopf, William R. Newman, and Matthew D. Eddy, 135–57. Osiris 29. Chicago: University of Chicago Press, 2015.

———. "Fermentation, Phlogiston and Matter Theory: Chemistry and Natural Philosophy in Georg Ernst Stahl's *Zymotechnia fundamentalis*." *Early Science and Medicine* 7 (2002): 53–57.

———. "Georg Ernst Stahl's Alchemical Publications: Anachronism, Reading Market, and a Scientific Lineage Redefined." In *New Narratives in Eighteenth-Century Chemistry*, edited by Lawrence M. Principe, 23–43. Dordrecht: Springer, 2007.

Clericuzio, Antonio. *Elements, Principles and Corpuscles: A Study of Atomism and Chemistry in the Seventeenth Century*. Dordrecht: Kluwer Academic, 2000.

———. "'Sooty Empiricks' and Natural Philosophers: The Status of Chemistry in the Seventeenth Century." *Science in Context* 23 (2010): 333.

Coleby, Leslie J. M. "Studies in the Chemical Work of Stahl." PhD diss., University College, University of London, 1938.

Debus, Allen G. *The Chemical Philosophy: Paracelsian Science and Medicine in the Sixteenth and Seventeenth Centuries*. 2 vols. New York: Science History Publications, 1977. Reprint, Mineola, NY: Dover Publications, 2002.

———. *Chemistry and Medical Debate: Van Helmont to Boerhaave*. Canton, MA: Science History Publications, 2001.

———. *The French Paracelsians: The Chemical Challenge to Medical and Scientific Tradition in Early Modern France*. Cambridge: Cambridge University Press, 1991.

Eklund, Jon Bledge. "Chemical Analysis and the Phlogiston Theory, 1732–1772: Prelude to Revolution." PhD diss., Yale University, 1971.

Helmont, Jean Baptiste van. *Ortus medicinae; id est, Initia physicae inaudita* . . . . Amsterdam: Ludovicum Elzevirium, 1648.

Jacob, Margaret C. *Strangers Nowhere in the World: The Rise of Cosmopolitanism in Early Modern Europe*. Philadelphia: University of Pennsylvania Press, 2006.

Kant, Immanuel. *Critique of Pure Reason*. London: Henry G. Bohn, 1855.

Kim, Mi Gyung. *Affinity, That Elusive Dream: A Genealogy of the Chemical Revolution*. Cambridge: MIT Press, 2003.

Lémery, Nicolas. *A Course of Chymistry: Containing the Easie Manner of Performing Those Chymical Medicines* . . . . 4th ed. London: Printed for A. Bell, 1720.

Kant, Immanuel. *Critique of Pure Reason*. London: Henry G. Bohn, 1855.

Macquer, Pierre-Joseph. *A Dictionary of Chemistry: Containing the Theory and Practice of That Science: Its Application to Natural Philosophy, Natural History, Medicine, and Animal Economy*. Paris: P. Fr. Didot, 1778.

———. *Elements of the Theory and Practice of Chemistry*. Edinburgh: Printed for A. Millar, 1758.

Metzger, Hélène. *Newton, Stahl, Boerhaave, and Chemical Doctrine: Translated with Supplementary Notes*. Translated by H. W. Laven and Larry J. Murphy. Hamilton, Ont.: Huxley, 2006.

Newman, William R. *The Summa Perfectionis of Pseudo-Geber: A Critical Edition, Translation and Study*. Leiden: Brill, 1991.

Pagel, Walter. *Paracelsus: An Introduction to Philosophical Medicine in the Era of the Renaissance*. 2nd rev. ed. Basel: S. Karger, 1982.

Powers, John C. "Chemistry without Principles: Herman Boerhaave on Instruments and Elements." In *New Narratives in Eighteenth-Century Chemistry*, edited by Lawrence M. Principe, 45–61. Dordrecht: Springer, 2007.

Principe, Lawrence M. *The Aspiring Adept: Robert Boyle and His Alchemical Quest*. Princeton, NJ: Princeton University Press, 1998.

Rappaport, Rhoda. "Rouelle and Stahl—The Phlogistic Revolution in France." *Chymia* 7 (1961): 73–102.

Siegfried, Robert. *From Elements to Atoms: A History of Chemical Composition*. Transactions of the American Philosophical Society, n.s., 92. Philadelphia: American Philosophical Society, 2002.

Shackelford, Jole. *A Philosophical Path for Paracelsian Medicine: The Ideas, Intellectual Context, and Influence of Petrus Severinus (1540–1602)*. Copenhagen: Museum Tusculanum Press, 2004.

Smith, Pamela H. *The Business of Alchemy: Science and Culture in the Holy Roman Empire*. Princeton, NJ: Princeton University Press, 1994.

Stahl, Georg Ernst. *Ausführliche Betrachtung und zulänglicher Beweiss von den Saltzen . . . .* Halle: Wäysenhauses, 1723.

———. *Opusculum chymico-physico-medicum . . . .* Halae Magdeburgicae: Typis & Impensis Orphano Trophei, 1715.

———. *Specimen beccherianum, sistens, fundamenta, documenta, experimenta, quibus principia mixtionis subterraneae, et instrumenta naturalia atque artificialia demonstrantur*. Lipsiae: Weidmaniana, 1738 [1703].

———. *Traité du soufre, ou remarques sur la dispute qui s'est élevée entre les chymistes, au sujet du soufre, tant commun, combustible ou volatil, que fixe, &c*. Paris: Didot, 1766.

———. *Zufällige Gedancken und nützliche Bedencken über den Streit, von dem so genannten Sulphure, und zwar sowol dem gemeinen, verbrennlichen, oder flüchtigen, als unverbrennlichen, oder fixen*. Halle: Waisenhaus, 1718.

Teich, Mikulas. "Interdisciplinarity in J. J. Becher's Thought." In *Johann Joachim Becher (1635–1682)*, edited by Gotthardt Frühsorge and Gerhard F. Strasser, 145–60. Wiesbaden: Harrassowitz, 1993.

Venel, Gabriel François. "Principes." In *Encyclopédie ou Dictionnaire raisonné des sciences, des arts et des métiers*, 13:375–75. Paris: Briasson, David l'ainé, le Breton, Durand, 1751–65.

White, John H. *The History of the Phlogiston Theory*. London: Edward Arnold, 1932.

Willis, Thomas. *Dr. Willis's Practice of Physick Being the Whole Works of That Renowned and Famous Physician . . .* London: Printed for T. Dring, C. Harper, and J. Leigh, 1684.

Part Two

Regional Contexts and Communities of Texts

Chapter 6

"If they are not pages that cure, they are pages that teach how to cure."

The Diffusion of Chemical Remedies in Early Modern Spain

Mar Rey Bueno

When Allen G. Debus published his article on the diffusion of Paracelsianism in Spain in 1998, it marked the first attempt within the international community of historians of science to discuss chemical medicine in early modern Spain.[1] Relying on earlier work of José María López Piñero,[2] Debus noted the absence of a Paracelsian tradition proper to Spain and attributed it primarily to the famous royal order promulgated by Philip II (1527–98) in 1559 that prohibited Castilian students from leaving Spain to further their education at European universi-

1. Debus, "Paracelsus and the Delayed Scientific Revolution in Spain."

2. López Piñero, "Química y medicina en la España de los siglos XVI y XVII"; and López Piñero, "Paracelsus and His Work in the 16th and 17th Century Spain." Compare, too, López Piñero, *Ciencia y técnica en la sociedad española de los siglos XVI y XVII.*

ties. It had been assumed that this order had the immediate effect of rendering Spanish science backward, therefore absenting Spain from the Scientific Revolution,[3] and of effecting more than a century-long retardation of the field in Spain. In this light, if, as Debus maintained, Paracelsus had at some point prevailed in Spain, it had been right when the rest of Europe was abandoning his ideas, that is to say, at the end of the seventeenth century. It was then that the supposed birth of Spanish chemical discourse took place within a group known as the *novatores*. In addition to defending the use of chemical remedies and practices for the treatment of illnesses, the *novatores* were also given to denouncing the backwardness of Spanish science and calling for the creation of new institutions where the doctrines of a new science could take hold.[4]

This was the narrative I encountered two decades ago when I began my research on the therapeutics practiced in Spanish courts during the sixteenth and seventeenth centuries, research based on archival documentation that, little by little, brought to light a completely different interpretation from the one prevailing when Debus wrote his important article.[5] The archives began to reveal a Philip II (who ruled from 1556 to 1598) significantly different from the historical figure portrayed by hostile contemporaries as a Catholic at all costs, who persecuted anything resembling modernity. Documents associated with his reign, in fact, showed him to be an enthusiast of the *ars separatoria*.[6] This passion led him, among others things, to establish as many as three distillation laboratories at three of his residences—the one at the Escorial becoming perhaps the most lavish in all of Europe[7]—and to institutionalize the profession of distiller (specialist in making chemical remedies)—by integrating it into the group of the physicians, surgeons, and pharmacists charged with maintaining his health.[8] Philip had, among his *protomédicos*, outstanding experts in the *ars separatoria* such as Francisco de Valles (1524–92), author of a treatise intended to teach the most appropriate methods of this art,[9] and Lorenzo Cózar (ca. 1540–ca. 1592),

3. See Padgen, "Reception of the 'New Philosophy' in Eighteenth-Century Spain"; and Goodman, "Scientific Revolution in Spain and Portugal."

4. López Piñero, *Introducción de la ciencia moderna en España*.

5. Rey Bueno, *Señores del fuego*.

6. Rey Bueno and Alegre, "Renovación de la terapéutica real"; and Rey Bueno and Alegre, "Destiladores de Su Majestad."

7. Rey Bueno, "*Mayson pour distiller des eaues*," 26–39.

8. Rey Bueno and Alegre, "La ordenación normativa de la asistencia sanitaria."

9. Valles, *Tratado de las aguas destiladas*.

author of the unique Paracelsian treatise published in sixteenth-century Spain[10] and creator of the chair *De remediis secretis* at the University of Valencia.[11] Philip also saw to it that pharmacists followed procedures in preparing distilled waters that prevented poisonings associated with the use of lead alembics.[12]

If this research resulted in the need to revise our understanding of science during Philip's reign, then an analogous revision is in order regarding the apparent resurgence of chemistry during the reign of Charles II (1665–1700), a resurgence, contrary to what Debus posited, that appears to have had nothing to do with a delayed reception of Paracelsus's doctrines. The *novatores* were not, *sensu stricto*, Paracelsians. To so characterize them reduces too severely the scope of their philosophico-scientific interests. It is true that they shared with the Paracelsians both a discourse in defense of applying chemical remedies to the treatment of disease and a denunciation of those whom they called traditionalist physicians. The latter had won university chairs, were the primary physicians of illustrious patients, and were viewed as having largely ignored the chemical literature of the preceding century and a half.[13] However, a reading of the hundreds of *opúsculos* exchanged between traditionalists and *novatores* over a forty-year period reveals that the charge of ignorance levied against the traditionalists was not as accurate as their detractors would have had us believe. Moreover, each side used the same information to attack the other. This paradox suggests that perhaps the *novatores* were not as advanced nor were the traditionalists as stagnant as conventional wisdom would have had it. In fact, it seems evident that the *novatores* decided to embrace chemical medicine in order to achieve one clear objective: to accede to the positions of power held by the traditionalists.[14]

Falling within two different centuries, the reigns of Philip II and Charles II point simply to two stages characterized by a greater *visibility* of chemical practice in early modern Spain. In contrast to the apparent lack of interest in the subject that seemed to characterize the first part of the seventeenth century, the second

10. Cózar, *Dialogus veros medicinae fontes indicans*. Cózar was initially studied by López Piñero in *El Dialogus (1589) del paracelsista Llorenç Coçar*.

11. López Terrada, "Llorenç Coçar."

12. Rey Bueno, "El informe Valles."

13. López Piñero, "Juan de Cabriada"; and López Piñero, "*Carta filosófica, médico-chymica*."

14. López Pérez and Rey Bueno, "Instrumentalización de la espagiria." Recently, John Slater has revisited the constituent document of the *novator* movement, the *Carta filosófica, médico-chymica* (Valencia, 1687) written by the physician Juan de Cabriada. In his masterful article, "Rereading Cabriada's *Carta*," Slater stressed the rhetoric of the *Carta*, an aspect fundamental to understanding the true interests that motivated Cabriada and, similarly, the majority of the writings of the *novatores*.

half of the sixteenth century (when Philip II reigned), and the last two decades of the seventeenth century (under the rule of Charles II) actually witnessed a prodigious number of publications and related announcements concerning chemistry. Since the principal protagonists of both periods were physicians directly linked to the court, it could be assumed that chemistry was exclusively a royal fashion, another whim of monarchs fascinated by the collection of rarities and curiosities. One could also link the health problems of both monarchs with an interest in all sorts of therapies, chemical therapies being just one type among many.[15] The choice of Flemish and Neapolitan distillers during the reign of Philip II and of Neapolitan spagyrists during that of Charles II seemed to support a theory that the absence of experts in the *ars chemical* in Spain forced the monarchy to look for professionals in other European kingdoms within the Hispanic sphere of influence. The investigations of Miguel López Pérez and José Rodríguez Guerrero have shown that this theory, while valid, is incomplete; chemical practice linked to medicine had great currency in early modern Spanish society throughout both centuries.

With this in mind, this chapter—in this volume in honor of our teacher, Allen Debus—aims to show the influence of *chimica*[16] in sixteenth- and seventeenth-century Spain, not by exploring the great pronouncements of great historical figures but by examining chemical practices in Spanish everyday life. That, of course, is an ambitious plan, especially given both a time frame that spans some one hundred and fifty years and the wealth of historical material that has come to light in the last fifteen years. For these reasons, it is best to be modest and to offer only the most relevant aspects of those practices in broad brushstrokes, in the hopes of demonstrating that Spanish chemical practice did in fact exist during the sixteenth and seventeenth centuries. It not only existed, but it must also be meaningfully incorporated into the historical narrative in order properly to understand early modern European science.[17]

15. Rey Bueno, *El hechizado*.

16. "Chymica" or "chimica" was the word used by early modern Spaniards to encapsulate a set of practices used for separating the pure from the impure in any substance in order to get the best remedy. *Chimica* was thus practiced in concert with traditional alchemy. In fact, *chimica* draws from alchemy and recaptures the essence of its doctrine, in such a way that *chimica* was understood as divided into two parts. One was devoted to making medicines from animals, vegetables, and minerals by separation of the active part. The other was exclusively devoted to the transmutation of metals, also called "alchymia," "alchemia," "crisopeya," "metalurgia," "arte aurífera," and "arte hermética." For more analysis of this terminology, see Newman and Principe, "Alchemy vs. Chemistry"; and Abbri, "Alchemy and Chemistry."

17. On the need to incorporate Spain into the narrative of the Scientific Revolution, see the studies of

Peddlers of Secret Panaceas[18]

In April of 1598, the Royal Court of the *Protomedicato,* the highest health-regulatory institution within the Hispanic monarchy, granted permission for the commercial production of what was styled a "white dust solution of gold's quintessence [*polvos blancos solutivos de la quinta esencia del oro*]." A medicinal remedy made and sold by Alessandro Quintilio, a Roman citizen residing in Madrid, it was one of many chemically derived products that claimed to be effective against all diseases. Quintilio, who published at least three different books on his numerous treatments,[19] gained almost instant notoriety in the cosmopolitan city of Madrid, the capital of an empire that spanned the globe and the residence of people of all nationalities. His patients included Genoese ambassadors, apostolic notaries, Augustinian friars, royal officers, acknowledged necromancers, and royal distillers, among others.

Quintilio came to light as a historical figure worthy of note thanks to the research of López Pérez, which provided new insights into the diffusion of chemical medicines in seventeenth-century Spain.[20] López Pérez's work was complemented by Rodríguez Guerrero's finding that Quintilio was in fact not the true inventor of the white dust solution. Rather, the miraculous panacea was invented by the Veronese physician Vittorio Algarotti (1553–1604), creator of a great commercial network that extended to several European cities and had brokers in charge of distributing his products. Quintilio was merely Algarotti's agent in Spain.[21]

Slater and Prieto, "Was Spanish Science Imperial?"; and Eamon, "'Nuestros males no son constitucionales, sino circunstanciales.'"

18. The point of departure of this section is Eamon's pioneering study, *Science and the Secrets of Nature,* which sets the interpretive foundation for this type of turbulent literature, situated halfway between guild secrets and the first intimations of the modern experimental method.

19. A copy of the first *relación* does not survive. The second edition, entitled *Relación y memoria de los maravillosos efetos* [sic] *y notables prouechos que han hecno y hacen los polvos blancos solutiuos de la quinta esencia del oro, que compone y haze el author Alexandro Quintilio para embiar con los mismos a las Indias. . . . Va añadido a esta segunda impression el privilegio de Su Magestad,* was printed in Madrid in 1609. The third edition is entitled *Relación y memoria de los maravillosos efetos* [sic] *y notables prouechos que han hecno y hacen los polvos blancos solutious de la quinta esencia del oro, que compone y haze el autor Alexandro Quintilio para embiar con los mismos a las Indias . . . Va añadido a esta tercera impression la Memoria del Priuilegio de Portugal* (Madrid: Luis Sánchez, 1616).

20. López Pérez discovered the third edition of the printed book (Madrid, 1616) stored in the Biblioteca Nacional de España (BNE), R/10576. This book lacks a front cover, preliminary material, and final conclusion. While searching for other surviving copies in Spanish libraries, he found one in the library of the University of Zaragoza (78–52) that includes a calcographic portrait of Quintilio taken from the 1609 edition. See López Pérez, "Alquimia, terapéutica y sanidad en la España," 578–628; and López Pérez, *Asclepio renovado,* 199–232.

21. Rodríguez Guerrero, "Primera gran red comercial."

The *polvos blancos solutivos* headed the list of tremendously popular chemical panaceas in seventeenth-century Spain, among them, the quintessence of the *manna solutivo* of Antonino Polizzi, the medicinal oil of *germanía* of Giuseppe Balsam, the quintessence of rosemary's balsam of Cosme Novella, the solutive pink sugar of Francisco Ruiz Zapata, the quintessence of the chemical sun of Matías de Beinza, the water of life of Luis de Alderetey Soto, and the arcane of Fra Buenaventura Angeleres.[22] Precise references to all of these may be found in publications and manuals authored by their inventors and written to attract clients through their professional and authoritative tone.[23]

The quintessence of the *manna solutivo* and the medicinal oil of *germanía* were, like the *polvos blancos solutivos*, Italian inventions exported to Spain. The first, made from a mixture of three resins (escamonea, bdellium, and Arabic gum) and combined with an alcohol (a quintessence derived from wine), was the creation of Antonino Polizzi, a physician at the Sicilian Inquisitorial Court.[24] Its sale was authorized in 1595 by the general *protomédico* of the kingdoms of Naples and Sicily, both territories then under Hispanic rule, and it arrived at the court in Madrid early in the seventeenth century via a secretive adept unwilling to reveal its composition.[25] The distribution of the medicinal oil of *germanía*,

22. Rodríguez Guerrero, "Vendedores de panaceas alquímicas."

23. As Rodríguez Guerrero asserted, his objective was to mix denotative and connotative information: "On the one hand, they give information about the product they want to sell; they show the qualities of the product and they invite purchase, noting the places allowed to sell it. On the other hand, the medication is associated with feelings of happiness, gratitude, and satisfaction that together form the testimonials of healed patients [*Por un lado informan del producto que se pretende vender, muestran sus cualidades e invitan a un acto de compra posterior dejando indicados los lugares dispuestos para tal fin (denotación); por otro, se asocia el medicamento con los mensajes de felicidad, agradecimiento y satisfacción que conforman las credenciales de pacientes sanados (connotación)*]"; Rodríguez Guerrero, "Primera gran red comercial," 20.

24. Politius, *De quinta essentia solutiva*.

25. According to the Toledan pharmacist Gervasio de Barrionuevo, "The owner of the secret did not want to reveal it to anyone, not to doctors, not to court pharmacists. Indeed, he only wanted to keep his secret within his group. He gave his secret to one of his family members who guarded it closely in this city. It did surprise (it did not scare me, and it has always been the mother of admiration) because he sometimes named it distilled manna, other times clarified, and others Quintessence. He did not noise it around . . . Moreover, he kept it as if he were guardian of the divine laws" (cuyo secreto por lo curioso no quiso su dueño por sus razones revelarló a ninguno, así Médicos, como Pharmacéuticos de la Corte; si sólo quiso que entre los de su facción se quedasse. Pasando pues dicho secreto a esta ciudad por un deudo suyo, guardó el dicho el mismo sigilo, sirvió de novedad [mas no me espanto, que siempre ha sido madre de la admiración] porque unas veces la apellidaba manna distillado; otras, clarificado; otras, Quinta esencia. No hizo poco ruido . . . mas en guardarlo parece que guardaba las leyes divinas); Gervasio de Barrionuevo, *Tratado sobre el láudano opiato de Iosepho Querzetano*. After the death of the last distributor, who settled in Spain, the secret was lost. Nevertheless, Barrionuevo explained that some physicians in Madrid still urged him to experiment in his pharmacy in the hopes of finding a possible substitute.

a creation of Giuseppe Balsam, was somewhat different in that Balsam, unlike Polizzi, chose *not* to put his business in the hands of others. After personally peddling his oil in the cities of Granada, Cordova, Seville, and Toledo, he established himself in Valencia, where he obtained sales licenses from the viceroy.[26]

Soon after the arrival of these Italian products, the number of Spanish inventors promoting their own remedies rose. To mention just two examples, Cosme Novella, a druggist at the Royal and General Hospital of Our Lady of Grace in Zaragoza, manufactured a quintessence of rosemary balsam,[27] while Francisco Ruiz Zapata, a physician at the University of Zaragoza, distributed his panacea under the name *azúcar rosado solutivo*.[28] By the late seventeenth century, three more alchemical panaceas had gained notoriety. The first of them, with the sonorous name "quintessence of the chemical sun," was the invention of Matías de Beinza, a Navarrese druggist trained in Madrid in the 1630s.[29] The last two—Luis de Alderete y Soto's water of life and Fra Buenaventura Angeleres's arcane—came to light in turbulent, end-of-the-century Madrid. Associated with Madrid high society, Luis de Alderete y Soto was the greater solicitor of the Royal Councils and Fra Buenaventura Angeleres served as general commissioner of the Conventual Minors of San Francisco.[30] These men lived in a time and place characterized by disputes in the form of *opúsculos* or medical duels in which defenders and detractors of chemical practice faced off.

Amusements of Idle Men

Angeleres and Alderete exemplify the interest in chemical practice in certain sectors of Spanish society during the sixteenth and seventeenth centuries. They were persons of high social status who found, in their experiments and chemical

26. Despite institutional support, Balsam had to face the city's College of Apothecaries, which did not hesitate to denounce him for practicing pharmacy without permission; López Terrada, "Prácticas médicas extraacadémicas."

27. Novella, *Tratado de la quintaesencia del Bálsamo de Romero*.

28. Although the Zaragozan College of Physicians and Surgeons opened a file on him in 1624 and eventually both ordered his expulsion and denounced the remedy, Ruiz Zapata obtained the support of physicians and pharmacists of the court, some of whom wrote prefaces to treatises in which Ruiz Zapata detailed the complete procedure for the manufacture and application of his remedy. Ruiz Zapata also left a detailed account of the trial in which he was involved. See Ruiz Zapata, *Discurso sobre la composición del azúcar rosado solutivo*.

29. De Beinza, *Discurso sobre los polvos universales purgantes*.

30. López Pérez and Rey Bueno, "Instrumentalización de la espagiria."

readings, both diversion and a way to differentiate themselves from their social peers. Another gentleman practitioner, the Aragonese Vincencio Juan de Lastanosa (1607–81), was described by his contemporaries as learned in *alchimia*. Lastanosa's palace at Huesca reportedly housed "the miracles of spagyric and hermetic chemistry,"[31] specifically a cabinet of curiosities, an alchemical laboratory, a garden with plants from both the Old and New Worlds, and a library of exemplars of universal knowledge.[32] With a true passion for quintessences and elixirs, Lastanosa was a virtuoso, curious and fascinated by the wonders of nature, who thus bore the distinguishing marks of power and wealth.[33] His patronage of several alchemists further testifies to his social status. One, the Neapolitan alchemist Nadal Baronio, worked in Lastanosa's palace for years elaborating a variety of chemical remedies, including the mythical potable gold.[34] Another, Diego de Bercebal (d. 1707), a Franciscan nurse interested in learning the chemical art, judged Huesca to be the best place to study in light of Lastanosa's enthusiasm for secrets and exquisite remedies.[35] In fact, an analysis of the alchemical content of Lastanosa's library reveals that, in addition to alchemical treatises written by artificers from many eras, he had a large selection of treatises exclusively dedicated to the preparation of chemical remedies.[36] While it is unknown whether Lastanosa himself ever elaborated the remedies and alchemical panaceas whose recipes he collected, it is clear that he prized manuscripts of secrets and wrote a translation into Castilian of *Les elemens de chymie* by the French pharmacist Jean Béguin (1550–1620).[37]

Juan Vázquez de Mármol, the general book censor during the reign of Philip II, shared Lastanosa's passion for collecting books of secrets. A prominent figure in Madrid society as well as a paleographer, publisher, and critic of texts, Vázquez de Mármol maintained relationships with the most outstanding humanists of the

31. Lastanosa, *Tratado de la moneda iaquesa*. We owe the discovery of Lastanosa's alchemical interests to López Pérez, who initially published his article "Lastanosa" in the e-journal *Panacea*, republished in the journal *Azogue* 5 (2002–2007). He later extended his work on Lastanosa as alchemist in "Anatomía del virtuoso" and "Alquimia y Vincencio Juan de Lastanosa."

32. Rey Bueno and López Pérez, *Gentleman, Virtuoso, Inquirer*.

33. Eamon, "Appearance, Artifice, and Reality."

34. Lastanosa himself described Baronio's sojourn in *Narración de lo que le pasó a don Vincencio Juan de Lastanosa a 15 de octubre del año 1552 con u religioso doctor y grave*, a manuscript preserved in New York City at the Hispanic Society of America; see HSA, B-2424, fols. 52r–79v.

35. We learn this from Manente's introduction to Bercebal's *Recetario medicinal espagírico*.

36. Moran, "Extracting the Virtues of Nature."

37. Rey Bueno, "Collector of Secrets."

sixteenth century.[38] His passion for the world of books went beyond a mere professional interest, however. In a handwritten notebook of his property conserved in Madrid in the National Library of Spain, he commented on his long-standing desire to see, read, and browse new books as well as on his care not to cross the line of what was prohibited by *Santa Madre*, the Catholic Church.[39] Next to Latin texts taken from diverse authors, books on Roman emperors, popes, and religious orders, and a Flemish-Spanish vocabulary, Vázquez de Mármol dedicated the largest portion of his notebook to recipes by one Alexis of Piedmont and other authors.[40] Among these recipes, for example, one aimed to eliminate the salty phlegm that was given to him by a servant of the archbishop of Toledo, another to make black ink, and yet another—made from quinces, cloves of spices, cinnamon, *almaciga*, and mint—to treat vomiting, *cámaras*, and lack of appetite.

Reading alchemical and hermetic texts as well as actually working in the laboratory was commonplace in Philip II's court. Activities directly promoted by the monarch were supplemented by private initiatives like those of the Flemish Giraldo París, a successful businessman who invested a substantial portion of his fortune in conducting alchemical experiments and acquiring treatises directly related to hermetic philosophy. In Madrid from 1570 on, París moved in diplomatic and courtly worlds, forming part of the retinue that in 1580 accompanied Philip II on his triumphal trip to Lisbon, the capital of the then recently inherited Portuguese empire. This trip secured for París the lucrative business rights to Portuguese pepper,[41] the financial rewards of which allowed him to settle in a mansion located behind the Hospital of the Passion in Madrid. There, he performed many chemical experiments and hosted spagyric gatherings of adepts.[42] París's prominent socioeconomic position aroused the envy of his enemies, who cited his interest in chemical experiments as well as his spagyric gatherings in bringing his case before the Inquisition. Although París was jailed by the inquisitorial court in August of 1603, the documentation of his trial confirms

38. Rolán and Suárez-Somonte, "Sobre los avatares de la edición en el humanismo español"; and Herrero, "Ciencia paleográfica hispano-latina en el siglo XVI."

39. BNE, ms. 9226, fol. 249.

40. *Recetarios acado de D. Alejo Piemontés y de otros autores, y de otras muchas recetas que me han dado.* Ibid., fols. 145r–225r.

41. With his associate Giovanni Battista Rovelasca, París controlled the monopoly between Malaca and Antwerp via Lisbon. An administrative adjudication beginning in 1587 allowed him to live on the proceeds; Thomas, "Flamencos en la península ibérica."

42. Rey Bueno, "Juntas de herbolarios y tertulias espagíricas."

that the charges dealt not with his alchemical activities but rather with his interpretations of various religious tenets—the virginity of Mary, the essence of the Trinity, the existence of purgatory, and the Israelites crossing the Red Sea—that he made from a chemical perspective.[43] As one witness put it in París's defense, however, he did not offer these interpretations "as a theologian, but as a natural philosopher who had studied alchemical enigmas."[44] París himself argued along similar lines in a memorial he wrote in his defense and in which he affirmed that his chemical practices had helped him understand many obscure passages of the Bible. As he explained, it was for that reason that he had philosophized about the possibility of using knowledge derived from experience to remove the stumbling blocks that separated different Christian nations and to fashion a dialectic weapon against the objections raised by Protestants, Jews, and Muslims. París thus used alchemy—an art uniquely able to imitate nature—as a language of mediation between nations and religions.[45]

It might seem that Philip II's chemical interests were simply mimicked by courtiers and close associates eager to emulate their monarch's tastes and desires. What is certain, however, is that similar cases are to be found in circles that thrived at other times and functioned in spheres removed from court. Consider, for example, Manuel Franco de Guzmán, a Castilian nobleman and resident of Valladolid. With important and lucrative properties distributed throughout Burgos, Franco de Guzmán spent much of his free time compiling alchemical treatises. A manuscript of his property conserved at the National Library of Spain provides a remarkable optic through which to judge Spanish tastes in alchemy during the second half of the sixteenth century. Next to pseudo-epigraphic medieval treatises attributed to famous Greek, Arab, or Latin authors, Franco de Guzmán also recorded aspects of theoretical alchemy and diverse procedures for making therapeutic remedies. Most prominent among the latter were two recipes for making potable gold taken from Philip Ulstadt (fl. 1530), a recipe to cure tongue ulcerations, and a fragment of *De secretis naturae liber* attributed to the author known as Ramón Llull (ca. 1232–ca. 1315).[46]

43. For the inquisitorial trial of Giraldo París, see AHN, sección Inquisición, legajo 100/21. See also Muñoz Calvo, *Inquisición y ciencia en la España moderna*, 44–57.

44. Muñoz Calvo, *Inquisición y ciencia en la España moderna*, 44–57. He was none other than Phelipe de Cortavilla, a Flemish pharmacist in charge of supplying medicines to royal servants.

45. I have revisited París's trial from a new interpretative perspective in "Mágicos prodigiosos y verdades acrisoladas."

46. BNE, ms. 7443. One of the most renowned alchemical compendia in the Spanish historical liter-

Spagyric Friars

Chemical practice within the Spanish nobility was not limited—as the evidence we have just reviewed might suggest—to the second half of the sixteenth and early seventeenth centuries. The practice of alchemy continued well into the seventeenth century. For example, Juan José de la Calle, knight of the Order of Santiago and *Oidor* at the Chancellery of Granada, was a great connoisseur of the chemical art, while Prince Don Vicente Gonzaga (1602–94), member of the Council of State that governed the destinies of the monarchy, routinely engaged in self cures using chemical remedies made in his own laboratory. Accounts of these men have come down to us thanks to Fra Andrés de Villacastín, a Hieronymite monk of the monastery of San Lorenzo de El Escorial, who was also an expert artificer and participant in chemical discourse of the late seventeenth-century court in Madrid.[47] Villacastín was one of many Spanish monks who, during the early modern era, read and practiced chemistry. His privileged position in the monastery of San Lorenzo (its Mansion of the Waters continued to operate throughout the seventeenth century[48]) and his knowledge of the art were sufficient reasons for Charles II to choose him as the *artifice* charged with certifying the properties of Alderete's Water of Life.

Fra Esteban Villa, another outstanding pharmacist of seventeenth-century Spain, directed the pharmacy of the Hospital of San Juan de Burgos. Associated with the Benedictine monastery there from 1616 to 1660, Villa was considered by his brethren to be an illustrious son of the order, a philosopher, and a theologian dedicated to the faculty of *espagiria* in which, it is said, he was

ature, this text, although discovered in the nineteenth century, has only recently been studied in detail. See Rodríguez Guerrero, "Manuscrito 7443 de la Biblioteca Nacional de España."

47. See Villacastín, *Chymica despreciada.*

48. We know this from testimonies offered by both pharmacist Juan del Castillo and the above-cited Matías de Beinza. Castillo, a Frenchman from Bordeaux who settled in Cádiz, learned how to distill quintessences at El Escorial. As he mentioned in the "Prologue to the Reader," "if there is something good [in his work] it should be attributed to . . . effort and desire. I have obtained the quintessence from the pharmacy of the Escorial, that is what I have most learned in Spain [*si algo de bueno hay* [in his work] *atribúyanlo . . . al trabajo y codicia que tomé en la botica del Escurial de las quintasesencias, que es lo más, y con mayor efecto he aprendido en España*]"; Castillo, *Pharmacopea universa medicamenta.* Beinza himself spent two years in El Escorial's pharmacy, where he studied the *chemical art* with pharmacist friar Fermín de San Jerónimo, "distinguished in his profession and admirable in chemistry [*insigne en su profesión y admirable en lo chímico*]"; Beinza, *Discurso sobre los polvos universales purgantes,* 1:60.

unsurpassed.[49] He was described in similarly glowing terms by the Mercedarian brother Bernardino de Palacios in his 1729 *History of the City of Burgos, Its Families and Its Holy Church*. According to Bernardino, Fra Esteban was a great chemical philosopher and an even better administrator of the monastery's pharmacy,[50] who had learned his art in a Galenic and spagyrical pharmacy equipped with a vegetable garden for the cultivation of medicinal herbs and a laboratory for making distilled waters and quintessences.[51] The pharmacy ultimately became one of the most famous in Castile, owing as much to the wealth of the medicinal herbs and drugs it purveyed as to the skill of its pharmacist friars. Both made the pharmacy the point of reference for patients and *sanadores* throughout the kingdom.[52] Fra Esteban's chemical ideas, moreover, are well documented in the five works on pharmacy that he published over a twenty-year period.[53] In them, he made it clear that his favored reference works were Evónimo Filiatro's *De remediis secretis*, Ulstad's *Coelum philosophorum*, Wecker's *Antidotarium*, and Diego de Santiago's *Arte separatoria*.[54] The skill, moreover, with which he prepared his glassware and stills as well as his flowers and roots for preliminary "digestions" prior to distillation[55] attested to his expertise as a practitioner. He devoted much space in his later works to discussions of alchemical secrets, fully accepting the use of chemical practices in medicine as well as the possibility that gold could be obtained by alchemical transmutation.[56]

Villa's successor as regent of the monastic pharmacy, Fra Esteban Núñez, continued the chemical activity of his teacher, combining the task of medicine

49. See *Directorio, regla y advertencias que se hacen a los abades que serán de este Real Monasterio de San Juan de Burgos*, RMSJB.

50. Jimeno, "Antigua y famosa botica."

51. Traditional studies on San Juan's pharmacy are Jimeno, "Antigua y famosa botica"; and Bonilla, *Las afamadas boticas burgalesas*. During a research trip to the city of Burgos in 2002, López Pérez and I found a whole series of manuscripts concerning this pharmacy—inquiries, inventories, and relations—that had remained uncatalogued at the El Archivo Municipal, Burgos (AMB). We analyzed these documents in López Pérez and Rey Bueno, *Miropolio general y racional de botica*, 15–24.

52. Palacios noted that most of the citizens of Burgos bought their medicines there. Even more interesting, royal physicians and pharmacists also bought remedies there that they could not find in Madrid; Jimeno, "Antigua y famosa botica," 72.

53. They are *Examen de boticarios* (Burgos, 1632); *Ramillete de plantas* (Burgos, 1636); *Libro de simples incógnitos en medicina* (Burgos, 1643); *Libro de las vidas de doze príncipes de la medicina* (Burgos, 1647); and *Segunda parte de simples incógnitos en la medicina* (Burgos, 1654).

54. Villa, *Examen de boticarios*, 4.

55. Villa, *Ramillete de plantas*, fols. 126v–127v.

56. Villa, *Libro de las vidas de doze príncipes de la medicina*, fols. 109r–119v.

maker with writing about the pharmaceutical art. Along with an official price list of medicines dating from 1680[57] and a short dictionary of pharmaceutical operations,[58] Núñez left an unfinished pharmaceutical treatise entitled *Miropolio general y racional de botica* and dated 1680, which was a compendium of the most common spagyric medicines in the pharmacies of his day.[59] Since his *Miropolio* remained unfinished, however, it provides only a fragmentary glimpse of the Spanish therapeutic landscape. Still, it covered what Núñez had learned not only from his teacher but also from the many authors whose books were reviewed in its pages. The latter range from the classics on distillation—Conrad Gesner (1516–65), Giambattista della Porta (1535?–1615), and Pietro Mattioli (1501–77)—to the specialists in spagyria—Joseph Duchesne (1544–1609), Angelo Sala (1576–1637), and Johann Schröder (1600–1664)—to the principal pharmacopeias of the Galenic and spagyric corpus.

One of Núñez's contemporaries, the priest and beneficiary of the parochial church of San Pablo of Zaragoza Juan de Vidós y Miró (1645–1710), was also an active and renowned practitioner of the spagyric art. Having received a papal exemption, he practiced spagyric medicine in Zaragoza beginning in 1669, with his fame ultimately spreading as far as Madrid, Salamanca, Murcia, Seville, Valencia, and Navarre. Apparently moved by the Christian charity that governed all his acts, Vidós decided to collect a lifetime of knowledge of the art and to publish it so that it could be consulted whenever a doctor was unavailable. The result was the 1691 *Primera parte de medicina y cirugia racional y espagírica* (or *First Part of Medicine and Rational Surgery and Spagyrics*), de facto the first chemical pharmacopeia published in Spain.[60]

57. *Traslado de la cédula real que Su Majestad el Rey Nuestro Señor Don Carlos Segundo mandó guardar, en quanto a los precios de medicinas, en veinte y siete días del mes de noviembre de mil y seiscientos y ochenta años, y se publicó en Madrid en catorce de diciembre del mismo año*. The Crown was forced to decree a general custom tax on 10 February 1680, owing to monetary deflation; Folch Jou and Muñoz Calvo, "Estudio preliminar sobre la tarifa de medicamentos en España en 1680."

58. Entitled *De operationibus pharmaceuticis,* it collected all practices carried out by the pharmacists and recorded not only a recipe of Pierre Potier's *Syrupusaureus* but also one for medical tape based on an original recipe of pharmacist friars. Both the *Tarifa* and *Diccionario* were incorporated at the end of the second edition of *Ramillete de plantas* (Burgos, 1646) and sold together. A version of the book is available at the BHUC, FOA 2890 (1).

59. The manuscript is in AMB, section Monasteriode San Juan, and was published with an introductory study in López Pérez and Rey Bueno, *Miropolio general y racional de botica*, 87–317.

60. Vidós, *Primera parte de medicina y cirugia.*

REMEDIES FOR THE POOR

Seventeenth-century practitioners like Vidós actually authored numerous writings that, under the premise of charity, offered simple, easy-to-make remedies for the poor. This kind of work had a long tradition with roots deep in the Middle Ages—as exemplified by the *Thesaurus pauperum* of Petrus Hispanus (fl. 13th century)—and immediate antecedents in the well-known books of secrets. This literature of charitable medicine, also known as domestic pharmacopeias, was penned mostly by physicians whose main objective was to provide patients with therapeutic knowledge independent of the work of pharmacists. These manuals generally discussed both how to recognize the medicinal properties of plants and how to carry out distillations designed to yield distilled waters, elixirs, and quintessences. The relationship, in fact, between distillation and remedies for the poor seems to have been common in Spain during the early modern era. Works on distillation techniques, such as *Discursos de la vía particular y verdadero modo de destilar compuestos* (or *Discourses of the Particular Way and True Mode of Compound Distillation*), written by pharmacist Juan de Castro Medinilla, even proposed a special distillatory apparatus for "the poor men who can enjoy perfect distilled waters and with less expense."[61] Vidós's 1691 *Primera parte de medicina y cirugia* represented an even more comprehensive work in this genre.

In his dedication to the "Charitable and Pious Reader," Vidós briefly described the circumstances that brought him to specialize in the making of chemical remedies. As he put it, he desired to help the most destitute, those who could not pay for the services of a university-trained physician or could not afford the costs of chemical compounds made by a licensed pharmacist. He explained that after having learned "the theory and practice of the Art of Medicine and Surgery," he began to distribute his remedies "without any gain to anyone who requested them."[62] The attention that his grateful patients drew to the efficacy of his medicines ultimately resulted in the publication of his pharmacopeia. Of the latter's pages, Vidós humbly remarked that "if they are not pages that cure, they are pages that teach how to cure; if they do not formally absorb the juice against the malady, they show and document it. These pages are all charity,

61. "Pobres gocen perfectas aguas distiladas i con menos costa." This is an unpublished manuscript dated Cordova, 1619 and preserved at BNE, ms. 4250; for the quotation, see p. 49.

62. "sin interés alguno a cuantos lo pedían." See Vidós, "Dedication to the Most Noble Realm of Aragon," in *Primera parte de medicina y cirugia*. The quotations that follow in this paragraph may also be found here.

all document. They are all remedy, all health; all Pharmacy of such a sweet and convenient price that the poorest subject will find medicine and health, without his poverty hindering him." This charitable approach was also evident in encomiums of the book, in which Vidós was presented as a medical Columbus who offered a new route to healing—ingenious and cheap—that found its purpose in relieving the suffering of the poor.[63]

Felipe Borbón, a contemporary of Vidós, was a graduate with a bachelor's degree in medicine and surgery from the University of Huesca and the author of *Medicina doméstica, necessaria a los pobres y familiar a los ricos* (or *Domestic Medicine Necessary for Poor People and Rich Families*), a book intended to instruct the poor on easy ways to concoct the remedies needed to alleviate ailments. Borbón especially emphasized "chemical laxatives for the poor [*purgantes chimicos para los pobres*]," remedies that supposedly fought ailments quickly and affordably.[64] Somewhat later in his book, he also incorporated a brief description of the main chemical practices that poor men may use for extracting the medicinal virtues that the Omnipotence had placed in the plants.[65] What Borbón had in fact done in this section of his book was to translate the work of the famous seventeenth-century French physician Philibert Guybert, author of a number of charitable books that were frequently reprinted and translated into the main European languages.[66] In his works, Guybert attacked pharmacists and their complicated compositions and so received severe criticism from that sector, not only in France but also in Spain. These assaults motivated Pedro Gutiérrez de Arevalo, a pharmacist in Madrid, to write a response to Guybert's first charitable book aimed at defending Spanish therapeutics against the criticisms of the French doctor.[67]

63. "sino son hojas que curan, son hojas que enseñan a curar; sino embeben formalmente el jugo contra el achaque, lo enseñan y documentan. Son hojas todas caridad, todas documento. Todas remedio, todas salud; todas Botica de tan dulce y acomodado precio, que el más pobre regnícola hallará medicina y salud, sin que le estorbe su pobreza"; see Vidós, "The Censure of Fra Luis Pueyo and Abadia, Chair of Philosophy at the University of Zaragoza and Provincial of the Carmelites in Aragon," in *Primera parte de medicina y cirugia.*

64. Borbón, *Medicina doméstica*, 11–17.

65. Ibid., 31–37. Specifically, Borbón aimed to provide an "easy method to extract the spirit, oils, salt, tincture, from the roots, woods, grains, seeds, leaves, and flowers by the chemical art [*método fácil para sacar el espíritu, el aceite, la sal, la tintura, el extremo fijo de las raíces, madera, granos, simientes, hojas y flores por Arte Chímica*]."

66. These works bore titles such as *Médecin charitable* (Anvers, 1624), *Apothicaire du médecin charitable* (París, 1625), and *Oeuvres charitables de Philber Guybert* (París, 1629). See Ramsey, "Popularization of Medicine in France, 1650–1900"; and Albou, "Histoire des *Oeuvres charitables* de Philibert Guybert."

67. See Gutiérrez de Arevalo, *Práctica de boticarios.*

WOMEN'S WORK

The so-called *Book of the Prior*, a book that falls somewhere between the genre of books of secrets and technical agricultural manuals, may also be situated within this tradition of domestic pharmacopeias. Written by Miguel Agustín (1560–1630), prior of the Order of San Juan of Jerusalem in Perpignan, it was originally published in Catalan in 1617. Translated into Castilian in 1625, it was reedited more than twenty times between 1625 and 1785.[68] Judging from its title, the work would seem to have been the product of an exercise in compiling knowledge drawn from many authors, old and new. A detailed reading of the text, however, reveals that Agustín's book was actually a literal translation of *L'agriculture et la maison rustique* (or *Agriculture and the Rustic House*) by Charles Estienne (1504–64) and Jean Liebault (1535–96).

Initially published in Paris in Latin in 1545 as *Praedium rusticum*, Estienne's *L'agriculture* attempted to popularize ancient knowledge of botany, medicine, horticulture, and agronomy. Translated into French in 1564, the same year in which its author died, it was extended in successive editions by Jean Liebault, Estienne's son-in-law and a great believer in the world of secrets.[69] Liebault's interests led him to incorporate into the text the *Bref discours de la distillation des eaux*, and the *Bref discours sur la distillation des huiles et quintes essences* (or *Brief Discourse on the Distillation of Waters* and *Brief Discourse on the Distillation of Oils and Quintessences*, respectively), which included detailed illustrations of furnaces, distillatory apparatus, stills, and various techniques. All of this material was reproduced in Agustín's translation and incorporated into the second chapter of the first book devoted to "the secrets of the office and duties of the mother, head of the farmhouse, and how she must teach her daughters and servants."[70] That chapter concludes with an account of the remedies for the diseases of a farmhouse family as detailed by Estienne and Liebault in their work. The two French authors assumed that the woman of the house would be in charge of making medicinal remedies

68. Agustín, *Libro de los secretos de agricultura*. Its title in Catalan is *Libre dels secrets de agricultura, casa rústica y pastoril*.

69. His interest in the world of secrets led him to publish *Quatre liures des secrets de medicine et de la philosophie chimique* (Paris: Chez Jacques du-Pauys, 1573), a translation into French of Gesner, *Evonymus Conradi Gesneri . . . De remediis secretis liber secundus* (Tiguri, 1569). This, the second part of Gesner's book of secrets, was published by his disciple Gaspar Wolf after his master's death.

70. "los Secretos de la condición y oficio de la madre de familia de la casa de campo, y cómo debe adoctrinar a sus hijas y criadas"; Agustín, *Libro de los secretos de agricultura*, 32.

for the ailments of all those in her household. Although it is not mentioned explicitly, it would seem that the chapter was inspired by the work of the Spanish humanist Juan Luis Vives (1493–1540).

Vives has been recognized as the major voice responsible for defining the role and education of women in sixteenth-century Europe, a characterization based on the popularity of a single work, his *De Institutione foeminae christianae* (or *On the Education of Christian Women*). Written in 1523 at a crucial transition in Vives's life, the text was dedicated to his countrywoman Catherine of Aragon (1485–1536), consort of the English King Henry VIII (1491–1547), and was regarded as a model for the education of her daughter, Princess Mary (1516–1558). By 1600, it had appeared in more than forty editions in English, Dutch, French, German, Italian, and Spanish as well as in the original Latin. In Book 2, "On Married Women," in the ninth chapter, "On How She Should Behave at Home," Vives explained that

> since the care of the inhabitants of the house falls upon the woman, she will keep remedies on hand for common and almost daily maladies and will have them ready in a larder so that she may attend to her husband, small children, and the servant when required and will not have to send for the doctor often and buy everything from the apothecary. I should not wish that a woman dedicate herself to the art of medicine or have too much confidence in it. I advise her to be familiar with the remedies for frequent and everyday illness, like coughs, catarrhs, itching, colic, loose bowels or constipation, intestinal worms, headache or aching eyes, slight fevers, dislocations, burns, slight cuts, and similar ailments that occur almost daily for trivial reasons. Add to this regulation of the daily diet, of greatest importance for the maintenance of good health, what should be consumed, what avoided, when and in what amount. She can learn this skill from the experience of other prudent matrons rather than from the advice of some nearby physician, or some simple handbook on that subject rather than from big, detailed medical tomes.[71]

This advice was similar to that found in the *Libro del Prior*, a book that also incorporated valuable knowledge of distillation for the housewife. Although distillation could have been construed as an art limited to trained philosophers and alchemists, the manual held that because it offered so many benefits to those who knew it, the female head of the household should carry out distillations not only for her own household but also "to help all her neighbors in their illnesses,

71. Vives, *Education of a Christian Woman*, 263.

the same way some Lords in their charity distil waters and other liquors as efficacious remedies to help the poor."[72] In this way, women joined men in the ranks of those adept at distillation with their knowledge of the proper temperatures and techniques to use in furnaces, of cooking glasses and other tools used for distilling, of how to prepare materials before distillation, and, of course, of the medicinal virtues of the resulting waters.

This brings us into territory just now being explored by historians of science, namely, housewifery and its associated natural philosophy during the early modern period. Over the past two decades, it has been shown that women played a fundamental role in domestic medicine, not only as midwives—work traditionally assigned to the female universe—but also as experts in the production of remedies used to cure the illnesses most common in the domestic sphere.[73] Women performed these duties in concert with a number of others—such as perfumers, aestheticists, and hygienists—charged with bodily cleanliness and adornment through the care of skin, hair, and teeth.[74] All of this knowledge came with a cornucopia of recipes and prescriptions that were not only transmitted verbally but also compiled into recipe books that were passed down from mother to daughter as treasures of quotidian knowledge.[75] Written by women, these books contained both medical and culinary recipes and so serve as registers of the ideals and practices of domestic work, physical health, and sustenance. Recipe collections stand at an historically significant intersection between the practical sciences of the body (which are represented in anatomies, herbals, midwives' manuals, and medical handbooks) and the mechanical arts (prominent in manuals of instruction for navigation, geometry, surveying, and metallurgy, among others). In style and content, food recipes and health remedies reflect the knowledge traditions and

72. "para socorrer a todos sus vecinos en las enfermedades, del modo y manera que vemos ser cosa y usanza caritativa de algunos Señores, que destilan aguas y otros licores que son eficaces remedios para valer y ayudar a los pobres." Agustín, *Libro de los secretos de agricultura*, 241.

73. Martínez Crespo, "Mujer y medicina en la Baja Edad Media."

74. Cabré i Pairet, "Cosmética y perfumería en la Castilla bajomedieval."

75. Among the oldest surviving manuscripts are *Manual de mugeres en el qual se contienen muchas y diversas reçeutas muy buenas*, BPP, mss. 834, with a critical edition by Martínez Crespo. The Biblioteca Nacional de España has three sixteenth-century manuscripts that join pharmacy, cosmetology, and cooking: *Livro de receptas de pivetes, pastilhas elvvas perfumadas y conserbas* (BNE, mss. 1462); *Las recetas experimentadas para diversas cosas* (BNE, mss. 2019), and *Recetas y memorias para guisados, confituras, olores, aguas, afeites, adobos de guantes, ungüentos y medicinas para muchas enfermedades* (BNE, mss. 6058). These recipe books have been studied by Pérez Samper in "Recetarios de mujeres y para mujeres."

scientific developments of the early modern period, standing as unacknowledged companions of the experimental texts of the "new science."

Rather Than a Conclusion, a Travel Journal

This chapter may be understood as part of a historiographical current that, over the past few years, has attempted to rescue early modern Spanish science from the systematic neglect to which it had largely been subjected by the international community of historians of science. This neglect goes back to the seventeenth century when Spain became the "villain" that English, French, and German historians needed in order to complete their "melodramas of modernity."[76] Fortunately, this is a situation that has been changing due to the work of, among others, Jorge Cañizares, Antonio Barrera, Daniela Bleichmar, and María Portuondo.[77] Such scholars have begun to demonstrate the pioneering character of the Spanish scientific community in grappling with the New World as well as the natural world and in making geographical and cosmographical advances. Four decades have now elapsed since Allen Debus's groundbreaking studies—on what he defined as the "chemical philosophy"—forced historians to reframe the traditional narrative of the Scientific Revolution and redirect their exclusive focus on mathematics and astronomy.[78] Something similar is happening now. As William Eamon has affirmed, "perhaps it is time to refocus and reassess the questions that we are asking about early modern Iberian science and, in particular, about Iberia's role in the Scientific Revolution."[79]

Spain was not a country that turned its back on novelties. In fact, it might be asserted that many activities considered novel in the current historiography (alchemy and patronage, women and chemical practice, chemical remedies and everyday life) were firmly embedded in Spanish practices of the sixteenth and seventeenth centuries. Incorporating Spain into the discourse of early modern science is thus a way of retrieving the early modern experience itself. Indeed, as Eamon remarked, "[a] world without Spain is certainly not the world that early modern Europeans thought they were living in."[80]

76. Eamon, "'Nuestros males no son constitucionales,'" 14.

77. See Cañizares, *How to Write the History of the New World*; Cañizares, "Iberian Science in the Renaissance: Ignored How Much Longer?"; Cañizares, *Nature, Empire, and Nation*; Barrera, *Experiencing Nature*; Bleichmar et al., *Science in the Spanish and Portuguese Empires, 1500–1800*; and Portuondo, *Secret Science*.

78. Debus, *Chemical Philosophy*.

79. Eamon, "'Nuestros males no son constitucionales,'" 21.

80. Ibid., 16.

Works Cited

Archives

AHN = Archivo Histórico Nacionàl, Madrid, Spain.

AMB = El Archivo Municipal, Burgos, Spain.

BHUC = Biblioteca Histórica de la Universidad Complutense, Madrid, Spain.

BNE = Biblioteca Nacional de España, Madrid, Spain.

BPP = Biblioteca Palatina, Parma, Italy.

HSA = Hispanic Society of America, New York, NY.

RMSJB = Real Monasterio de San Juan de Burgos

Printed Sources

Abbri, Ferdinando. "Alchemy and Chemistry: Chemical Discourses in the Seventeenth Century." *Early Science and Medicine* 5 (2000): 214–26.

Albou, Philippe. "Histoire des *Ouevres charitables* de Philibert Guybert." *Histoires des sciences médicales* 32 (1998): 11–26.

Agustín, Miguel. *Libro de los secretos de agricultura, casa de campo y pastoral.* Barcelona: En la imprinta de Juan Piferrer, 1722.

Barrera, Antonio. *Experiencing Nature: The Spanish American Empire and the Early Scientific Revolution.* Austin: University of Texas Press, 2006.

Barrionuevo, Gervasio de. *Tratado sobre el láudano opiato de Iosepho Querzetano y su elixir apropiado: Con el secreto del manna solutivo.* Toledo: Por Agustín de Salas Zaço, 1684.

Beinza, Matías De. *Discurso sobre los polvos universales purgantes.* Bayona: Antonio Fauvet, 1680.

Bercebal, Diego. *Recetario medicinal espagírico.* Zaragoza, 1713.

Bleichmar, Daniela, Paula DeVos, Kristin Huffine, and Kevin Sheehan, eds. *Science in the Spanish and Portuguese Empires, 1500–1800.* Stanford: Stanford University Press, 2009.

Bonilla, Alejo Arnaiz. *Las afamadas boticas burgalesas de los Hospitales de San Juan y San Julián y San Quirce.* Burgos: Burgos, 1981.

Borbón, Felipe. *Medicina doméstica, necessaria a los pobres y familiar a los ricos: Transcrita del médico caritativo con algunos remedios de otros autores. Con escolios en las materias y afectos que se tratan, así chirúrgicos como medicos.* Zaragoza: Domingo Gascón, 1686.

Cabré i Pairet, Montserrat. "Cosmética y perfumería en la Castilla bajomedieval." In *Historia de la ciencia y de la técnica en la Corona de Castilla,* vol. 2, *Edad Media,* by Marcelino V. Amasuno and Luis García Ballester, 772–79. Valladolid: Junta de Castilla y León, 2002.

Cañizares, Jorge. *How to Write the History of the New World.* Stanford: Stanford University Press, 2001.

———. *Nature, Empire, and Nation: Explorations of the History of Science in the Iberian World.* Stanford: Stanford University Press, 2006.

———. "Iberian Science in the Renaissance: Ignored How Much Longer?" *Perspectives on Science* 12 (2004): 86–125.

Castillo, Juan del. *Pharmacopea universa medicamenta in officinis pharmaceuticis usitata complectens et explicans.* Gadibus: Apud Ioannem de Borja, 1622.

Cózar, Lorenzo. *Dialogus veros medicinae fontes indicans.* Valentiae: Apud Petrum Patricium, 1589.

Debus, Allen G. *The Chemical Philosophy: Paracelsian Science and Medicine in the Sixteenth and Seventeenth Centuries.* 2 vols. New York: Science History Publications, 1977.

———. "Paracelsus and the Delayed Scientific Revolution in Spain: A Legacy of Philip II." In *Reading the Book of Nature: The Other Side of the Scientific Revolution,* edited by Allen G. Debus and Michael T. Walton, 147–61. Kirksville, MO: Sixteenth Century Journal Publishers, 1998.

Eamon, William. "Appearance, Artifice, and Reality: Collecting Secrets in a Courtly Culture." In *The Gentleman, the Virtuoso, the Inquirer: Vincencio Juan de Lastanosa and the Art of Collecting in Early Modern Spain,* edited by Mar Rey Bueno and Miguel López Pérez, 127–43. Newcastle, UK: Cambridge Scholars Publishing, 2008.

———. "'Nuestros males no son constitucionales, sino circunstanciales': The Black Legend and the History of Early Modern Spanish Science." *Colorado Review of Hispanic Studies* 7 (2009): 13–30.

———. *Science and the Secrets of Nature: Books of Secrets in Medieval and Early Modern Culture.* Princeton, NJ: Princeton University Press, 1994.

Folch Jou, Guillermo, and Sagrario Muñoz Calvo. "Estudio preliminar sobre la tarifa de medicamentos en España en 1680." *Boletín de la Sociedad Española de Historia de la Farmacia* 112 (1977): 197–204.

Gonzalez Rolán, Tomás, and Pilar Saquero Suárez-Somonte. "Sobre los avatares de la edición en el humanismo español: Acercamiento a la actividad del granadino Juan Vázquez del Mármol como corrector general y crítico textual." *Cuadernos de filología clásica* 4 (1992): 23–37.

Goodman, David. "The Scientific Revolution in Spain and Portugal." In *The Scientific Revolution in National Context*, edited by Roy Porter and Mikulas Teich, 158–78. Cambridge: Cambridge University Press, 1992.

Gutiérrez de Arevalo, Pedro. *Práctica de boticarios: Guía de enfermeros. Remedios para pobres*. Madrid: María de Quiñones, 1634.

Herrero, Carmen Olmedilla. "La ciencia paleográfica hispano-latina en el siglo XVI." Edición y valoración de las Abreviaturas de Juan Vázquez Del Mármol. *Cuadernos de Filología Clásica. Estudios Latinos* 4 (1993): 191–232.

Jimeno, Pascual Domingo Jimeno. "La antigua y famosa botica del Hospital de San Juan Evangelista, de Burgos." *Anales de la Academia Nacional de Farmacia* 3 (1934): 62–83.

Lastanosa, Vincencio Juan de. *Tratado de la moneda iaquesa y de otras de oro y plata del Reyno de Aragón*. Zaragoza: Dedication of Diego Vincencio Vidania, 1681.

López Pérez, Miguel. "Alquimia, terapéutica y sanidad en la España de los Austrias Menores." PhD diss., Universidad Complutense de Madrid, Facultad de Geografía e Historia, 2001.

———. "La alquimia y Vincencio Juan de Lastanosa." In *Vincencio Juan de Lastanosa (1607–1684): La pasión de saber*, 177–83. Exhibition catalog. Huesca: Instituto de Estudios Altoaragonenses, 2007.

———. "Anatomía del virtuoso: Coleccionismo y melancolía en la figura de Vincencio Juan de Lastanosa." *Argensola* 115 (2007): 167–85.

———. *Asclepio renovado: Alquimia y medicina en la España moderna (1500–1700)*. Madrid: Corona Borealis, 2003.

———. "Lastanosa, la alquimia y algunos helmoncianos aragoneses." *Azogue* 5 (2002–2007): 139–50.

López Pérez, Miguel, and Mar Rey Bueno. "La instrumentalización de la espagiria en el proceso de renovación: Las polémicas sobre medicamentos químicos." In *Los hijos de Hermes: Alquimia y espagiria en la terapéutica española moderna*, edited by Javier Puerto et al., 281–347. Madrid: Corona Borealis, 2001.

———. *Miropolio general y racional de botica (1680) de Fray Esteban Núñez (Estudio introductorio, transcripción y notas)*. Burgos: Colegio Oficial de Farmacéuticos, 2003.

López Piñero, José María. "La *Carta filosófica, médico-chymica* (1687) de Juan de Cabriada, punto de partida de la medicina moderna en España." *Asclepio* 17 (1965): 207–14.

———. *Ciencia y técnica en la sociedad española de los siglos XVI y XVII*. Barcelona: Labor, 1979.

———. *El Dialogus (1589) del paracelsista Llorenç Cóçar y la cátedra de medicamentos químicos de la Universidad de Valencia (1591)*. Valencia: Cátedra e Instituto de Historia de la Medicina, 1977.

———. "Juan de Cabriada y las primeras etapas de la iatroquímica y de la medicina moderna en España." *Cuadernos de historia de la medicina española* 2 (1963): 129–54.

———. *La introducción de la ciencia moderna en España*. Barcelona: Ariel, 1969.

———. "Paracelsus and His Work in 16th and 17th Century Spain." *Clio medica* 18 (1973): 113–41.

———. "Química y medicina en la España de los siglos XVI y XVII: La influencia de Paracelso." *Cuadernos de historia de la medicina española* 11 (1972): 3–40.

López Terrada, María Luz. "Llorenç Coçar: Protomédico de Felipe II y médico paracelsista en la Valencia del siglo XVI." *Cronos* 8 (2005): 31–66.

———. "Las prácticas médicas extraacadémicas en la ciudad de Valencia durante los siglos XVI y XVII." *Dynamis* 22 (2002): 85–120.

Martínez Crespo, Alicia. *Manual de mugeres en el qual se contienen muchas y diversas reçeutas muy buenas*. Salamanca: Ediciones Universidad de Salamanca, 1995.

———. "Mujer y medicina en la Baja Edad Media." *Hispania* 54 (1994): 37–52.

Moran, Bruce T. "Extracting the Virtues of Nature: Spagyric Remedies and Chemical Metaphors in the Library of Don Vincencio Juan de Lastanosa (1607–1681)." In *The Gentleman, the Virtuoso, the Inquirer: Vincencio Juan de Lastanosa and the Art of Collecting in Early Modern Spain*, edited by Mar Rey Bueno and Miguel López Pérez, 144–56. Newcastle, UK: Cambridge Scholars Publishing, 2008.

Muñoz Calvo, Sagrario. *Inquisición y ciencia en la España moderna*. Madrid: Editora Nacional, 1977.

Newman, William R., and Lawrence M. Principe. "Alchemy vs. Chemistry: The Etymological Origins of a Historiographic Mistake." *Early Science and Medicine* 3 (1998): 32–65.

Novella, Cosme. *Tratado de la quintaesencia del Bálsamo de Romero*. Zaragoza: Juan de Lanaja, 1606.

Olmedilla Herrero, Carmen. "La ciencia paleográfica hispano-latina en el siglo XVI: Edición y valoración de las Abreviaturas de Juan Vázquez del Mármol." *Cuadernos de filología clásica* 4 (1993): 191–232.

Padgen, Anthony. "The Reception of the 'New Philosophy' in Eighteenth-Century Spain." *Journal of the Warburg and Courtland Institutes* 51 (1988): 125–40.

Pérez Samper, María de los Ángeles. "Recetarios de mujeres y para mujeres: Sobre la conversación y transmission de los saberes domésticos en la época moderna." *Cuadernos de historia moderna* 19 (1997): 121–54.

Politius, Antonio. *De quinta essentia solutiva atque brevi epylogo componendorum medicamentorum, cum aliquibus phylosophiae ac medicinae problematibus: Libri duo.* Panormi: Ioannem Baptistam Maringum, 1613.

Portuondo, María. *Secret Science: Spanish Cosmography and the New World.* Chicago: University of Chicago Press, 2009.

Ramsey, Matthew. "The Popularization of Medicine in France, 1650–1900." In *The Popularization of Medicine, 1650–1850,* edited by Roy Porter, 97–124. London: Routledge, 1993.

Rey Bueno, Mar. "The Collector of Secrets: Potable Gold, an Italian Alchemist and a Nurse Soldier in Lastanosa's Laboratory." In *The Gentleman, the Virtuoso, the Inquirer: Vincencio Juan de Lastanosa and the Art of Collecting in Early Modern Spain,* edited by Mar Rey Bueno and Miguel López Pérez, 157–71. Newcastle, UK: Cambridge Scholars Publishing, 2008.

———. *El hechizado: Medicina, alquimia y superstición en la corte de Carlos II (1665–1700).* Madrid: Corona Borealis, 1998.

———. "El informe valles: Los desdibujados límites del arte de boticarios a finales del siglo XVI (1589–1594)." *Asclepio* 56 (2004): 243–68.

———. "Juntas de herbolarios y tertulias espagíricas: El círculo cortesano de Diego de Cortavila (1597–1657)." *Dynamis* 24 (2004): 243–67.

———. "Mágicos prodigiosos y verdades acrisoladas: Inquisición, magia, experiencia y conocimiento en el siglo XVII español." *Colorado Review of Hispanic Studies* 7 (2009): 49–66.

———. "*La mayson pour distiller des eaues* at El Escorial: Alchemy and Medicine at the Court of Philip II, 1556–1598." In *Health and Medicine in Hapsburg Spain: Agents, Practices, Representations,* edited by Teresa Huguet-Termes, Jon Arrizabalaga, and Harold J. Cook, 26–39. London: Wellcome Trust Centre for the History of Medicine at UCL, 2009.

———. *Los señores del fuego: Destiladores y espagíricos en la corte de los Austrias españoles.* Madrid: Corona Boreales, 2002.

Rey Bueno, Mar, and María E. Alegre. "Los destiladores de Su Majestad: Destilación, espagiria y paracelsismo en la corte de Felipe II." *Dynamis* 21 (2001): 323–50.

———. "La ordenación normativa de la asistencia sanitaria en la corte de los Habsburgo españoles (1515–1700)." *Dynamis* 18 (1998): 341–75.

———. "Renovación de la terapéutica real: Los destiladores de Su Majestad, maestros simplicistas y médicos herbolarios de Felipe II." *Asclepio* 53 (2001): 23–51.

Rodríguez Guerrero, José. "El manuscrito 7443 de la Biblioteca Nacional de España: Identificación de su origen, autor y contenidos." *Azogue* 5 (2002–2007): 57–69.

———. "La primera gran red comercial de un medicamento chymico: Vittorio Algarotti y su quinta esencia del oro medicinal." *Azogue* 6 (2008–2009): 12–67.

———. "Vendedores de panaceas alquímicas entre los siglos XVI y XVII." *Azogue* 5 (2002–2007): 90–99.

Rolán, T. Gonzáles, and P. Saquero Suárez-Somonte. "Sobre los avatares de la edición en el humanismo espanõl." Acercamiento a la actividad del granadino Juan Vázquez del Mármol como corrector general y crítico textual. *Cuadernos de Filología Clásica. Estudios Latinos* 4 (1992): 23–37.

Ruiz Zapata, Francisco. *Discurso sobre la composición del azúcar rosado solutivo, defendiendo las ordinaciones reales y la del collegio de médicos y cirujanos, atendiendo al beneficio común de la salud*. Zaragoza: Por Pedro Verges, 1625.

Slater, John. "Rereading Cabriada's *Carta*: Alchemy and Rhetoric in Baroque Spain." *Colorado Review of Hispanic Studies* 7 (2009): 67–80.

Slater, John, and Andrés Prieto. "Was Spanish Science Imperial?" *Colorado Review of Hispanic Studies* 7 (2009): 3–10.

Thomas, Werner. "Los flamencos en la península ibérica a través de los documentos inquisitoriales (siglos XVI–XVII)." *Espacio, tiempo y forma* 4 (1990): 167–95.

Valles, Francisco. *Tratado de las aguas destiladas, pesos y medidas de que los boticarios deben usar por nueva ordenanza y mandato de su Majestad y su real consejo*. Madrid: Luis Sánchez, 1592.

Vidós y Miró, Juan de. *Primera parte de medicina y cirugia racional y espagirica, sin obra manual de hierro ni fuego, purificada con el de la caridad, en el crisol de la razón y experiencia, para alivio de los enfermos: Con su antidotario de rayzes, yerbas, flores, semillas, frutos, maderas, aguas, vinos etc. Medicinales, que usa la Medicina racional y espagirica: y la farmacopea donde se explican el modo y composición de los remedios con el uso, dosis y aplicacion de ellos*. Zaragoza: Gaspar Tomás Martinez, 1691.

Villa, Esteban. *Examen de boticarios*. Burgos: Pedro de Huydobro, 1632.

———. *Libro de simples incógnitos en medicina*. Burgos: Colegio Oficial de Farmacéuticos, 1643.

———. *Libro de las vidas de doze príncipes de la medicina*. Burgos, 1647.

———. *Ramillete de plantas*. Burgos: [Pedro] Gomez de Baldivielsso, 1636.

———. *Segunda parte de simples incógnitos en la medicina*. Burgos: Colegio Oficial de Farmacéuticos, 1654.

Villacastín, Andrés de. *La chymica despreciada: D. Luys de Alderete y Soto perseguido. Defendida y defendido . . . con las doctrinas de los médicos griegos, árabes y latinos, así los príncipes, como los clásicos de sus escuelas*. Granada: Imprenta de la Santísima Trinidad, 1687.

Vives, Juan Luis. *The Education of a Christian Woman: A Sixteenth-Century Manual*. Edited and translated by Charles Fantazzi. Chicago: University of Chicago Press, 2000.

Chapter 7

Prescriptions of Alchemy

Two Austrian Medical Doctors and Their Alchemical Manuscripts

Anke Timmermann*

Allen G. Debus once summarized his motivations for working in the history of alchemy and medicine with a logic reminiscent of a syllogism. "[D]ebates over chemistry and chemical medicine in the sixteenth and seventeenth centuries," he said, "played a very important role in the development of a new science. Science is an essential part of the world we live in. We cannot understand where we are today without a knowledge of the history of science."[1] Debus's oeuvre not only put Paracelsian iatrochemistry on the map in the history of science—and back in its place in early modern history—but also set a precedent for current

* I would like to thank Karen Parshall, one of the editors of this volume, for her excellent suggestions for and kind assistance with the polishing of this chapter. The research for this publication has received funding from the European Union Seventh Framework Programme (FP7-PEOPLE-2009-IIF Marie Curie Action), project "AlchemVienna," project number 252918.

1. Debus, interview by Bohning. See also Debus, "Essay Review: Alchemy and the Historian of Science: Elias Ashmole."

scholarly debates about Paracelsian influences. Today, the history of alchemy/chemistry and medicine has intermingled and moved into periods and areas beyond Paracelsus and his followers.

The geographical region now covered by Austria provides a particularly interesting yet little-known example of the interaction between early modern alchemy and medicine. A central part of the Holy Roman Empire in the early modern period, the region—divided into various dukedoms—enjoyed autonomy in many public and legal sectors but in others fell under the authority of the imperial court in Vienna. It also experienced intellectual, linguistic, and cultural exchange with its immediate geographical neighbors. Most important for medical professionalization was the University of Vienna, which had been founded in 1365 both to surpass the rival universities in Prague and Krakow and to supersede the Italian universities. It regulated not only the examination of medical doctors but also the practice of pharmacies long before similar developments took place in other parts of Europe. Alchemy formed a natural part of the pharmaceutical and intellectual interests of doctors in Central Europe thanks to its highly developed mining industry, sophisticated practical crafts, flourishing trade along the Danube, and active circulation of alchemical knowledge through books, materials, and equipment. The history of alchemy and medicine in the archduchy of Austria and the Habsburg territories thus presents a rich combination of institutional and national factors, academia and craft, experimentation and regulation.[2]

This chapter investigates the lives and works of two doctors: Wolfgang Kappler (1493–1567), an apothecary doctor at Krems on the Danube in the archduchy of Austria, and Nicolaus Pol (ca. 1470–1532), a physician at the imperial court of Innsbruck in the Tyrol. Both showed a keen interest in alchemy and its medical uses. Their surviving books—substantial remnants of an extensive library in the case of Pol and a single manuscript reflective of Pol's influence in the case of Kappler—not only provide evidence of their practice and experiences as writers, readers, and practitioners of medicine but also inform our growing knowledge about the relationship between alchemy and medicine in the early modern period.

2. There has been some recent work on these themes. See, for example, Soukup, *Chemie in Österreich*; and Soukup and Mayer, *Alchemistisches Gold, Paracelsistische Pharmaka*. The history of the Medical Faculty of Vienna has not been fully realized. See, however, Horn, "Examiniert und Approbiert."

WOLFGANG KAPPLER (1493–FEBRUARY 1567)

At the age of thirty-four, Wolfgang Kappler settled as a physician and apothecary in the town of Krems on the Danube, about fifteen leagues (seventy kilometers) outside of Vienna. Krems had a population of some four thousand at the time, comprised mostly of craftsmen whose work was closely connected with trading along the Danube. Kappler had received his medical education at the philosophical-medical Collegium artium liberalum et physicorum in San Giovanni/Bragora and in Venice, and subsequently held appointments as municipal physician (*Stadtphysicus*) in Brno and physician in Znojmo (both now in the Czech Republic), about 130 and 90 kilometers north of Vienna, respectively. He had thus gained extensive experience in different parts of Central Europe—both as a medical practitioner and in the manufacture of remedies—by the time he arrived in Krems. There, he soon became an influential physician, held positions on the town council, and generally showed a forceful disposition reminiscent of another physician born in 1493, Paracelsus. It is Kappler's personality that has largely determined his historical image.[3]

Three prominent episodes in Kappler's life particularly tie in with his interest in alchemy. First, on being offered the position in Krems, he demanded that from that moment on his pharmacy would be the only one authorized to sell medical remedies to physicians and the public. Not surprisingly, his blatant attempt to control the local medical marketplace did not sit well with other medical practitioners in the city.[4] Both Kappler and the city of Krems referred to and called upon Viennese experts—as well as their regulations of pharmaceutical practice—in the ensuing, extensive quarrel about the details of this arrangement.[5] Fundamentally a clash between money and power on the one hand, and the safety and quality of remedies on the other, this episode brought Kappler's experience in writing prescriptions and producing remedies, alchemical ones included, to bear in establishing professional standards in Krems.

3. Ottner, "Die Streitbare Natur des Doktor Wolfgang Kappler," esp. 79. Kappler's vita is outlined in Kühnel, "Kremser Apotheker," esp. 16–22.

4. Ottner, "Die Streitbare Natur des Doktor Wolfgang Kappler," 79–80; and Kühnel, "Kremser Apotheker," 17.

5. Ottner, "Die Streitbare Natur des Doktor Wolfgang Kappler," 80ff. For a more detailed discussion of medical and pharmaceutical regulations at Krems, see ibid., 82ff.; and Ottner, *Dem gemeinen wesen zum besten* (on Kappler, see esp. 84–89). For Kappler's use of Viennese experts in the evaluation of his pharmacy, see Kühnel, "Kremser Apotheker," 19.

Another episode from Kappler's life, or rather from a literary commemoration of his botanical and pharmaceutical expertise, was inspired by a supposed encounter between Kappler, a colleague named Dr. Christoff Schaffner, and Paracelsus during the latter's Bohemian travels. Their debate about medical remedies survives in the form of a contemporary poem in a manuscript by Kappler's son-in-law Matthias Koch, who later followed in Kappler's footsteps as an apothecary in Krems.[6] Even if it is not based on an actual event, the poem is noteworthy for its association of Kappler and Paracelsus through both their conception of *materia medica* and their temperaments.

The final noteworthy incident concerns Kappler's possible contact with alchemical experimentation, which presents itself in the person of Christoph von Trenbach (d. 1552), canon of Passau and priest at Kirchberg am Wagram. Kappler lent Trenbach a substantial amount of money, which Trenbach never repaid. So great was Kappler's dismay over this debt that he criticized Trenbach posthumously, even though the latter's brother had offered to make amends.[7] Exactly what the original purpose of the loan was remains unclear, but it is possible that Trenbach borrowed the sum with the intention of building an alchemical laboratory.[8] It is known that a laboratory at Oberstockstall, a manor in the Kirchberg region and thus in Trenbach's vicinity, was in existence in the final years of Kappler's life. It is also well documented that Trenbach had connections with the manor and its lords, thus putting him, and by association Kappler, in close contact with both the site of and individuals engaged in alchemical experimentation.

The picture of Wolfgang Kappler that emerges from these anecdotes is one of a medical doctor and apothecary whose Italian training and extensive experience opened doors for him in the city of Krems. A wealthy and learned man who lent substantial sums of money to influential contemporaries, Kappler had ambitions for social status that surpassed his possibilities. Most importantly, as a sixteenth-century physician, Kappler developed his pharmaceutical enterprise at precisely the time that Paracelsian theories began to permeate the disciplines he studied.

6. Ottner, "Die Streitbare Natur des Doktor Wolfgang Kappler," 84–86; and Soukup, *Chemie in Österreich*, 222–24. Also, a laudatory poem on Kappler, composed by Matthias Cornax, his contemporary and fellow royal physician to Ferdinand I, can be found on the inner cover of ÖNB MS 11410.

7. Ottner, "Die Streitbare Natur des Doktor Wolfgang Kappler," 81; and Soukup, *Chemie in Österreich*, 222n590.

8. This is explored in Soukup and Mayer, *Alchemistisches Gold, Paracelsistische Pharmaka*, 10. See also Osten, *Das Alchemistenlaboratorium von Oberstockstall.*

Kappler's Manuscript: ÖNB MS 11410

While Kappler's public image has largely been shaped by the reports of others, his actual activities and interests—among them, active engagement with alchemy and alchemical remedies—are recorded in the sole surviving manuscript directly attributable to him. Sporting Kappler's colorful and bold coat of arms as an *ex libris*, ÖNB MS 11410 forms part of the collection of the Austrian National Library and is entitled *Practica pro menbris* [sic] *corporis humani / a capite usque ad pedes et primo de / capitis ad sectionibus*.[9] A folio volume of some 438 leaves, it contains hundreds of notes on illnesses and their remedies. These, as the copyist's preface explains, were gathered, tried, and tested during the daily experiences of Kappler, whom he described as a doctor and apothecary ("dealer in spices") with a strong commitment to constantly improving his pharmaceutical practice.[10] A precautionary note declares, however, that any textual mistakes are the copyist's and not Kappler's, while a quotation on the art of writing clearly and concisely (borrowed from the *Poetria nova* of Geoffrey of Vinsauf [fl. 1200]) concludes the preface.

Contrary to what the text's preface might suggest, most of the remedies gathered in Kappler's manuscript are not original. An astute reader's bit of marginalia elucidates this on the verso page. Kappler's volume, in fact, represents a copy of a manuscript originally written by Nicolaus Pol, physician at the court of Innsbruck, and finished by Pol's son, Johann.[11] When Johann died in 1536 (just four years after his father), the city of Znojmo, Kappler's former place of residence, acquired the manuscript. Kappler managed, with difficulty but through the influence of his financial stakes in the municipality, to convince the city to loan him the manuscript so that he could have a copy made. The copy that he commissioned, probably in the late 1530s, is the manuscript discussed here, ÖNB MS 11410, while Pol's original is now, unfortunately, lost.[12] Pol's life and extensive library (which will be discussed in detail below) thus provide a

9. Ex libris and coat of arms of *Practica for the Parts of the Human Body*. A detail is reproduced in Kühnel, "Kremser Apotheker," unnumbered plate after 16; and the entire page in Osten, "Das Alchemistenlaboratorium von Oberstockstall," esp. 339. The title can be found on fol. 1r of the manuscript.

10. ÖNB MS 11410, fol. IIIr. See also Academia Caesarea Vindobonensis, *Tabulae codicum*, 320.

11. ÖNB MS 11410, fol. IIIv. A transcription and early mention of Kappler in relation to Pol may be found in Ferrari, "Doctor Nicolaus Pol," esp. 119n3.

12. The year 1554, mentioned variously if not originally in the manuscript, is an unlikely date for the manuscript's composition for various reasons. See Kühnel, "Kremser Apotheker," 18; and Soukup, *Chemie in Österreich*, 222n590.

context within which to view and consider Kappler's sole surviving manuscript; they shaped its content before Kappler and his son-in-law, Matthias Koch, annotated and expanded the volume during the course of their daily routine of mixing and selling remedies to patients.[13]

Read from front to back, the manuscript leads the casual browser through the range of illnesses that Kappler encountered and anticipated in his medical practice. As was customary in *materia medica*, and as detailed in the volume's title, the complaints and their corresponding remedies are arranged from head to toe, starting with cures for headache and epilepsy, nerve pain and worms of the teeth, fainting, and dropsy, and followed by treatises on urine and on the heart, blood, and veins. Later sections contain recipes for laxatives, solutions for complications in childbirth and gynecological issues, and treatments for the relief of fistulas and sciatic pain. While there are some cures for boils, there are otherwise few remedies applicable to the legs and feet. Tracts on the plague, instructions for the proper cleaning of instruments of various sorts of materials, culinary recipes, and a collection of varia (many added by later hands) occupy a relatively large portion of the manuscript. Kappler's annotations record additional recipes, emendations of Pol's preparations, notes and personal notices (such as the wedding of his daughter and the feast provided for the occasion), and extensive lists of particular regimens.

Both the manuscript in its original form, that is, in the form in which Kappler first encountered it, and the heavily annotated version that has come down to us, show Kappler's ongoing engagement with certain methods, equipment, substances, and approaches that bridge medicine, alchemy, and pharmacy. Also in evidence is Kappler's knowledge of medical and alchemical literature, both traditional and contemporary. He referenced not only authorities like Arnold of Villanova (ca. 1238–ca. 1310) but also local and international colleagues, most of whom have been largely forgotten by historians of science. Title lines of individual remedies often purport to be "ex alchimico," occasionally employ alchemical symbols for substances such as mercury, and include methods such as distillation used in the alchemical laboratory. The manuscript, through its incorporation of alchemical terminology, literature, and methods, thus represents a key reference work for Kappler's medico-alchemical practice.

13. See also Kühnel, "Kremser Apotheker," 19–20. Kappler's script dominates the final folios of the manuscript beginning on fol. 426.

Nicolaus Pol (ca. 1470–October/November 1532)

The owner, possibly even the writer of Kappler's original source, Nicolaus Pol, had left his own legacy in medical and bibliophilic circles by the time Kappler acquired and copied his manuscript. A member of the generation preceding Kappler's, Pol received his medical training before the birth of Paracelsus, at the time Arnold of Villanova's works were gaining new popularity, and in an age when printing presses had begun to produce medical and other scientific books. Many of those books found their way into Pol's library.

Pol was raised and educated around the court of Innsbruck, then the imperial residence of the Tyrol and the intellectual and geographical nexus of Bavaria and the Italian city-states. Entering the service of Duke Sigismund (1427–96) at the age of seventeen, Pol appears several times in the court books as "distiller" before being sent off for further education. Unfortunately, the exact nature and extent of his tasks in that post are not documented.[14] After returning from his university education, probably in nearby Italy, Pol served as a physician at the Innsbruck court. Duke Sigismund's effort to employ "scores of physicians" as well as astronomers and his own apothecary, and to seek out alchemists' services likely influenced Pol's approach to medicine and science.[15] Pol would also have had access to Sigismund's library, which attracted visitors particularly for its extensive medical holdings. From 1495 on, however, Pol was in the employ of Maximilian I, who transformed Innsbruck into a center of political and cultural activity in the Holy Roman Empire. There Pol was surrounded by other royal physicians, who may have educated him further in matters both medicinal and bibliophilic.[16]

From 1513 to 1520, while already an established physician advancing towards middle age, Pol cultivated Veit Bild (1481–1529), a Benedictine

14. A letter of Pol's on his early life is quoted in Bachmann, "Dr. Nikolaus Poll," 410. Bachmann's theories on Pol's university education cannot be confirmed. Pol's vita is detailed more fully in Ferrari, "Doctor Nicolaus Pol," 115ff.; Lehmann, "Ein Deutscher auf der Suche"; and Fisch, *Nicolaus Pol Doctor 1494*. Here, Fisch's observation on Pol's religious book acquisitions and potentially clerical occupation remain to be substantiated; ibid., 26–27. See also Tenge-Wolf, "Nikolaus Pol und die Llull-Handschriften."

15. Tenge-Wolf, "Nikolaus Pol und die Llull-Handschriften," 267, with reference to an unpublished dissertation that I have not been able to access: Margarete Ortwein, "Der Innsbrucker Hof zur Zeit Sigismunds des Münzreichen."

16. Fisch, *Nicolaus Pol Doctor 1494*, 13, mentions particularly Ulrich Ellenbog. See Goldschmidt, "Hieronymus Muenzer and Other Fifteenth-Century Bibliophiles," 504–5.

monk and humanist at an Augsburg monastery, as a scientific correspondent. It is telling that Bild addressed Pol as "physician and astrologer most expert."[17] During these years, Pol's library grew significantly, but he had yet to contribute to it as an author in print. It was not until 1535, three years after his death and eighteen years after he had completed the text, that Pol's tract on the guaiac treatment for syphilis was finally published. Exactly why he investigated this popular topic remains unclear, although he may have been part of a commission sent to Spain by Cardinal Lang in 1517, and the text may have resulted from that excursion.[18] Apart from occasional visits to Venice, this trip is the only tentative trace of Pol's physical movement beyond Innsbruck.[19] While his life leaves fewer and thus less obvious traces than does Kappler's, Pol nevertheless emerges as an intelligent, well-educated physician with a deep curiosity about nature and its workings. He was a man whose profession perfectly matched his intellect.

Pol's Library: Its Books and Manuscripts

It is perhaps solely because of his extensive library that Nicolaus Pol, otherwise just one of many court physicians and medical practitioners in the Tyrol, has been the subject of scholarly research over the course of the past century. He may have owned as many as 1,350 volumes,[20] almost half the number of items accumulated by John Dee (1527–1609) about a half century later and a large multiple of the number of volumes on the shelves of most contemporaneous physicians. How he secured the money to buy his books, given his modest salary, remains a mystery.[21] Although it is unknown where Pol may have stored his

17. Fisch, *Nicolaus Pol Doctor 1494,* 19–24, quote at 20.

18. Pol, *De cura morbi gallici.* And see Fisch, *Nicolaus Pol Doctor 1494,* 16, 37–48, 50ff.

19. Fisch, *Nicolaus Pol Doctor 1494,* 17.

20. Today, books and manuscripts identifiable as from Pol's collection number close to five hundred volumes, scattered across various libraries around the world. A brief if not necessarily fully reliable sketch of the fate of Pol's holdings after his death can be found in Fisch, *Nicolaus Pol Doctor 1494,* 34–36. See also Baum, "Katalanische Philosophen," 615; and the foreword in *Medical Books from the Library of Dr. Nicolaus Pol.* I would like to thank archivist and museum registrar Jennifer Nieves at the Dittrick Medical History Center for making this catalogue available to me. Pol appears as "Legator Bibliothecae Anno 1494" in the Innichen monastery donor catalogue. Regarding Innichen, see Tenge-Wolf and Kühebacher, "Nikolaus Pol und die Stiftsbibliothek von Innichen," 53; and Kühebacher, *Kirche und Museum des Stiftes Innichen,* 218.

21. Estimations of the total number of books originally in his possession vary; the given estimate is based on Fisch, *Nicolaus Pol Doctor 1494,* 137ff. For private libraries of physicians in England, a useful point of comparison in the absence of a comparative study on Pol's compatriots is Jones, "Medical Libraries." On Pol's

books in Innsbruck or whether he kept them in a private or public setting, it is likely, judging by the different contemporary hands in their marginal notes, that colleagues and visitors at court had access to them.[22]

For an erstwhile book collector at the turn of the sixteenth century, Innsbruck was ideally located between two cities, Augsburg and Venice, with renowned publishing industries. Both would have been accessible to Pol, and books from both cities were well represented in his library.[23] For the acquisition of specific, rare editions, however, Pol enlisted the help of others. For instance, he asked his correspondent Veit Bild to purchase for him the translation by the French humanist Jacques Lefèvre d'Etaples (ca. 1455–1536) of works by the Syrian scholar John of Damascus (ca. 675–ca. 750). Bild also ordered books for Pol in Augsburg, Frankfurt, and France, and delivered titles of new publications with them.[24]

Others supplied Pol with information on where and how to find items on his growing list of desiderata. If Pol was indeed the compiler of a list of locations and owners of Llullian works found in one of his manuscripts, his informants included scholars, clerics, and physicians across the Germanic lands from Trier in the Rhineland to Gurk in Carinthia (now southern Austria) as well as in Italy (Bologna, Venice, and elsewhere) and Paris. Many of the book repositories listed are those of individual clerics in a host of monasteries.[25]

Pol's acquisition efforts were clearly methodical and successful. For example, his manuscripts now at Innichen include fourteen fifteenth-century copies of writings by and about Llull (primarily Llull's *Ars magna* or *Ars generalis ultima* and its short version, the *Ars brevis*) with a considerable number in duplicate versions. Covering as much as a quarter of all known Llullian works, Pol's collection is considered the most significant still extant.[26] And his knowledge of

---

funding of the library, see Tenge-Wolf and Kühebacher, "Nikolaus Pol und die Stiftsbibliothek von Innichen."

22. For comparison, consider John Dee's library, its use and visitors; Sherman, *John Dee*, 29–52.

23. Fisch, *Nicolaus Pol Doctor 1494*, 29.

24. Lehmann, "Ein Deutscher auf der Suche," 234–35.

25. ÖNB MS 5510, fol. 224r, in Lehmann, "Ein Deutscher auf der Suche," 238–39 (German translation) and 239–40 (Latin original).

26. The manuscripts carry shelf marks 17–23 (VIII b. 8–14), 25 (VIII b. 16), 28 (VIII c. 3), 33 (VIII c. 8), 36 (VIII c. 11), 38 (VIII c. 13), 40 (VIII d. 1), and 43 (VIII d. 4); MS 21 dates from the sixteenth century. They are described in Balaguer, "Los códices Lulianos," 314–40. See also Tenge-Wolf, "Nikolaus Pol "; Tenge-Wolf and Kühebacher, "Nikolaus Pol und die Stiftsbibliothek von Innichen," 53–54; and Kühebacher, *Kirche und Museum des Stiftes Innichen*, 223. Out of the seventy-four texts included in the manuscripts, only two appear to be medical in nature.

and access to works by other authors and on other subjects is equally impressive. Approximately half of his surviving incunabula and a third of his sixteenth-century holdings are medical or broadly scientific in nature.[27]

What, however, can Pol's library tell us about his attitudes toward alchemy, medicine, and pharmacy? Today, his books and manuscripts often carry either or both of two distinguishing marks. One is an ownership mark, "Nicolaus Pol doctor 1494," on the inner cover or flyleaf verso. This was most likely added at the library of the monastery at Innichen (now San Candido, Italy) that came to house a substantial part of his collection soon after his death; the year 1494 probably marks the date of Pol's bequest.[28] The other is a large majuscule letter inked onto the lower, and sometimes upper, cut, a shelf mark[29] that was visible when books were placed, with the back up and the side cut down, onto shelves for storage. These shelf marks offer an opportunity to investigate how Pol kept, arranged, and thought about the content of his books.

While an analysis of the shelf marks of all extant Pol holdings would be ideal, many items are no longer readily accessible, while others appear largely irrelevant. For example, the above-mentioned Llullian manuscripts are interspersed with religious tracts and have only limited medical relevance; none of them stems from the pseudo-Llullian alchemical corpus. By contrast, the seven manuscripts now held at the Austrian National Library represent a wider range of Pol's interests and strongly emphasize alchemy and medicine.[30] As a somewhat random selection from Pol's collection, they provide a counterpoint to Kappler's pharmacopoetic manuscript. These seven Pol manuscripts, supple-

27. Pol's printed book holdings are described in Fisch, *Nicolaus Pol Doctor 1494*, 25–48, with a list of 467 books and manuscripts (see pp. 137–235). Although Fisch's list is considered "preliminary and tentative" by Fisch himself (p. 25), his remains the most comprehensive description of Pol's books. For the Innichen/San Candido holdings (not included by Fisch), see Balaguer, "Los códices Lulianos," esp. 311ff. The part of Pol's collection now at the Dittrick Medical History Center at Case Western Reserve University is described in *Medical Books from the Library of Dr. Nicolaus Pol*, the 1929 catalogue mentioned above.

28. On the current holdings of the Innichen monastery and museum, see Kühebacher, *Kirche und Museum des Stiftes Innichen*, 218; and Tenge-Wolf, "Nikolaus Pol und die Llull-Handschriften," 273. For various, often dubious theories on the matter of the significance of the date in Pol's ownership mark, see Balaguer, "Los códices Lulianos," 312; "Introduction," in *Medical Books from the Library of Dr. Nicolaus Pol*; Ferrari, "Doctor Nicolaus Pol," 113; Goldschmidt, "Hieronymus Muenzer and Other Fifteenth-Century Bibliophiles," 506; Fisch, *Nicolaus Pol Doctor 1494*, 14, 30; and Tenge-Wolf, "Nikolaus Pol und die Llull-Handschriften," 265. See also the appendix below.

29. The term "shelf marks" used here is modern. Pol refers to a passage in a book marked with such a sigil as "fo. 109 segno G," hence the letters are plainly markers (marginal note in MS 5487, fol. 13v).

30. For the history of these (and other) manuscripts' journey to Vienna (in part via Castle Ambras in the Tyrol) and their absorption into the Viennese imperial library, see Gottlieb, *Die Ambraser Handschriften*.

mented by information about thirty-three printed books now housed at Case Western University, prompt the following reflections.

The shelf marks appear to have been added to Pol's printed books and manuscripts in the late fifteenth or early sixteenth century, that is, during his lifetime and most likely by Pol himself. If Pol organized his collection of more than a thousand books via shelf marks, what was his ordering principle? By using the twenty-four letters of the Latin alphabet, Pol clearly did not intend to assign unique shelf marks to individual books, a method first introduced in monastic libraries in the fourteenth century.[31] Rather, Pol's system followed the convention of earlier medieval libraries, where shelf marks were repetitive and the librarian's knowledge was key to the retrieval of individual items.

In practice, this system was successful for Pol, as contemporary marginal notes referring to other volumes by shelf mark confirm. For example, ÖNB MS 5489, fol. 65v reminds the reader to "vide in libro M" (see book *M*), fol. 274 for more information about *sal anatron*, and, indeed, the corresponding passage can be found in a codex marked *M*, now ÖNB MS 5230 (fol. 274r in the old pagination, now fol. 257r). The identification of the correct volume out of several marked *M* was, however, the reader's responsibility and a task that was not necessarily straightforward. Consider the fact that six volumes in the sample of seven manuscripts and thirty-three books under examination here carry the letter *F*, with no distinction made between manuscripts and printed books. If Pol's library numbered 1,350 items and even if shelf marks were distributed evenly, each letter would have occurred at least twenty-six times. Interestingly, too, some books in Pol's library actually remained unmarked. As a result, it is impossible to trace from the surviving evidence any further distinction by press or shelf position indicative of a more sophisticated classification system.

These kinds of uncertainties and inconsistencies suggest that drawing historical conclusions based on Pol's library classification scheme—whatever it may have been—needs to be done with caution. As Richard Sharpe has so aptly put it, "It is rarely possible to understand what marks mean if we have only the evidence of those in surviving books. When we have other evidence [such as library catalogues and finding aids], it becomes apparent that marks of the same

31. Sharpe, "Accession, Classification, Location," 281, 284–87. In the following paragraphs, information about shelf marks in the Dittrick Medical History Center collection is taken from *Medical Books in the Library of Dr. Nicolaus Pol.*

type may in reality have very different signification."[32] The evidence provided by the shelf marks, contents, and marginal notes of the forty items considered here from Pol's library nevertheless allow for some interesting observations.

First, Pol's shelf marks function in part as running numbers—perhaps marked in order of acquisition—and cluster similar items within the same subject group. Thus three Avicenna volumes, all large folios on medicine published in 1498, carry the shelf marks *F*, *G*, and *H*.[33] Two of Pol's other medical volumes, also works by Avicenna carrying shelf marks *D* and *I*, may have flanked the trio of 1498. This implies that shelves may have been assigned a priori—in anticipation of the acquisition of related books—so that Pol could reasonably add consecutive shelf marks to several, associated books once a critical mass had been achieved. Books bound together and assigned a single shelf mark, such as three pharmaceutical-medical volumes by different authors and all published in Venice in 1495 and 1496, provide further evidence for this.[34] The fact, however, that some items remained unbound—some with and some without a shelf mark[35]—casts doubt on how thoroughly this grouping of similar items may have applied across the library. Regardless, these items suggest distinct efforts at classification in Pol's library, even if his actual categories cannot be reconstructed.

Second, Pol's regular use of marginal notes to reference shelf marks provides further evidence of the purposeful arrangement of his library. Together, Pol's notes and library created a network of knowledge. Books opened up the world of literature to a Pol largely isolated in Innsbruck; his methodic organizational and cross-referencing system advanced his knowledge of contemporary science, including medicine and alchemy. Pol's was a working library and book repository with substantial scientific holdings and ambitions of comprehensiveness and usability.

32. Sharpe, "Accession, Classification, Location," 280.

33. Avicenna, *Canonis de medicina libri I, III et IV* (Lyons: Jean Trechsel and [completed by] Jean Clein, 1498).

34. Shelf mark C: Quiricus de Augustis, *Lumen apothecariorum cum certis expositionibus noviter impressum* (Venice: Joannes et Gregorius de Greboriis, 22 September 1495); Joannes de Ketham, *Fasciculus medicinae* (Venice: Joannes et Gregorius de Gregoriis, 15 October 1495); and Joannes Jacobus de Manliis, *Luminare maius* (Venice: Bonetus Locatellus, for Octavianos Scotus, 28 May 1496).

35. For example, shelf mark F: Haly Abbas, *Liber medicinae dictus regalis, trans. Stephanus, ed. Antonius Vitalis* (Venice: Bernardinus Ricius, Novariensis, for Joannes de Nigro, 25 September 1492) (unbound); no shelf mark: Iocabus Foroliviensis (Giacomo della Torre), *Expositio in primum librum canonis Avicennae*; and Hugo Senensis (Ugo Benzo), *Quaestio de malitia complexionis diversae* (Pavia: Christophorus de Canibus, for Nieronymus de Durantibus, 7 May 1488).

Finally, the implications of Pol's library organization for his acquisition, storage, and use of *alchemica* are small but significant. While shelf marks for *alchemica* are the same as those for medical works and other subjects, it is still possible that they occupied a separate shelf, perhaps mixed with other items of special interest.[36] However, such a separation would have been motivated by Pol's intentions for the books' usage and not necessarily reflective of his treatment of alchemy as a special subject. How, or indeed if, Pol classified alchemy and medicine is simply not apparent from the surviving shelf marks alone. [37]

The Practice of Medicine and Alchemy through the Optic of Pol's Working Library

So, what if anything can manuscript and book collections tell us about the history of alchemy *cum* medicine? That is the question that the remainder of this chapter will investigate through a consideration of the engagement of both Pol and Kappler with alchemical texts.

Current trends in historiography address the divide between manuscripts as written evidence and the actual practice of a science or craft.[38] In the context of the present chapter, this would suggest two questions: What use did Pol and Kappler actually make of their alchemical reading in their daily practice as doctors, that is, in fashioning their prescriptions and producing their remedies? What did they actually do?[39]

The motivation for their everyday practice is easily inferred from their professional commitments to take care of the sick. Indeed, the recipes for waters and salves, pills and powders listed in their manuscripts attest to that. Their

36. Compare John Dee's "internal," private part of his library. See Sherman, *John Dee*, 33.

37. Early modern libraries such as Trinity College Library at Cambridge were in the habit of storing *alchemica* with *medica*, the latter a term used for scientific books in general. Gaskell, *Trinity College Library*, esp. 9, 23, 86–90, 128.

38. For example, Principe and Newman discussed early approaches to the history of alchemy and their limits, desiderata, and future perspectives in "Some Problems with the Historiography of Alchemy." More recent developments include a focus on the history of alchemy in *Isis*, including, among others, Newman's "What Have We Learned from the Recent Historiography of Alchemy?" See also Martinón-Torres, "Some Recent Developments in the Historiography of Alchemy." Interdisciplinary approaches to a joint history of alchemy and medicine have also been presented at various conferences such as the meeting on "Alchemy and Medicine from Antiquity to the Enlightenment" held in Cambridge (UK) on 22–24 September 2011.

39. This very question was recently asked for the general history of alchemy in Principe, "Alchemy Restored," 310.

alchemical recipes, however, cover the spectrum from medications incorporating metalline substances to *chrysopoeia*. For example, ÖNB MS 5315, a miscellany completed prior to Pol's acquisition of it, gathers instructions in the space of a few folios for the preparation of potable gold, a corrosive water for the separation of gold and precious stones, a metalline water, and "Pil[u]le Imperiales" (imperial pills).[40] Other manuscripts provide instructions for making *aquae vitae* and quintessences, not always medical in their intended application.[41] As for the medically relevant recipes, ingredients in the collections of both men would have been at home not just in an alchemical laboratory but also in a kitchen or in Kappler's famous garden or apothecary shop.

What role, then, did the texts themselves, alchemical or more traditionally pharmaceutical, play in the doctors' approaches to their practice? It seems that neither Pol nor Kappler changed or questioned the wording or content of either medicinal or alchemical texts in their copies or annotations. Rather, they contributed additions and cross-references to the texts. They viewed their task as textual exegesis (Pol) or the practical realization (Pol and Kappler) of preparations selected from a variety of sources rather than as a critical engagement with any individual text. Consider, for example, this alchemically inclined medical preparation that was added in the margins of Kappler's manuscript to a section on pain medication. Entitled "Oleum tartari cum sale nitro" (Oil of tartar with saltpeter), the recipe, Kappler noted, had been obtained from one "P. Neplachio In Brun" (P. Neplachio at Brno).[42] Many other remedies added by Kappler to this reference work for his daily practice also testify to his effort to collect and preserve recipes from doctors of repute. Such collecting activity would likely not have stopped at a textual level. The gathered recipes would have been tested and then added to the repertoire of remedies Kappler expected to—or actually did—prepare. Patients' complaints and their demand for certain remedies would have informed the types of cures Kappler retrieved and recorded.

Annotations like this one in Kappler's text actually mirror the growth of Pol's library on a smaller scale and in the medium of a single manuscript. The practical inclination of Kappler's collecting activities is perhaps particularly pertinent for plague remedies, given the outbreak of epidemics in his day. Indeed, while the manuscript as copied from Pol devoted ample space to plague treat-

40. ÖNB MS 5487, fol. 46v; MS 5315, fols. 99r–108v.
41. ÖNB MS 11410, fols. 211v, 355ff.
42. Ibid., fol. 59v.

ments, Kappler added, as proof of efficacy, that they originated at the Medical Faculty of Vienna.[43] In this context, both his and Pol's pursuit of knowledge was cumulative, connected with the manufacture and application of remedies, and motivated partly by impending epidemics and partly by the medical marketplace. Alchemically informed recipes were part of that picture. And, where Pol created structure in his library through the organization of books, Kappler fit his textual acquisitions into the received structure of Pol's manuscript.

The motivation for both doctors' work in the library, in preparation for and development of their daily practice, reveals itself in their personal notes and manuscripts. As a case in point, Pol's manuscripts—and especially his cross-references—uncover his library as a place of active research. They reflect his targeted interest in applied alchemical texts, regardless of whether they had direct applications to medicine.

As mentioned above, for example, an annotator of ÖNB MS 5489 (fol. 65v) directed the reader to the codex marked *M* (now labeled ÖNB MS 5230) that then led to a series of technical recipes involving *sal anatron*.[44] ÖNB MS 5230, a miscellany produced in the early sixteenth century, bears the imprints of several annotating hands and contains many alchemical symbols, very distinctive manicules, and generally evidence of extensive usage that bridges the library and the laboratory. Another cross-reference by the same annotator appears in ÖNB MS 5487 beneath an alchemical recipe—complete with alchemical symbols as shorthand for substances—that is illustrated with the drawing of a vessel in ashes.[45] The recipe opens with a reference to the contemporary Viennese bishop Jurij Slatkonja (Georg von Slatkonia [1456–1522]), which in turn directs the reader to a volume with the shelf mark *G*, fol. 109, perhaps a passage in MS 5485 on the red elixir (fol. 109r) or a marginal note referring to space left for a drawing (fol. 109v). The uniformity of the script in these cross-references, their connection with the shelf marks, and their paleographical dating (in connection with the documented times of Pol's ownership) suggest that it was indeed Pol himself who read these volumes so carefully and with an eye for alchemical content.

Taken together, Pol's marginal cross-references also attest to the fact that this was an ongoing exercise. References in Cod. 5230, fol. 274r, for example, were clearly written at different sittings. His connective, extensive, and dynamic

43. Ibid., fol. 393v.
44. ÖNB MS 5230, fol. 274.
45. ÖNB MS 5487, fol. 13v.

reading of *alchemica* must therefore have influenced his understanding of medical remedies and their manufacture as well as his book-collecting activities over a long period of time.

Kappler's manuscript demonstrates how this active research was in turn applied in the practical occupation with remedies by both Pol and Kappler. Kappler's volume exhibits the label "ex alchimico" as well as alchemical terminology in a staggering range of sections, in those on epilepsy and other head-related seizures, on urine, on miscarriages, on sciatic pain, and elsewhere.[46] The cure for seizures, for example, employs vitriol under the alchemical code word "leo viridis." Some recipes use common ingredients available at Kappler's pharmacy and elsewhere, while others, often not distinguished as alchemical in the title, nevertheless call for ingredients or methods more closely related to the alchemical laboratory. Since the phrase "ex alchimico" only appears in the main body of Kappler's manuscript, and not in his or later annotations and additions, it probably stemmed from Pol's manuscript, the prototype for Kappler's text. Pol's remedy titles thus actually reference the origin of the recipes. Those *ex alchimico* were taken from an alchemical manuscript or book; others were referred to by the name of the author or compiler. These titles fulfill a function similar to that of Pol's cross-references; they enable the reader to trace the source of a particular remedy. Pol, then, used the term "alchemy" in his library as a reference, but like his shelf marks, it did not necessarily form a subject category or carry additional meaning.

It seems, then, that once applied to medical purposes, alchemy and medicine were amalgamated in Pol's and hence Kappler's manuscript. Pol did not distinguish alchemy from other subjects in its use and function. His need for distinction was linguistic; it aided the navigation of his library for the identification of sources. Nor did Kappler introduce a separation of alchemy from medicine in his notes. Kappler's manuscript thus represents the very conjoining of medicine and alchemy that has been emblematic of the birth of modern pharmacy. For Pol and Kappler, the separation of alchemy and medicine represents a preliminary step, one confined to the act of reading and sorting information. In practical applications, however, a dichotomy between them was at odds with the living, applied, interacting organizational principles in effect in Pol's library as well as with Kappler's concept of remedies.

46. See, for example, "Ex alchimico contra sciatica & omnem dolorem iuncturarum," ÖNB MS 11410, fol. 187r. Medications for suppressing impending miscarriage are on fol. 153 and epilepsy features around fol. 30v.

Pol's and Kappler's manuscripts also illuminate their actual, hands-on involvement with alchemy through their forays into the world of equipment and objects. The manuscripts of both men not only called for practical measures including distilling processes, putrefaction, and heating by *bain marie,* but they also contain actual drawings of furnaces and distilling apparatus.[47] Two instances, in particular, are noteworthy.

The first is Kappler's sketch of a *Rosenhut,* a distilling apparatus used in a recipe for a universal medicine as well as in many other instances. His sketch is simple and carries a single caption. Written into a square marked at the bottom front of the furnace, near the enclosed fire that provides heat for the evaporation of liquids, that caption specified the location of the *Luft Looch,* the air hole necessary for the regulation of heat.[48] Such a specification would have been of interest only to a reader who actually aimed to understand the workings of the furnace and who might have sought to use or even to build a similar apparatus. Such distilling equipment might also have come in useful in Kappler's pharmaceutical preparations.

The second example, a manuscript of Pol's written during his lifetime, is more complex. It depicts a "Faul Hayntz" (or Fauler Heinz/Piger Henricus) furnace. This sketch is much more elaborate both in its execution and in its captions.[49] The written description details the symmetric arrangement of the furnace, its openings, the positions of hob-like places on which several vessels can be heated at once, and some measurements relative to its construction that allow for an advantageous and adjustable distribution of heat. There are also detailed depictions of parts of other apparatus generally found in manuscripts aimed solely at those interested in alchemy. Notes and possibly even the sketch of the Fauler Heinz oven were most likely added by Pol himself and are consistent with Pol's interest—reflected in his other manuscripts and annotations—in the practicalities of working with substances over heat.[50] It would seem, then, that Pol's collecting efforts and intricate knowledge were not confined to books and manuscripts. Equipment and skills were additions, now lost, to his comprehensive literary approach to medicine, pharmacy, and alchemy.

47. ÖNB MS 11410, fol. 40r. Sketches of apparatus appear in ibid. fols. 249v, 300r, 377v–378r, 434r, and elsewhere.

48. Ibid., fol. 434r.

49. ÖNB MS 5487, fol. 11v, "Furnus Faul Hayntz." This manuscript also contains texts attributed to Ramón Llull and John of Rupescissa.

50. Unfortunately, Pol's own hand cannot be identified with any certainty. Tenge-Wolf and Kühebacher, "Nikolaus Pol und die Stiftsbibliothek von Innichen," 54, present a hand they consider likely to be Pol's.

Isolated in a single manuscript, Pol's notes also provide evidence of a categorization of alchemical items not apparent from his library organization. At the research stage of his work, Pol investigated alchemy seriously, in isolation from its immediate intersections with medicine and together with its practical applications. He was neither an "armchair alchemist" nor a physician pragmatically and unquestioningly adopting alchemically relevant methods in his production of remedies. His thoroughness, already apparent in the magnitude of his library, also extended to his exploration of alchemy as a subject and practice.

It should now be clear why Pol's manuscript was so valuable to Kappler that he used forceful persuasion to gain access to it. Pol's lost original of Kappler's manuscript was a record of Pol's personal reading notes. As an extraction of useful remedies from the medical and alchemical collections of his library, and written by a physician in possession of extensive knowledge of alchemy, Pol's digest of recipes certainly would have been more beneficial to Kappler and his practice than any one of the books Pol might have originally read. At a time when Pol's library was already being dispersed, this manuscript preserved—for Kappler's purposes—that library's most practical and valuable parts.

Moreover, Kappler's elaborated manuscript provided evidence of Pol's sense of how knowledge should be organized, at the same time that it reflected the contents of Pol's library and both doctors' approaches to remedies. It thus reveals the evolution of an early modern understanding of *materia medica*, medicine, and alchemy that in many ways defies modern classifications. As objects and as stores of historical knowledge, Kappler's manuscript and Pol's library support the contention that medicine and alchemy were very much intertwined and integrated in the early modern period.

Conclusion

Wolfgang Kappler's manuscript and Nicolaus Pol's library are emblematic of their times and yet as unique as their owners. Kappler, a doctor and apothecary whose manuscript tells of his everyday practice, seemed to approach remedies pragmatically. His manuscript presents a mixture of medicine, alchemy, and the knowledge of apothecaries. Pol, whose manuscript served as the basis for Kappler's, had an extensive library and lived in a courtly context of scholarly exchange. His library is evidence of his structured collecting activities and intensive study of a wide variety of texts and practices—including the alchemical—for purposes that extended beyond his work as a doctor. The work in both library and medical practice of Pol and Kappler is marked by a complex interplay of categorization and creativity. Their books both inform and form the development of their knowledge, a combination of alchemy and medicine that would become second nature to doctors and pharmacists of later generations.

This description of Pol, Kappler, and their books may appear straightforward to modern readers. Both types of Renaissance reader of *alchemica* are familiar to us from existing accounts of early modern libraries, books, manuscripts, and their users.[51] However, the lives and works of Pol and Kappler were conceived of differently just a few decades ago. Early scholarship on Pol was motivated by the bibliophilic interests of an Italian and a Spanish researcher, picked up by a German with similar interests in 1941, and exhausted by the end of the same decade with the publication of a definitive study. Research on Kappler started and ended in 1961 as part of a survey of doctors and apothecaries at Krems.[52] The early bibliographic emphasis, especially in scholarship on Pol, and the subsequent scholarly silence are not unusual for the Germanic countries, where the history of science developed differently from its English counterpart.

Perhaps particularly obvious to the modern reader is the absence of a discussion in the earlier literature of Pol and Kappler's practice of pharmacy and their conception of illness and the efficacy of remedies. Harry Kühnel's failure to mention the frequent references to alchemy in Kappler's manuscript, for example, reflected not so much a personal blind spot as the state of the discipline in his day.[53] By contrast, the joint history of alchemy and medicine that is currently emerging affords a wider, more inclusive perspective on issues both raised and ignored in the history both of medicine and of alchemy. An interdisciplinary, contemporary perspective not only questions but also adds to our picture of the early modern medical marketplace and its practices. It necessarily rests on prior scholarship in the two subdisciplines of the history of science, but ideally selects and surpasses the approaches used in them. It is to be hoped that the history of alchemy *cum* medicine will one day effect the same natural amalgamation it has observed in its historical objects.

51. I have discussed this previously in Timmermann, "Doctor's Order."

52. See Kühnel, "Kremser Apotheker." Early publications on Pol are, in chronological order, Balaguer, "Los códices Lulianos"; Ferrari, "Doctor Nicolaus Pol"; Lehmann, "Ein Deutscher"; and Fisch, *Nicolaus Pol Doctor 1494*. Recent publications, all considering individual aspects (Pol's Llulliana and Kappler's role in the development of medical regulations in Krems) are Traninger, "Bildgebende Verfahren"; Ottner, "Die Streitbare Natur des Doktor Wolfgang Kappler"; and Tenge-Wolf, "Nikolaus Pol und die Llull-Handschriften."

53. Kühnel, "Kremser Apotheker," 20.

APPENDIX: HANDLIST OF POL'S MANUSCRIPTS AT THE AUSTRIAN NATIONAL LIBRARY

This list, taking into account the history and organization of Pol's library as described in the main part of the chapter, consolidates and corrects extant information on the origins, contents, and physical features of all known Pol manuscripts now held at the Austrian National Library.[54]

E: MS 5509

Volume on alchemy. Bavaria or Austria,[55] mid-fifteenth century (prior to Pol's lifetime). 218 x 147 mm, 258 folios. Binding: Vienna, 1755.

Texts authored by or attributed to authorities, ancient (Hermes Trismegistus, Geber) and medieval (Boethius, Thomas Aquinas, Roger Bacon, Albertus Magnus); to alchemist-physician Arnold of Villanova (at least one a unique copy),[56] Ramón Llull and others, as well as a significant number of *anonyma*. Theoretical alchemy and recipes.

MS 5509 only shows the letter *E* on both the upper and lower cut, but not Pol's ownership note. The letter on the upper cut is missing its vertical stroke due to rebinding. A historical description of the manuscript in a previous binding confirms Pol's ownership. Some folios cut from MS 5509 can now be found in MS 5510.[57] Since the manuscript was rebound at the

54. Unless otherwise noted, information is taken from the Austrian National Library catalogue (ÖNB-HANNA-Katalog; http://aleph.onb.ac.at/F?func=file&file_name=login&local_base=ONB06, accessed 10 January 2012), a digital combination of information of the following standard sources: Gentilotti, *Catalogus*; Menhardt, *Verzeichnis*; Unterkircher, *Die datierten Handschriften*; and Academia Caesarea Vindobonensis, *Tabulae codicum*. A table of most Pol manuscripts at the Austrian National Library, their ownership marks, dates of their bindings, and letters on cut may be found in Höller, *Neues Licht auf den Kongreß der Goldmacher*, 58. See also Lehmann, "Ein Deutscher auf der Suche," 236–37. Contrary to information given in Menhardt, *Verzeichnis*, 1116, MSS 5002 and 5239 do not show any clear evidence of Pol's ownership. The letters present on the cut of MS 5239 are rather small majuscules at the top of the side cut, *A* and *E*, which may also be more recent than the fifteenth or sixteenth century.

55. Geographical origins refer to modern territories rather than historical ones.

56. Glorieux, *Répertoire des maîtres en théologie*, Nr 211 (dd): Arnold of Villanova's letter to the pope.

57. Höller, *Neues Licht auf den Kongreß der Goldmacher*, 59 and 55, with reference to Gentilotti, *Catalogus manuscriptorum codicum latinorum*; and Menhardt, "Die altdeutschen Ambrasiani der Österreichischen Nationalbibliothek," 56–57. Höller, *Neues Licht auf den Kongreß der Goldmacher*, 52–59, contains a detailed description of the manuscript; her publication also provides an edition of a German alchemical tract, fols. 249r–253v.

Viennese imperial library, this may have happened at any point between the addition of the shelf mark and 1755.

F: MS 5487[58]

Volume of Llull tracts and alchemy. Vienna(?), end of the fifteenth century (contemporary with Pol's early medical career). 219 x 150 mm, 178 folios. Binding: Vienna, undated.

Llullian works including the *Testamentum* and commentaries; John of Rupescissa on Llull; and some anonymous *alchemica*, including recipes.

Pol's ownership note may be found in the manuscript as well as the shelf mark *F* (on the lower cut only). Possibly part of the Fugger library in the sixteenth or early seventeenth century, via humanist Johannes Schöner, who owned various scientific manuscripts now also kept at the Austrian National Library.[59]

G: MS 5485

Volume of Llull tracts and alchemy. Bavaria or Austria, second half of the fifteenth century (probably contemporary with Pol's early medical career). Octavo, 157 folios. Binding: southwest Germany, predating Pol's acquisition of the manuscript.[60]

Llulliana including the *Liber quintae essentiae* and tracts on mercury; John of Rupescissa's *Liber de conderatione quintae essentiae*; anonymous vernacular recipes.

The ownership note appears in the manuscript, the letter *G* on the lower cut only. This is one of the manuscripts later absorbed into the library of Castle Ambras near Innsbruck before it ended up in Vienna.

I: MS 5489

Volume on alchemy, recipes, and experiments. Middle West of Germany, end of the fifteenth century (contemporary with Pol's early medical career). Octavo, 99 folios. Binding: fifteenth century.

58. Both Menhardt, *Verzeichnis*; and Höller, *Neues Licht auf den Kongreß der Goldmacher*, fail to include this volume in their lists of Pol manuscripts.

59. Lehmann, "Ein Deutscher auf der Suche," 237.

60. The HANNA catalogue, s.v. "Signatur: 5485," accessed 10 January 2012, dates the binding to pre-1494, mistaking Pol's ownership mark for an acquisition date.

Albertus Magnus's *Semita recta*; several anonymous recipes, including alchemical ones as well as recipes for ink and colors; mostly in Latin.

Letter *I* appears on the lower cut only. Pol's ownership note is also present.[61]

M: MS 5230

Volume on alchemy. Vienna(?), second half of the fifteenth and sixteenth century (during Pol's lifetime). 225 x 155 mm, 411 folios. Binding: southern Germany, early sixteenth century.

Due to the erroneous assumption that the ownership mark, "Nicolaus Pol doctor 1494," contains the manuscript's date of acquisition, it has been proposed that the manuscript changed hands soon after Pol acquired it. In spite of the presence of a number of annotating hands, none of which can be identified, it may be the case that Pol himself left the following note on fol. 10v: "Item hic liber fuit michi presentatus ab Ottone Kappenmayster per filium sororis uxoris eius iuvenem anno 1503 curr. 20. die Novembris" (This book of Otto Kappenmayster's was given to me by the young son of his sister-in-law this year, 1503, on the twentieth day of November). Nothing is known about Otto Kappenmayster.

Various and diverse texts authored by or attributed to authors, ancient (Avicenna, Geber, Virgil), medieval (for example, Thomas Aquinas, with particular emphasis on Ramón Llull and Arnold of Villanova), and many others; some (such as texts by one Kochperger and Fridericus de Nüremberga) may be contemporary with Pol. Those in the vernacular include recipes by one Petrus of Prague and one Wilhelm von Ghauch;[62] another, anonymous treatise appears to be in Low German.[63] In addition to the ten titled *anonyma* listed in the catalogues, this manuscript contains countless anonymous alchemical tracts on alchemical theory and practice, most in Latin.

Pol's ownership note appears in the manuscript, the letter *M* on the lower cut only, although remainders can be spotted on the upper cut.

61. The HANNA catalogue mistakes the letter on the cut for an *F*; see s.v. "Signatur: 5489," accessed 6 January 2012. It is corrected to an *I* in Höller, *Neues Licht auf den Kongreß der Goldmacher*, 58; and Lehmann, "Ein Deutscher auf der Suche," 237.

62. Keil, "Wilhelm von Ghauch," 535, discusses the recipes for copper in this codex on fols. 41v–42r. See also Keil and Halbleib, "Zwei alchemische Kupferprozesse," 101–5.

63. Lehmann, "Ein Deutscher auf der Suche," 236, remarks upon an owner of this manuscript of Low German origin and his annotations.

N: MS 5510

Volume on alchemy. Vienna, Bavaria, Swabia, and/or Middle Eastern Germany, fifteenth and beginning of the sixteenth century (during Pol's lifetime). 220 x 150 mm, 302 folios. Binding: southern Germany, early sixteenth century.

Most texts are anonymous or unattributed and provide advice, instructions, or memory aids for alchemical theory and practice. Among the attributed items are tracts by Christophorus Parisiensis, Ramón Llull, and one Iohannes, as well as Arnold of Villanova, Gratian, Thomas Aquinas, and Heinrich von Mügeln. Most texts are in Latin, with some German and Italian items.

The letter *N* appears on the lower cut only, the ownership note in the manuscript. The volume incorporates some folios from MS 5509.[64] This is one of the manuscripts absorbed into the Castle Ambras library before it came to Vienna.

Z: MS 5315[65]

Volume of texts by Arnold of Villanova and *medica*. Unknown location, mid-fifteenth century (prior to Pol's lifetime). 295 x 215 mm, 272 folios. Original binding.

Hands present in the manuscript can also be found in astronomical manuscripts: one possibly written in Salzburg in the second half of the fifteenth century, another written in Vienna, and a third and fourth written in Vienna or Klosterneuburg, all dating from the mid-fifteenth century.[66] Also, the flyleaves containing Pol's ownership note (see below) were taken from an Austrian manuscript from the mid-twelfth century.[67] It seems likely, then, that it is of Austrian origin.

Collection of texts by or attributed to Arnold of Villanova with a medical inclination as well as authorities in the natural sciences and medicine as varied as Albertus Magnus, Petrus de Abano, and Costa ben Luca, and various *anonyma* on medicine and alchemy.

For this manuscript, the ownership note is on a fragment, which was detached from the main manuscript and rebound with other materials, now MS Ser. n. 298. Letter *Z* on lower cut only.[68]

64. Höller, *Neues Licht auf den Kongreß der Goldmacher*, 55.

65. Both Menhardt, *Verzeichnis*; and Höller, *Neues Licht auf den Kongreß der Goldmacher*, fail to include this volume in their lists of Pol manuscripts.

66. Kratochwill et al., *Die Deutschen Handschriften des Mittelalters*, 342.

67. Mazal and Unterkircher, *Katalog der abendländischen Handschriften*, 103.

68. Catalogue HANNA mistakenly lists the letter as an *N* (s.v. ,"Signatur: 5315," accessed 6 January 2012).

Works Cited

Archives

HANNA = Austrian National Library catalogue, ÖNB: http://aleph.onb.ac.at/F?func=file&file_name=login&local_base=ONB06 (accessed at time of article creation).

ÖNB = Austrian National Library (Österreichische Nationalbibliothek), Vienna, Austria.

Interview

Debus, Allen G. Interview by James J. Bohning at Deerfield, IL, 29 March 2007. Philadelphia: Chemical Heritage Foundation, Oral History Transcript #0365. In American Chemical Society, Division of History of Chemistry. "Allen G. Debus (1926–2009)." Online http://www.scs.illinois.edu/~mainzv/HIST/awards/Dexter%20Papers/DebusDexterBioJJB.pdf.

Printed Sources

Academia Caesarea Vindobonensis. *Tabulae codicum manu scriptorum praeter graecos et orientales in Bibliotheca Palatina Vindobonensi asservatorum*. 10 vols. Vienna: Gerold, 1864–99.

Bachmann, Hanns. "Dr. Nikolaus Poll, Hofarzt zu Innsbruck." *Veröffentlichungen des Museums Ferdinandeum* 27/29 (1946–49): 409–18.

Balaguer, Jordi Rubió. "Los códices Lulianos de la Bibliotheca de Innichen (Tirol)." *Revista de filología española* 4 (1917): 303–40.

Baum, Wilhelm. "Katalanische Philosophen in Tiroler Klöstern." *Der Schlern: Monatszeitschrift für Südtiroler Landeskunde* 58, no. 10 (1984): 612–21.

Debus, Allen G. "Essay Review: Alchemy and the Historian of Science: Elias Ashmole." *History of Science* 6, no. 1 (1967): 128–38.

Ferrari, Luigi. "Doctor Nicolaus Pol, la collegiata di S. Candido ed i susi incanaboli." *Atti del Reale Istituto Veneto di scienze, lettere ed arte* 96 (1937): 109–69.

Fisch, Max H. *Nicolaus Pol Doctor 1494: With a Critical Text of His Guaiac Tract*. Edited and translated by Dorothy M. Schullian. New York: Cleveland Medical Library Association, 1947.

Gaskell, Philip. *Trinity College Library: The First 150 Years*. Cambridge: Cambridge University Press, 1980.

Gentilotti, Giovanni Benedetto. *Catalogus manuscriptorum codicum latinorum Bibliothecae Palatinae Vindobonensis VII: Codices medici*. Vienna: Österreichische Nationalbibliothek, 1721.

Glorieux, Palémon. *Répertoire des maîtres en théologie de Paris au XIIIe siècle.* 2 vols. Paris: J. Vrin, 1933–34.

Goldschmidt, Ernst Philip. "Hieronymus Muenzer and Other Fifteenth-Century Bibliophiles." *Bulletin of the New York Academy of Medicine* 14, no. 8 (1938): 491–508.

Gottlieb, Theodor. *Die Ambraser Handschriften: Beitrag zur Geschichte der Wiener Hofbibliothek.* Leipzig: M. Spirgatis, 1900.

Höller, Daniela. *Neues Licht auf den Kongreß der Goldmacher: Edition und Kommentar von ÖNB 5509 (fol. 249r–253v).* Vienna: Praesens, 2007.

Horn, Sonia. "Examiniert und Approbiert: Die Wiener medizinische Fakultät und nicht-akademische Heilkundige in Spätmittelalter und Früher Neuzeit." PhD diss., University of Vienna, 2001. Online at http://www.meduniwien.ac.at/typo3/fileadmin/Josephinum/Diss_Horn_UB_pdf.pdf.

Jones, Peter Murray. "Medical Libraries." In *The Cambridge History of Libraries in Britain and Ireland,* edited by Elisabeth Leedham-Green and Teresa Webber, 1:461–71. Cambridge: Cambridge University Press, 2006.

Keil, Gundolf. "Wilhelm von Ghauch (Wilhelmus de Ple[s]e)." *Würzburger Medizinhistorische Mitteilungen* 18 (1999): 535.

Keil, Gundolf, and Marianne Halbleib. "Zwei alchemische Kupferprozesse des Wilhelm von Ghauch." *Nova Acta Paracelsica* n.s. 16 (2002): 101–5.

Kratochwill, Dagmar, Annemarie Mühlböck, Peter Wind, and Gerold Hayer. *Die Deutschen Handschriften des Mittelalters der Erzabtei St. Peter zu Salzburg.* Vienna: Verlag der Österreichischen Akademie der Wissenschaften, 1982.

Kühebacher, Egon. *Kirche und Museum des Stiftes Innichen: Begleiter und Führer bei der Betrachtung der Kulturdenkmäler und Kunstwerke des ältesten Tiroler Stiftes.* Bozen: Verlagsanstalt Athesia, 1993.

Kühnel, Harry. "Kremser Apotheker und Ärzte des Mittelalters und der Frühen Neuzeit." *Mitteilungen des Kremser Stadtarchivs* 1 (1961): 9–32.

Lehmann, Paul. "Ein Deutscher auf der Suche nach Werken des Raymundus Llullus." *Zentralblatt für Bibliothekswesen* 58, no. 7/8 (1941): 233–40.

Martinón-Torres, Marcos. "Some Recent Developments in the Historiography of Alchemy." *Ambix* 58, no. 3 (2011): 215–37.

Mazal, Otto, and Franz Unterkircher. *Katalog der abendländischen Handschriften der Österreichischen Nationalbibliothek, "Series nova" (Neuerwerbungen).* Vienna: G. Prachner, 1963.

Medical Books from the Library of Dr. Nicolaus Pol: Court Physician to the Emperor Maximilian I. London: Maggs Bros., 1929.

Menhardt, Hermann. "Die altdeutschen Ambrasiani der Österreichischen Nationalbibliothek." In "Festschrift für Josef Stummvoll, Alois Kisser, Ernst Trenkler zum 50. Geburtstag dargebracht von Kollegen, Freunden und Mitarbeitern," edited by Michael Stickler, 56–57. Special issue, *Das Antiquariat* 8 (1952).

———. *Verzeichnis der altdeutschen literarischen Handschriften der österreichischen Nationalbibliothek*. Vol. 2. Berlin: Akademie Verlag, 1961.

Newman, William R. "What Have We Learned from the Recent Historiography of Alchemy?" *Isis* 102, no. 2 (2011): 313–21.

Ortwein, Margarete. "Der Innsbrucker Hof zur Zeit Sigismunds des Münzreichen: Ein Beitrag zur Geschichte der materiellen Kultur." PhD diss., University of Innsbruck, 1936.

Osten, Sigrid von. *Das Alchemistenlaboratorium von Oberstockstall: Ein Fundkomplex des 16. Jahrhunderts aus Niederösterreich*. Monographien zur Frühgeschichte und Mittelalterarchäologie 6. Innsbruck: Wagner, 1998.

———. "Das Alchemistenlaboratorium von Oberstockstall." In *Sein & Sinn, Burg & Mensch: Niederösterreichische Landesausstellung 2001*, edited by Falko Daim and Thomas Kühtreiber, 315–44. St. Pölten: Niederösterreichisches Landesmuseum, 2001.

Ottner, Christine. *Dem gemeinen wesen zum besten: Verwalten von Krankheit und Gesundheit in Krems an der Donau und Österreich unter der Enns (1580–1680)*. St. Pölten: Selbstverlag des NÖ Instituts für Landeskunde, 2003.

———. "Die Streitbare Natur des Doktor Wolfgang Kappler oder Der Arzt als Apotheker: Zur Entwicklung des Apothekergewerbes in Krems/NÖ am Beginn der frühen Neuzeit." In *Sozialgeschichte der Medizin—Stadtgeschichte und Medizingeschichte*, edited by Susanne C. Pils and Sonia Horn, 78–85. Vienna: Verlagshaus Thaur, 1998.

Pol, Nicolaus. *De cura morbi gallici per lignum guaycanum*. Venice: G. Paduano and V. Ruffinelli, 1535.

Principe, Lawrence M. "Alchemy Restored." *Isis* 102, no. 2 (2011): 305–12.

Principe, Lawrence M., and William R. Newman. "Some Problems with the Historiography of Alchemy." In *Secrets of Nature: Astrology and Alchemy in Early Modern Europe*, edited by William R. Newman and Anthony Grafton, 385–431. Cambridge, MA: MIT Press, 2001.

Sharpe, Richard. "Accession, Classification, Location: Shelfmarks in Medieval Libraries." *Scriptorium* 50 (1996): 279–87.

Sherman, William Howard. *John Dee: The Politics of Reading and Writing in the English Renaissance*. Amherst: University of Massachusetts Press, 1997.

Soukup, Rudolf Werner. *Chemie in Österreich: Bergbau, Alchemie und frühe Chemie. Von den Anfängen bis zum Ende des 18. Jahrhundert*. Vienna: Böhlau, 2007.

Soukup, Rudolf Werner, and Helmut Mayer. *Alchemistisches Gold, Paracelsistische Pharmaka: Laboratoriumstechnik im 16. Jahrhundert.* Vienna: Böhlau, 1997.

Tenge-Wolf, Viola. "Nikolaus Pol und die Llull-Handschriften der Stiftsbibliothek San Candido/Innichen." In *Ramon Llull und Nikolaus von Kues: Eine Begegnung im Zeichen der Toleranz,* edited by Ermenegildo Bidese et al., 261–86. Turnhout: Brepols, 2005.

Tenge-Wolf, Viola, and Egon Kühebacher. "Nikolaus Pol und die Stiftsbibliothek von Innichen." *Der Innicher,* 5 February 2005, 51–54.

Timmermann, Anke. "Doctor's Order: An Early Modern Doctor's Alchemical Notebooks." *Early Science and Medicine* 13, no. 1 (2008): 25–52.

Traninger, Anita. "Bildgebende Verfahren: Ramon Llull, Giordano Bruno, die Illumination des Gedächtnisses und die Bibliothek des Nicolaus Pol in Innichen/San Candido." In *Archäologie der Phantasie: Vom "Imaginationsraum Süditrol" zur longue durée einer, "Kultur de Phantasmen" und ihrer Wiederkehr in der Kunst de Gegenwart,* edited by Elmar Locher and Hans Jürgen Scheuer, 127–44. Innsbruck: Studienverlag, 2012.

Unterkircher, Franz. *Die datierten Handschriften der Österreichischen Nationalbibliothek.* Vienna: Verlag der Österreichischen Akademie der Wissenschaften, 1969–76.

Chapter 8

The Chemical Philosophy and Kabbalah

Pantheus, Khunrath, Croll, and the Treasures of the Oratory and the Laboratory

Michael T. Walton*

Although the majority of Plato's works were unavailable in the Latin Christian West prior to the later fifteenth century, his ideas were not completely unknown.[1] Augustine, Boethius, and Macrobius had transmitted knowledge of his doctrines, especially those of the soul and one god, which seemed to presage Christian dogma. Plato was regarded as the exemplar of the virtuous pagan, and medieval philosophers referred to him as Moses Atticus. His ideas

* Michael Walton died in 2013, before the publication of this volume and his chapter in it could be realized.

1. This study grows out of the work of Walter Pagel and Allen G. Debus, who demonstrated the need to broaden our understanding of the context of the development of early modern natural philosophy. It is also an elaboration of the themes raised in my book, Walton, *Genesis and the Chemical Philosophy*. I thank Dane Daniel for reading this essay and suggesting improvements.

had also been developed in the Hellenistic world into the philosophical religion of Neoplatonism. Influenced by the mystery cults, Neoplatonism purported to offer true seekers of wisdom and knowledge a way to move beyond the symbols in the temples of worship to soul-knowledge of the essence of Being.[2]

The works of Plato and the philosophical mystics became widely available in Greek and Latin in western Europe in the fifteenth and sixteenth century. Their study created Renaissance Neoplatonists, who viewed the ancient wisdom traditions—including Neopythagorianism, gnosticism, and hermeticism, no matter how much they had been adulterated—as forming a coherent whole. Taken together with Christian scripture, those traditions revealed how God formed and sustained the world. Such knowledge conferred upon its possessors not only intellectual understanding but also the power to influence nature, that is, to practice white magic or to obtain the philosopher's stone.

Jewish mysticism, which purported to have been revealed to Moses at Sinai, also had roots in the Hellenistic world. Hidden knowledge and secret words gave the "kabbalist" access to mystical experience and power over nature. The word "kabbalah" appears frequently in sixteenth- and seventeenth-century texts, where it is usually spelled "Cabala." Kabbalah complemented the Renaissance Neoplatonic tradition, and it was soon adopted and Christianized. Although most Christian kabbalists were primarily concerned with learning hidden religious truths, some believed they could find material truths as well. As these Christian Neoplatonists moved from textual studies to the laboratory, kabbalah became part of the chemical philosophical tradition.

This chapter will explore the use of the Christianized kabbalah in the chemical understanding of nature. Chemical kabbalah will be seen to be a subcategory of Christian kabbalah. This fact is best appreciated by analyzing aspects of the Christian kabbalah that were taken into chemical thought.

2. Ancient Neoplatonism reached its apex with the work of Plotinus, Porphyry, and Iamblichus in the third and fourth centuries of the Christian era.

GIOVANNI PICO DELLA MIRANDOLA, JOHANNES REUCHLIN, AND CORNELIUS AGRIPPA

The Renaissance iteration of Christian kabbalah, adopting and adapting Jewish ideas into the service of magic and Christian theology, was initiated by Giovanni Pico della Mirandola (Pico) (1463–94) and taken up by Johannes Reuchlin (1455–1522). In the kabbalah, Pico believed he had found the origin of the wisdom transmitted by Pythagoras and Plato; the Jewish texts, however, also revealed the incomplete nature of the Neoplatonic tradition. Although his own study was based on relatively few Jewish sources, translated into Latin, he had access in these to quotations from the primary kabbalistic works of *Zohar* and *Sefer Bahir* and from treatises by the mystics Menachem Recanati (1250–1310) and Abraham Abulafia (1240–91).[3] Among the quotations was Recanati's statement that the divine names and attributes revealed by kabbalah were for the worshipper "the tools an artisan uses to fulfill his task."[4] Kabbalistic tools were useful, if not essential, in gaining a proper understanding of scripture and restoring the pure knowledge given to Moses at Sinai.

Pico read the translations of Jewish texts through the lens of Christian tradition. He was especially attracted to magical doctrines involving the divine name, which, properly understood, supported the truth of the doctrine of the Trinity. Multiples of three, therefore, were preferable to other numerical configurations. For example, he preferred the *nine* angelic hierarchies of Dionysius Areopagite to the *ten* mentioned in his Latin copy of the *Corona nominis boni* by Abraham Axelrod of Cologne (fl. 1240).[5] Divine names, including the ineffable name (*nomen ineffabile*), were keys to understanding the creation, holy emanations (*sefiroth*), and Christian theology. Moreover, knowledge of divine names conferred power on religious adepts.[6]

In Pico's work, prophecy and its clear perception of reality and the future depend on divine illumination. The twentieth "Cabalistic Conclusion" of *900*

3. Wirszubski, *Pico della Mirandola's Encounter*. Wirszubski has identified four Latin manuscripts of Jewish texts, translated by Flavius Mithridates, as the probable sources for Pico's kabbalistic theses. Wirszubski's study remains a fundamental source on Pico's kabbalah. For a modern edition, with an English translation and commentary see Farmer, *Syncretism in the West*.

4. Recanati, *Commentary on the Daily Prayers*, 2:388.

5. Wirszubski, *Pico della Mirandola's Encounter*, 22–23.

6. Ibid., 23. See Pico della Mirandola, *900 Theses*, in *Syncretism in the West*, trans. Farmer, 346–47.

Theses refers to the mystical teaching concerning the light of the mirror.[7] The association of mirrors and divine illumination could be found in both Christian and Jewish sources. St. Paul used the images of seeing the divine glory through "a glass darkly" (I Cor. 13:12) and beholding the glory of the Lord "in a glass" (II Cor. 3:18). Jewish tradition and the kabbalah also associated mirrors with mystical illumination. The Talmudic tractate *Sukkah* mentions the *aspeklaria* or mirror of divine perception, and Simeon bar Yohai, as the putative author of the *Zohar*, used the *aspeklaria* to understand the mystical-magical nature of the Tetragrammaton, the four-letter name of God, YHVH.[8]

How kabbalistic insight and magic worked is glimpsed in the writing of Pico's teacher Johannes Alemano (ca. 1435–post 1504), who taught that a successful kabbalist should follow the example of Moses, who "had precise knowledge of the spiritual world, which is called the world of the Sefiroth and the world of divine names, or the world of letters." "Channels" connected the spiritual world to the material world, and Moses, as a kabbalist, was able to manipulate and direct the flow of powers between the higher spiritual world and the lower material world through "thoughts and prayers."[9] The worshipper thereby united the material and spiritual worlds.

Johannes Reuchlin's first work on kabbalah, *De verbo mirifico* (1494), dealt with divine names. He had carefully read Pico's work, but he also consulted primary Jewish sources, as illustrated in his letter to Rabbi Jacob Margoles of Nüremberg (ca. 1497) requesting kabbalistic materials.[10] In *De verbo mirifico,* Reuchlin presented the kabbalah as a source, if not *the* source, for the philosophical ideas that had been transmitted from Moscus [Moses?] the Phoenician through Plato, Philo, and Porphyry, to his own era's Neoplatonism. All true philosophy descends from Moses and the prophets, and the holy tongue of Hebrew remains the repository of God's words of power. Indeed, Hebrew influenced other languages, which assimilated some of its magical powers. These vestiges of Hebrew are especially strong in the letters of Semitic languages, such as Chaldean (Aramaic).[11] Reuchlin's discussion of the background of true philosophy

7. Wirszubski, *Pico della Mirandola's Encounter*, 37–38; and Pico della Mirandola, *900 Theses*, in *Syncretism in the West*, trans. Farmer, 354–55.

8. Babylonian Talmud, *Sukkah*, 45b. For a brief but excellent discussion of the *Zohar's* actual authorship, see Liebes, *Studies in the Zohar*.

9. MS Oxford 2234, fol. 8b, translated by Idel in *Kabbalah*, 204.

10. Maimon, *Germania Judaica*, 3:1037.

11. *De verbo mirifico*, bk. 1, in Reuchlin, *De verbo mirifico*, 29 (page numbers refer to the facsimile edition).

introduced his primary subject, the revelation of the wonder-working word and other divine names.[12]

It was clear to Reuchlin that the scriptures taught that all things were created by God, the Tetragrammaton, *yod he vav he* (by means of wisdom [*hochma* in Hebrew]).[13] God identified himself to Moses as "Ehieh, alef he yod he" or "I will be," using three letters of the four-letter name of God, and a word most potent.[14] The Tetragrammaton, *yod he vav he* mispronounced as "Jehova" and so prominent as the name of the divine in the Hebrew scriptures, can also be glimpsed in the Tetrad of Plato and Pythagoras.[15] In kabbalah, the holy four-letter name created the world and introduced the ten *sephiroth*, the hypostatized divine emanations embodying individual godly powers.[16] The four-letter name also underlay the philosophical concepts represented by the Orphic muses and the gods Dionysus, Apollo, and Venus. The four qualities, the four elements, and the four aspects of the human soul—sense, imagination, fantasy, and intelligence—mirror the most holy name. Whereas Pico's kabbalah seems to prefer triads, for Reuchlin the world is filled with fours.[17]

Reuchlin noted that God had said, "Fiat lux," and there was light. The Tetragrammaton is the Creator and the Light of the World.[18] The "Four-Letter Name," "Wisdom," and the "Logos" of the Gospel of John each refer to the pre-incarnate Jesus.[19] The holy four-letter name is the concentrated form of the seventy-two-letter, wonder-working name of God, called in Hebrew "Semhamaphoras."[20] Kabbalah taught that there is no higher power.[21] The proper use of the Tetragrammaton could create, heal, and alter human nature. For example, the use of just one letter of the Tetragrammaton would work wonders. The letter *h* (*he*)

12. In the dedication to his second work on kabbalah, *De arte cabalistica* (1517), Reuchlin related how he was recovering the true Pythagorean philosophy from kabbalistic treatises (see the last page of his dedication to "Sanctissimo Leoni Decimo"). He also discussed the power of angelic names extensively in book 3. Paul Ricci's partial Latin translation and summary of Joseph Gikatilla's *Sha'arei Orah, Portae Lucis* (Augsburg, 1516) opened a direct Hebrew source for Christian scholars.

13. Reuchlin, *De verbo mirifico*, 57.

14. Ibid., 56.

15. Ibid., 64, 66.

16. Ibid., 67.

17. Ibid.

18. Ibid., 79.

19. Ibid., 81.

20. Ibid., 66, 86. Recanati, *Commentary on Daily Prayers*, argued that because the Jews could not invoke God by his *Semhamaphoras* name, their prayers were often not fulfilled.

21. Reuchlin, *De verbo mirifico*, 90–91.

of the four-letter name was added to the name of Abram, making him Abra*ham*, the exalted father.[22] The Tetragrammaton could also be used to control the angels.[23] So powerful was the Tetragrammaton that it was normally forbidden to utter it. Only once a year, on Yom Kippur, and then only in the Holy of Holies of the Temple under rigorous circumstances, did the high priest utter this name of God.[24]

De verbo mirifico's clear and persuasive presentation of kabbalah attracted both theoretical philosophers and seekers of natural magical powers. Natural magicians treasured the potent letters, the words they formed, and the insights they gave into the meaning of scriptures. Jacques Lefevre d'Etaples (ca. 1455–1536), Symphorien Champier (1471–1538), and John Dee (1527–1609) were students of the kabbalah and readers of *De verbo mirifico*, but of those who followed the path marked by Pico and Reuchlin, the most influential in propagating the natural magic-nature philosophy may have been Cornelius Agrippa von Nettesheim (1486–1535).[25] His *De occulta philosophia*, written in 1510 but not published in its entirety until 1533, popularized kabbalah among Christians, especially the doctrines of name magic.[26] Agrippa had not merely read *De verbo mirifico*, he had lectured on the work at the University of Dole in 1509,[27] and in book 3 of *De occulta philsophia*, he discussed the natural virtues and processes of the created world and revealed the magic by which they might be controlled. Divine names were crucial instruments of that magical control, and in the spirit of Reuchlin's work, he discussed the Tetragrammaton and the name "Eheia."

De occulta philosophia became a prominent source for name-magic in the sixteenth century. As a compendium of theory and magical practice, it lent itself to the use of scholars who themselves did not wish to delve into kabbalah's more recondite and less accessible sources. This was true of John Dee and even more so of practical chemical philosophers, who were focused on laboratory results.[28]

22. Ibid., 94.

23. Ibid., 96.

24. Ibid., 103.

25. Copenhaver, "Lefevre d'Etaples." For a discussion of Dee, Pico, and Reuchlin, see Walton, "John Dee's *Monas Hieroglyphica*," 116–17; and Walton and Walton, "Geometrical Kabbalah of John Dee and Johannes Kepler," 46.

26. Tyson, "Life of Agrippa," xviii, xxxi, xxxiii, in Agrippa, *Three Books of Occult Philosophy*. See also Nauert, *Agrippa and the Crisis of Renaissance Thought*.

27. Tyson, "Life of Agrippa," xviii, xxxi, xxxiii, in Agrippa, *Three Books of Occult Philosophy*.

28. See Walton, "John Dee's *Monas Hieroglyphica*," 116–17.

GIOVANNI AGOSTINO PANTHEUS

Because the alchemical-chemical art was believed to be intimately connected with both the spiritual and the material worlds, it did not take long for kabbalistic interpretations of the elements, of fire analysis, and of the alchemical creation to develop. The merger of kabbalah with traditional alchemy and its focus on transmutation is seen in the *Voarchadumia* of the sixteenth-century Venetian priest and alchemist Giovanni Agostino Pantheus.[29] Pantheus shared Pico and Reuchlin's belief that kabbalah had been revealed to ancient magi, but his interest in the art seems to have lain in more practical applications than theirs.[30]

Pantheus was inspired by the possibility of transmuting metal by using words. He published a treatise, the *Transmutation,* on the subject in 1518. In that work, Pantheus used or referred to the Tetragrammaton, Greek and Hebrew letters, and traditional alchemical texts such as the Emerald Table and *Turba philosophorum.*[31] At that time, he seems to have viewed alchemy as at least a quasi-kabbalistic study.

The *Transmutation* was censured by ecclesiastical authorities who opposed the study and practice of alchemy. Pantheus addressed the censure in his *Voarchadumia,* rejecting alchemy as a false art. He distinguished the kabbalah of the *Voarchadumia,* which contained the true and religiously correct art of transmutation, from vulgar alchemy. Pantheus argued that *Voarchadumia* cleared up the "alchemical" confusions of his *Transmutation,* correcting them with a well-developed transmutational kabbalah.

Pantheus was explicit in his preface that *Voarchadumia* is actually opposed to alchemy. He presented it as a distinct transmutative discipline.[32] This cabala of metals was a liberal art that genuinely transmuted metals.[33] Alchemy, on the other hand, did not make real silver and gold but only things with their appear-

29. Pantheo, *Voarchadumia contra alchimiam.*

30. Paracelsus also shared this belief. See his *Concerning the Nature of Things,* in *Hermetic and Alchemical Writings,* ed. Waite, 1:161.

31. Thorndike, *History of Magic and Experimental Science,* 5:539–40, discusses Pantheus's works. See also Thorndike, "Alchemy during the First Half of the Sixteenth Century," 30–32.

32. "Authoris intentio de Voarchadumica professione, contra Alchimiam, Arte distincta ab Archimia & Sophia"; Pantheo, *Voarchadumia contra alchimiam,* 7.

33. "Voarchadumia est ars liberalis, virtute praedita sapientiae occultae, non auara, non vana, possibilis verissima, necesaria, & consequenter perqirenda, quae metallorum. Cábala nuncupatur"; ibid., 12.

ance.[34] Pantheus explained that the art of *Voarchadumia* is not new but originated with Tubal-Cain.[35] The word itself was a Chaldean compound consisting of *Voarh*, meaning "gold," and *Mea a adumot*, the Hebrew for "from the two reds." In this compound word, *Aurum duarum rubearum*, there is a *caementationum*, "a perfect joining of the two."[36]

Pantheus regarded himself as the re-discoverer of this kabbalah of transmutation and of its revelation of the hidden qualities of spirits and bodies. The relationships of substances, which allow them to be purified, are known through alphabetic notations instilled into matter by the "omnipotent, eternal God." Pantheus taught that the elements—that is, fire, air, water, and earth—correspond to the Hebrew letters *lamed*, *kuph*, *gadic*, and *samech*. He added in his *Voarchadumia* that those letters also signify other Hebrew letters *zain*, *nun*, *mem*, and *iod*. He declared that "[g]old is created from nature's elements that are in the bowels of the Earth, purified by the arts of division and union or as said in Hebrew, Zahav niura mi Teuah-iesodii."[37]

Pantheus's *Voarchadumia* demonstrated that chemical processes were kabbalistic. His description of this true art, rooted in Hebrew, replaced fraudulent alchemy and guarded Pantheus from the criticisms directed against alchemical goldmakers. The art of *Voarchadumia* also influenced adepts such as John Dee, who owned and annotated a copy of the first edition. As a kabbalistic man, Dee referred to *Voarchadumia* in his *Monas hieroglyphica* (1564).[38] Gerhard Dorn

34. Ibid., 10.

35. Ibid.

36. "Sed iam ad verbum ipsum Voarchadúmia deuenientes, dicimus ipsum deriuari, ac denominari ab Auro ex duabus rubies, quod Chaldaeo idiomate componitur ab Voarh, Auro, particula Indica primitiua, & Mea à adumòt Hebraicè, ex duabus rubeis, quod Latine significat Aurum duarum rubearum, hoc est duarum caementationum perfectarum"; ibid., 11.

37. The text reads, "Secundo ea est tanquam regimen quoddam caelatū & secretum, quod per manus tātùm traditur filiis sapiētum sub luce (licet vulgò tenebris obuolutum) dispositionem, illuminationem, conuertionem, constrictionem, retentionem, metallificationem, purificationem, mutiplicationem, & proportionem, demonstrans naturalium ligaminum absconsi & abscondentis, animae & corporis, densi, & rari, diuini & humani, formae & materiae, fixi & volatilis, interioris & exterioris, metalli & petrae, mollis & duri, occulti & manifesti, puri & misti, artificio quodam mediante à domino aeterno omnipotenti instituto, sub Igne, Aere, Aqua, & Terra, vel magno arcano harum quatuor literarum, Lamed Kuph, Gadic, & Samech, quae in Voarchadúmia idem significant quod Zain, Nun, Mem, & Iod, quarum caracteres in quadruplici differentia apud Hebreos habentur. Nouissimi autem, & Vsuales sunt ii (de quibus paulò infra Schemata patebunt) qui in Voarchadúmia idem significant, quod Aurum creatum ex Natura elementari, id est in visceribus terrae, purificatúmque ob artificium Diuisionis, & Caementationis, ac Hebraicè sic denominatum, Zahav niurá mi Téuah-iesodii"; ibid., 12–13.

38. Josten, "Translation of John Dee's 'Monas Hieroglyphica,'" 137. For references to Dee and Pantheus,

(ca. 1530–84), a noted Paracelsian, also knew Pantheus's work.[39] As will be discussed below, another Paracelsian, Oswald Croll (ca. 1563–1609), used the term in his discussion of chemistry.

PARACELSIANISM

Pantheus's kabbalah of transmutation, despite his protestations to the contrary, was very much a part of the backward-looking alchemical tradition. Paracelsus's reformation of alchemy into the new chemical philosophy also made use of kabbalah. Theoretical Paracelsianism was deeply rooted in a chemical-kabbalistic understanding of Genesis. The biblical account of the creation provided the pattern for a chemical understanding of nature. Of special importance was the power of words as demonstrated in the Hebrew scriptures, where God creates by speaking, and in the Christian New Testament, especially the Gospel of John, where the Word, *Logos*, created the world. Chemical philosophers followed orthodox Christianity, identifying the *Logos* and the Tetragrammaton with Jesus.

Paracelsus taught that the word "fiat," used in the Vulgate's Genesis story of the creation, materialized as the prime matter of the created world. As he wrote in his *Opus paramirum*, "[s]ince the prime matter of the world was *fiat*, who will venture to explain the *fiat*? We have, however, some [basis] through the fire of Vulcan through which we can explain the first three [principles]."[40]

Creation by word and spirit was central to Paracelsian chemistry. Peter Severinus (1542–1602) declared that "[t]he Creator placed light and the seminal reasons for all things, by an incomprehensible magic, by virtue of that word and spirit that was moved over the waters."[41] Gerhard Dorn taught much the

see ibid., 88n27–28; Norrgren, "Interpretation and the Hieroglyphic Monad"; and Szulakow, "John Dee and European Alchemy," 26.

39. Kahn, "Les débuts de Gérard Dorn," 95–97. Kahn draws special attention to Dorn's borrowing of the idea of two elements from Pantheus.

40. Theophrast von Hohenheim [Paracelsus], *Opus paramirum* (1531), bk. 1, ch. 2, in *Sämtliche Werke I*, 1.9:48. The passage reads, "dieweil aber prima materia mundi fiat ist gewesen, wer wil sich unterstehen das fiat zu erklären? Nun aber etwas haben wir durch das feur vulcani, dadurch wir die drei ersten erklären." Pagel, "Prime Matter of Paracelsus," 118, translated the passage somewhat differently. His understanding and explication of Paracelsus's fundamental notions continue to inform Paracelsian scholarship, including my own; however, Pagel did not note the relationship of word-creation to the kabbalah. I have written on ideas of chemical creation in Walton, "Genesis and Chemistry in the Sixteenth Century." Forshaw also discussed the topic in "Alchemical Exegesis of Genesis."

41. "In his quator Naturis incorporeis, inanibus, vacuis, Lucem et seminales rerum omnium Rationes, incomprehensibili Magis imposuit Creator, virtue Verbi et Spiritus illius, qui super aquas ferebatur"; Severinus,

same thing in his *Physica genesis*: "The eternal word of God created the Heaven and the Earth."[42] Although scriptural, the magic word "fiat" is also kabbalistic. The inclusion of magic and kabbalah in chemical discourse offended many anti-Paracelsians, who believed such ideas to be foolish, if not heretical. Both Thomas Erastus (1524–83) and Andreas Libavius (1555–1616), for example, attacked Paracelsian magic. Erastus mocked Severinus's mention of incomprehensible magic,"[43] and Libavius found ridiculous the idea that the word "fiat" became corporeal.[44] In his customary acerbic prose, he accused Paracelsus of doing "impure magic."[45]

The kabbalistic nature of Paracelsian word-magic is not readily apparent, for its practitioners often used Latin words such as "fiat" and "creavit" rather than Hebrew words. Only when viewed in light of Reuchlin's comparison of the Latin and Hebrew terms does the connection with kabbalah become clear. Although the early Paracelsians preferred to use Latin words, chemical kabbalah using Hebrew terms did exist. The *Amphitheatrum sapientiae aeternae* (1595) of Heinrich Khunrath (ca. 1560–1605) is a learned and deeply kabbalistic work. Its subtitle illustrates this point—"true throne of the catholic trinity of Christian cabala divine magic and physical chemistry."[46] The essence of Khunrath's kabbalistic chemistry is glimpsed in the partial seventeenth-century English translation of his *Confessio de chao physico-chemicorum catholico* (1596).[47] Therein is presented a Paracelsian description of creation, using Hebrew words:

> But it is further necessary in the Catholike Chymicall Arte, yt we truly believe (ye know directly) & stedfastly hold that Ruach Elohim the Spirit of ye Lord wch in Gen: 1.2. did move upon the Water, is really become a body. . . . That

Idea medicinae philosophiae, 41–42. The translation is from Shackelford, *Philosophical Path for Paracelsian Medicine*, 218.

42. Dorn, *Liber de naturae luce physica* (1538), in Zetzner, *Theatrum chemicum*, 336 (my translation). The text reads, "Aliquid tamen fuit hoc principium, in quo Dei verbum aeternum creavit coelum et terram."

43. Erastus, *Disputationum de medicine nova* (1572), 104; translated in Shackelford, *Philosophical Path for Paracelsian Medicine*, 218. The text reads, "In his quator naturis incorporeis, inanibus, vacuis, seminales omnium rerum rationes incomprehensibili magis."

44. Libavius, *De universitate*, 78. I thank Bruce Moran for providing me with a copy of the text. The text reads, "Riduculum vero est verbum 'fiat' abiise in corpus illud materiale sensibile."

45. Libavius, *Appendix necessaria*, 3 (my translation). The text reads, "Sed Paracelsici hoc genus nimium dilatant, et *magiae impurae* praetexunt faciunt."

46. Khunrath, *Amphitheatrum sapientiae aeternae*. The text reads, "Solius verae christiano-kabalisticum, divino-magicum nec non physico-chymicum, tertriunius, catholicon."

47. British Library, MS. Ashmole 1459, 99–106. Online at http://www.levity.com/alchemy/khunconf.html.

> Spirit of the Lord was OR [Hebrew for "light"] preceeding of the Essence of the Spirit of the Lord (wch in ye beginning of the world was with the Lord) wch Gen: 1.2 moved upon the waters, wch Wisd: 1, 7 filled the whole world, and wch Wisd: 11, 26 is in all, even in ye most inward & sacredest Virgin womb and center of the Earth, the most misticall broodie [brood] Mother of the greate world. world [sic] bodily it is become a Corporall Salt of wisdome, tho little great world OR bred in the world, of the bodily essence & first matter yt is earth & water, the Parents, to wittt [*sic*] of ye great World.[48]

This type of word-kabbalah was crucial to the development of chemical theory and to a lesser degree to chemical practice. The Paracelsians worked in a world where word-magic was the primary mechanism of creation. Such magic revealed essential aspects of matter as it existed in the present world. Words retained power over objects and beings. Kabbalah, in its chemical iteration, was believed to be a pure philosophy of nature, untainted by pagan traditions. In the dictionary appended to the English translation of the pseudo-Paracelsus's *Three Books of Philosophy Written to the Athenians*, "cabala" was defined as "that most secret knowledge which the Hebrew Rabbins say was given by God with the Law of Moses."[49] Chemical kabbalah connected the Creator to his material world. As such, its use was an important introduction to the chemical philosophy. It gave sacred warrant to the practical work of the chemist, who was a servant of God in helping mankind.

OSWALD CROLL

Oswald Croll's *Basilica chymica* (1609) was among the most popular chemical texts of the early seventeenth century, especially its "Praefatio admonitoria." The "Praefatio," which was translated into both English and French, was essentially *the* introduction to Paracelsian chemistry. The work continued in print into the

48. Ibid.

49. *Three Books of Philosophy Written to the Athenians*, in Pinnell, *Philosophy Reformed and Improved*, 64. There are two chemical texts in the seventeenth century that are titled *Cabala chemica*: Francisco Kiesero [Franz Kieser], *Cabala chymica* (Muelhausen, 1606), and Johann Grashofer [Chortolassaeus], *Aus der cabala chemica* (Leipzig, 1658). Kieser's study is a Paracelsian discussion of the creation by *fiat*, the celestial creation of the soul's astral body, and the nature of the philosopher's stone. It contains no Jewish elements. His cabala is an interpretation of chemical symbols according to Paracelsus. Grashofer's text is, in essence, an epitome of Kieser's. They demonstrate the use of the word "cabala" to name an ancient interpretive tradition without reference to any Jewish doctrines. In this, they differ greatly from Croll and Khunrath.

1670s with at least eighteen editions published through 1658.[50] The preface is a detailed discussion of the doctrines central to Paracelsianism, including kabbalah. In his description of word-magic and the *voarachadumian* art, Croll offered a key to the connection of chemistry and kabbalah.

Croll asserted that adepts in the art had kept secret certain traditional chemical knowledge. He, however, chose to reveal these chemical mysteries, even though it would offend Hermetic philosophers. He believed that chemistry, like Christian doctrine, should be available to all seekers. "Cabala, magic, and Woarchadumia" were divine truths obtained in the "oratory and the laboratory."[51] Croll explained that *fiat* created *prima materia,* which in itself is unknowable except through the three principles into which all things are reducible in fire. The spagyrical art, or resolution by fire, demonstrates that matter in the world could be reduced to three elements—fluid (mercury), oil (sulphur), and solid (salt)—and no further:

> No body compos'd by Nature can by any dissolving skill be parted into more or lesse than *Three,* viz. Into *Mercury* or liquor, *Sulphur* or Oyle, and *Salt;* every created thing is generated and preserved in these three; For the Holy *Triunity* when it spake that *Triune* word FIAT created all things *Triune,* as in Spagiricall resolution is plainly to be seen. By the word FIAT (or Let there be) God produced the first matter, which is threefold in respect of the three Principles contained in the first, and afterward these three species are seperated [*sic*] into four divers bodies, or Elements.[52]

50. Partington, *History of Chemistry,* 2:173–74.

51. Croll, *Basilica chymica* (1611; reprint 1996), 6–7. The text reads, "Insurgent in me & irascentur graviter primo secretiores Philosophi Hermeteci, quib. potiora is haec arcane antea fuerunt ex parte perspecta: conquerentes injuriam sibi fieri, si illa, quae summis studiis & longo temporum intervallo conquisita, altoq. silentio hactenus compressa fuerunt, tàm subitò lucem videant, & cum hominum vulgo communicentur. . . . Verùm hi, quia verae Sapientiae heredes, & Regni Philosophici cives, clausâ invidiae fenestrà, Dei & proximi semper amantes sunt, aut saltem esse debent, sublato in altum capite & oculis Cabalisticis, non ignorabunt in verâ Cabalâ, Magiâ, & Woarchadumiâ longe potiores thesauros sibi Oratorii & Laboratorii benficio." Croll's English translator, Henry Pinnell (*Philosophy Reformed and Improved,* 8–9), rendered the passage this way: "The Mysticall Hermetick Philosophers who heretofore in part have pryed into these excellent secrets, will rise against me, and be very angry with me; complaining that they are wrong'd in those things, which with greatest industry and long spent time have been found out and hitherto concealed in deep silence; should so soon be brought to light, and made common to every one. . . . But in as much as these are heires of true wisedome, Citizens of the Philosophicall Kingdome, alwayes lovers of God and their neighbor, without envy, or at least ought so to be, whose heads being lift up on high, and their eyes Divinely enlightned, these shall know that in the true *Cabala,* Magick and *Woarchadumie,* there are laid up far better Treasures, to be got by them with the help of the Oratory and Laboratory."

52. Croll, *Basilica chymica,* in Pinnell, *Philosophy Reformed and Improved,* 32.

As noted above, *fiat* is a magical kabbalistic word. It was very much part of the early Christian kabbalah as taught by Reuchlin. Croll also seems to have been aware of a deeper kabbalah of creation similar to the doctrine of *tzimzum*, a Hebrew word that in kabbalah refers to God's contraction of himself to make space for the material world. Following God's *tzimzum*, a point of light burst forth creating the world. Croll referred to a first creation from a kabbalistic point, when the Word called light into existence from inchoate darkness, and all things moved from darkness to light: "Whence we may observe that all things in the first Creation were produced out of the DIVINE NOTHING, or invisible Cabalisticall Poynt, into something, which God did in a moment, for his works cannot be delayed by time: All things proceeded out of the invisible Darknesse, and were called out to the visible Light by the WORD speaking, and the Spirit cherishing."[53]

The idea of the "Word" as "Creator" was rooted in scripture. Proverbs 3:19 stated, "the Lord by wisdom hath founded the earth." In the Gospel of John "the Word [*Logos*] was with God and was God." Croll saw such texts as confirming the kabbalah, its cosmogony, cosmology, and magic. The Word was a manifestation of God and not simply a symbol. Croll identified the Word and its magical power with Jesus. The Hebrew name for Jesus, ISHUH (Latin for *Yeshuah*) was the "verbo mirifico." It and its powers were known and revealed in the kabbalah. Paracelsus called it the "Naturalem Divinum."[54] The wonder-working word was the foundation of the "magicall and cabalisticall art":

> [T]he very GATE of miracles in that *Only* Divine Name ISHUH in which all things are reckoned up and contained, that is he doth performe it in the WONDERFULL WORD by the Mind, Faith and Prayer, to wit, prayers

53. Ibid., 70. The kabbalistic doctrine of the expansion of a point of light was prominent in the *Zohar.* R. Isaac Luria and his school developed Zoharic kabbalah in the middle of the sixteenth century. It became popular at that time among Jews and was known to Christians. Isaiah Tishby has drawn attention to the account of Moses Isserles, the prominent late sixteenth-century rabbi: "Many ordinary people now jump at the opportunity to learn something of kabbalah, for it is a delight to the eyes. And they are particularly keen on the writing of the late [kabbalists] whose books reveal their ideas quite plainly, especially in our own time when kabbalistic works are printed, such as the Zohar, Recanati, and *Sha'arei Orah*"; Isserles, *Sefer Torah ha-Olah,* 3:4, in Lachower and Tishby, *Wisdom of the Zohar*, 3 vols., trans. Goldstein, 111n182. Croll was one whose "eyes delighted" in the kabbalah.

54. "Et Verus Cabalista (quem Paracelsus vocat Naturalem Divinum, qui Prophetis aequiparatur, & cujus mens unita & coaequata Deo facit omnia quae vult, vult autem quae ipse Deus) Dealiter supra Naturam, ficudiâ firmâ, & fide mirificâ miraculorũ Janua Unico Divinissimo Nomine ISHUH, in quo recapitulantur & continentur omnia, id est in Verbo Mirifico, per Mentem, idem & Orationem, preces scilicet in Spiritu & Veritate factas in momento exequitur"; Croll, *Basilica chymica*, 46.

> made in Spirit and in Truth. The New Birth is the Field of Caelestiall Physick which healeth with a word without Externall means: that one operation is in respect of God as the Artificer, and in respect of Man as the Instrument; every creature is at the beck and command of their faith who are men innocent and taught in the Law of the Lord, who are heard in all things whatsoever they pray for, witnesse *Elias, Elisha,* &c. By prayer in Faith we obtaine all things, I mean (not a lazy, sluggish prayer, but) a constant asking, seeking, knocking: by faithfull Prayer we ascend in a straite and most sure way to the highest Wisdome of Divine and humane Things; For in these Three principall Poynts also consisteth the whole Foundation of the Magicall and Cabalisticall Art, as appeareth by *Paracelsus* in his third book of the Signature of Things.[55]

Not only the divine name but also its letters contained magical potency. The living God imbued all aspects of his name with life-giving virtues:

> These things wisely and rightly considered, we shall not wonder that Almighty God could (and can) make men whole by the *Prophets* and True *Cabalists* with a word onely. God is a living God, the NAME also of the living God is lively, and so the Letters of the living Name are also lively: God liveth for himselfe, his Name liveth because of him, the Letters live by reason of the Name; as God hath life in himselfe, so hath he given to his Name to have life in it selfe, and the Name also to the Letters.[56]

God had instilled powers into letters at the creation. They were part of nature. Croll derived this understanding of "natural" letter-magic from Agrippa: "For Characters or constellated Names according to *Agrippa,* have no force from the Figureors Pronunciation, but by reason of the Vertue or Office which God or Nature hath ordained to such a Name or Character: There is no vertue, or power either in Heaven or Earth which descendeth not from God."[57]

As an adept, Croll sought to learn and apply the wonder-working word for the good of humanity. Through the Word's power, he came to understand the signatures of substances, that is, their healing power. A true chemist could use the chemical-kabbalistic art to purify and intensify the forces in matter. "True magicians" did not use black magic, but rather the knowledge of "signs and signatures" granted by God. Proper use of the powers of these signs and signatures honor God in this world.

55. Croll, *Basilica chymica,* in Pinnell, *Philosophy Reformed and Improved,* 88, translating from the above passage in the *Basilica.*

56. Ibid., 84.

57. Ibid.

> Great things have been affected by True *Magicians* (by whom I doe not mean Nicromancers or them of the Black Art) those accurate searchers out of Nature, by a *Word* and *Characters* or *Signes*, framed at a certaine time according to the power of Heaven, far from all superstition, which ariseth onely from ignorance, without any prophanation or scandall of the Divine Majesty, or any wrong to Faith and Religion; otherwise it were better for us always to be sick then [*sic*] to be cured with the dishonour of God.[58]

Croll's medicines were material vehicles containing hidden heavenly powers. Words and letters likewise were visible vessels created by *fiat*. The physician used medicines, words, and letters, which had originated in God's *fiat*, to heal naturally. For the physician,

> [m]edicines are visible bodies; Words are invisible bodies: whether the Hearb or Word healeth, it is by God the Naturall Vertue thereof, to wit, by the spirit of God made One with Nature by his Word FIAT. [There is a] Characteristicall Cure which affecteth Naturall operations by words pronounced, written, carved and hanged about the neck, by the caelestiall properties of the Stars through a marvelous Influence agreeing with our bodies.[59]

Croll's kabbalah was part of the religio-chemical philosophy that grew out of Paracelsus's writings. He, like his master, was a physician who focused his healing through a kabbalistic understanding of chemistry. He summed up his art as the "twofold physick." The Word of God made the visible and invisible worlds concordant: "There is a twofold Physick, Visible or created, and Invisible even the WORD of GOD: It is by the WORD of GOD therefore that any one whosoever is restored to his health, he that despiseth his WORD, despiseth Physick, and so on the contrary; for he that saith Physick is worth nothing, doth upon the matter affirm that there is no God."[60]

Pantheus's kabbalah of transmutation was a version of traditional alchemy, using Hebrew letters; chemically it looked back to gold-making. Croll's kabbalah, on the other hand, directed the adept to look for future discoveries. As a Paracelsian, Croll advocated a new chemical interpretation of the creation. Kabbalah provided both theory and methods useful to rethinking nature. Pantheus wanted to make noble metals, but Croll and other Paracelsians sought insights into the healing virtues that God had built into nature at the creation. The laboratory was

58. Ibid.
59. Ibid., 85.
60. Ibid., 172.

the place to study the invisible realm of the "Word of God." Croll and his fellows grasped that the proper entrance to the laboratory was through the oratory, the site of prayer, contemplation, and illumination.

Heinrich Khunrath

Heinrich Khunrath also regarded the oratory as the portal to the divine wisdom of chemistry. One of the best-known illustrations in Khunrath's *Amphitheatrum sapientiae aeternae* is of a laboratory containing an oratory in which a chemist kneels in prayer. This plate reveals Khunrath's chemical kabbalah and his understanding of how an adept sought and received the knowledge he could use to manipulate nature. The laboratory is surrounded with aphorisms. Many contain kabbalistic ideas combined with Christian theology and chemical processes. Connecting these seemingly disparate elements are the Tetragrammaton, the letters YHVH of the name of God. Light, fire, physical knowledge, and spiritual illumination flow from God's ineffable name. The first text states that an adept should study YHVH with complete heart, soul, and strength, as commanded in Deuteronomy 6:4–9.[61]

The texts on or near the oratory contain a mixture of Latin and Hebrew. The tent of the oratory is capped with the words, "sic omnia gratorium"; below that the adept is reminded that "Happy is [he] whose counsel is Jehova" (Felix cui YHVH a consiliis). The tent is further decorated with Khunrath's name and the Hebrew words for "wisdom" and "light." The seeker is also told, "When we are diligent, God himself will help" (Hoc agentibus nobis, aderit ipse Deus). From the nearby incense burner rise the words, "May the prayer-sacrifice be acceptable to God" (Ascendat at sumus, oratio, sacrificiu[m] Deo acceptabile). Each phrase is a piece in explaining the role of the oratory in Khunrath's "Christian-Cabalistic, Divine-Magical, Physical-Chemical" endeavor. Prayer and contemplation allowed passage to divine knowledge. For Khunrath, the Tetragrammaton is Jesus, that is, the Word, Light, and Wisdom. An adept had to offer sacrifices of prayer and diligence to receive the Light. Contemporaries clearly understood Khunrath's kabbalistic message. John Arndts (1555–1621),

61. Laboratory Plate, in Khunrath, *Amphitheatrum sapientiae aeternae*: "Homo toto corde, tota anima, omnibus viribus, et omni mente, divina gratia stimulante et movente (est enim Dei misericordis donum) studens yhvh agnoscere, se ipsum (cognitione sui) abnegare, et mundum (cognitione quoque luminis naturae) contemnere immundum."

in a commentary printed in 1611, explained the *Amphitheatrum* as a kabbalistic treatise. His interpretation of the laboratory scene stressed Khunrath's conception of the three lights of nature: natural, supernatural, and godly.[62]

Croll's discussion of the oratory was very much like Khunrath's. A chemist requires divine illumination to comprehend nature. One approaches the art through "cabala, magic, and *Woarchadumie*." Knowledge of those disciplines is acquired with the "help of the Oratory and the Laboratory." The oratory is the place where one's "eyes [are] divinely enlightened." Prayer is especially important in mastering the division of magic "called Gabalia."[63] The oratory is like the holy of holies in the ancient Jewish temple, for kabbalistic magic is supported by "TRUE PRAYERS made in Spirit and Truth, when God and the Created Spirit are united in the *Holy* of *Holyes*, when God is prayed unto in the internall Spirit, not with noyse of words, but in a sacred silence, without opening the mouth and groaning."[64]

Croll did not write of common or perfunctory prayer. He advocated an ecstatic communication that would unite the wisdom-seeker with God. Through this method, he could know all things: "[F]or by the ardent and devout intention of him that prayeth with Fear and Trembling, the Understanding or *Mind* flaming with a Religious love, is joyned to the separated Intelligences. For internall Prayer proceeding with abundance of affection from a Godly mind, and continued with a fervent desire, uniteth the mind with God, and learneth and knoweth all things of God."[65] The oratory was thus the chapel that served as an antechamber to the laboratory: "But who among many thousands is it, who while he seeketh this very knowledge by a certaine and subduing judgement and due means, to whom the Stars are so benign, that he can by anothers diligence and endeavor passe thorow the porch and the

62. An early anonymous commentary on the figures in the *Amphitheatrum* was published at the end of Khunrath's *De igne magorum* (1611). It was reprinted in 1783 and attributed to Johann Arndts. Arndts stressed the kabbalistic nature of the figures. He understood the kabbalah as the power of heaven. The first figure of the laboratory taught the kabbalah, the second natural magic, and the third chemistry.

63. Croll, *Basilica chymica*, in Pinnell, *Philosophy Reformed and Improved*, 9, 72.

64. "Veris precibus in Spiritu & Veritate factis, ubi in Sancto Sanctorum sit unio Dei & Spiritus, Creati, ubi Deus interno spiritu, non vi vocabulorum, sed in silentio sacro, absque oris apertione & anhelitu compellatur"; ibid., 72; and Croll, *Basilica chymica* (1611), 38–39.

65. "[S]iquidem per ardentem devòtamaque cum Timore & Tremore invocantis intentionem, Intellectus seu Mens religioso amore flàgrans, jungitur intelligentiis separatis. Oratio enim Interna expiamente ortanimio affectu, ardentique, desiderio continuata, unit mentem cum Deo, & omnia dificit & cogniscit ex Deo"; Croll, *Basilica chymica*, in Pinnell, *Philosophy Reformed and Improved*, 74; and Croll, *Basilica chymica* (1611), 39.

gate of the Heavens into the Oratory or Chappell of *Apollo*, and get to the top of the mount of Chymistry?"[66]

Both Khunrath and Croll knew that without the oratory the laboratory is a vain venue. Knowledge comes from God, the Creator. The chemist ultimately gains power over nature with the ability to heal from the divine, not from mere laboratory procedures:

> But that the glorious God and most blessed bestower of all graces hath reavealed and made it plaine to his faithfull wise ones, to such as feare and honor him, that they might understand, meditate upon and love his omnipotent goodnesse, and by glorifying him in his wonders and all his power and virtues, serving him without any blemish, vice or sin in his holinesse, and true Righteousnesse, they might see how much he hath done for men of good will: And so finally they that are inflamed with a most fervent love of piety and Gratitude might find him that is worthy of infinite thanks, who is infinite in mercy, whose most holy and Fearful name be blessed for ever and ever.[67]

Conclusion

The early modern interest in the kabbalah extended beyond the adaptation of Jewish doctrines to theological and general magical ends. Chemical thinkers embraced kabbalistic notions of creation by words and letters. Pantheus developed a kabbalistic alchemy whereby Hebrew letters represented matter (the four elements) and revealed the art of transmutation. Khunrath and Croll saw kabbalah as essential to their religio-chemical researches. Kabbalah and its magic connected the divine with the material world. The insights gained by righteous adepts in the oratory enabled them to use the laboratory to serve God and to heal their fellow men. Both the oratory and the laboratory were essential in that quest.

66. "Sed quotusquisque est, qui dum certo subactoque judicio, medysque, debitis cognitionem hanc ipsam, quaeret, Astra ita habeat benigna, ut per Caelorum vestibulum & portam ad Apollinis sacraria intrarepossit, & alienâ operâ montem Chymiae ascendere?"; Croll, *Basilica chymica*, in Pinnell, *Philosophy Reformed and Improved*, 184; and Croll, *Basilica chymica* (1611), 88.

67. "Sed Deum gloriosum, gratiarum largitorem optimum suis fidelibus sapientibus, & ipsum timentibus ac honorantibus id aperuisse & propalasse, ut suam omnipotentem bonitatem intelligant, recolant & ament, eumque in suis mir abilibus glorificando omnibusque viribus, ei sine aliqua sorditie, vitio aut peccato in Sancititate & Justitia requisita inserviendo videant, quantum ipse fecit pro hominibus bonae voluntatis: Et sic finaliter qui misericordia Infinitus est, gratis infinitis, ferventissimo Pietatis & Gratitudinis amore accensi, dignum inveniant, cujus santissimum & tremendum nomen sit benedictum in saecula"; Croll, *Basilica chymica*, in Pinnell, *Philosophy Reformed and Improved*, 185; and Croll, *Basilica chymica* (1611; reprint 1996), 88–89.

WORKS CITED

Agrippa, Henry Cornelius. *Three Books of Occult Philosophy*. Translated by James Feake and edited by Donald Tyson. St. Paul, MN: Llewellyn Publications, 1997.

Copenhaver, Brian P. "Lefevre d'Etaples, Symphorien Champier, and the Secret Names of God." *Journal of the Warburg and Courtauld Institutes* 40 (1977): 189–211.

Croll, Oswald. *Basilica chymica*. Frankfurt, ca. 1611. Reprint, Hildesheim: George Olms Verlag, 1996.

———. *Philosophy Reformed and Improved in Four Profound Tractates*. London, 1657.

Dorn, Gerhard. *Liber de naturae luce physica, ex genesi desumpta* (1583). In E. Zetzner, *Theatrum chemicum*. Strasbourg, 1659.

Erastus, Thomas. *Disputationum de medicine nova . . . pars altera*. Basel, 1572.

Farmer, Stephen A., ed. and trans. *Syncretism in the West: Pico's 900 Theses (1486): The Evolution of Traditional Religious and Philosophical Systems*. Medieval and Renaissance Texts and Studies 167. Tempe: Arizona Center for Medieval and Renaissance Studies, 1998.

Forshaw, Peter J. "Alchemical Exegesis of Genesis." In *The Word and the World: Biblical Exegesis in Early Modern Science*, edited by Kevin Killeen and Peter J. Forshaw, 111–36. Hampshire: Palgrave, 2007.

Idel, Moshe. *Kabbalah: New Perspectives*. New Haven: Yale University Press, 1988.

Isserles, Moses. *Sefer Torah ha-Olah*. 3 vols. Prague, 1570. In *The Wisdom of the Zohar: An Anthology of Texts*, 3 vols., edited by Yaruham Fishel Lachower and Isaiah Tishby, translated by David Goldstein, 111. Oxford: Oxford University Press, 1989.

Josten, Conrad H. "A Translation of John Dee's 'Monas Hieroglyphica' (Antwerp, 1564) with an Introduction and Annotations." *Ambix* 12 (1964): 84–221.

Kahn, Didier. "Les débuts de Gérard Dorn d'après le manuscript autographe de sa *Clavius totius Philosophiae Chymisticae* (1565)." In *Analecta Paracelsica*, edited by Joachim Telle, 59–126. Stuttgart: Franz Steiner Verlag, 1994.

Khunrath, Heinrich. *Amphitheatrum sapientiae aeternae*. Hanau, 1609.

———. *De igne magorum*. Leipzig, 1783.

Libavius, Andreas. *Appendix necessaria syntagmatis arcanorum chymicorum*. Frankfurt, 1615.

———. *De universitate . . . iuxta historiam hexämeri mosaici*. Frankfurt, 1610.

Liebes, Yehuda. *Studies in the Zohar*. Translated by Arnold Schwartz, Stephanie Nakache, and Penina Peli. Albany: State University of New York Press, 1993.

Maimon, Ayre. *Germania Judaica*. Vol. 3, *1350–1517*. Tübingen: Mohr, 1987.

Nauert, Charles G. *Agrippa and the Crisis of Renaissance Thought*. Urbana: University of Illinois Press, 1965.

Nettesheim, Agrippa von, Heinrich Cornelius, Donald Tyson, and James Freake. *Three Books of Occult Philosophy*. St. Paul, MN: Llewellyn, 1993.

Norrgren, Hilde. "Interpretation and the Hieroglyphic Monad: John Dee's Reading of Pantheus' *Voarchadumia*." *Ambix* 52 (2005): 217s46.

Pagel, Walter. "The Prime Matter of Paracelsus." *Ambix* 9 (1961): 117–35.

Pantheo, Giovanni Agostino. *Voarchadumia contra alchimiam: Ars distincta ab alchimia et sophia: Cum additionibus, proportionibus numeris et figuris opportunis Joannis Augustini Panthei, veneti sacerdotis*. Venice, 1530.

Paracelsus. *The Hermetic and Alchemical Writings of Paracelsus*. Vol. 1, *Hermetic Chemistry*. Edited by Arthur Henry Waite. London: James Eliot and Co., 1894.

[Paracelsus] Theophrast von Hohenheim. *Opus paramirum* [1531]. In *Sämtliche Werke I: Medizinische, naturwissenschaftliche und philosophische Schriften*, edited by Karl Sudhoff, pt. 1, vol. 9. Munich: O. W. Barth, 1925.

Partington, James R. *A History of Chemistry*. 4 vols. London: Macmillan, 1961.

Pico della Mirandola, Giovanni. *900 Theses*. In *Syncretism in the West: Pico's 900 Theses (1486): The Evolution of Traditional Religious and Philosophical Systems*, edited and translated by Stephen A. Farmer. Medieval and Renaissance Texts and Studies 167. Tempe: Arizona Center for Medieval and Renaissance Studies, 1998.

Pinnell, Henry. *Philosophy Reformed and Improved in Four Profound Tractates*. London, 1657.

Recanati, Menachem ben Benjamin. *Commentary on the Daily Prayers: Flavius Mithridates' Latin Translation, the Hebrew Text, and an English Version*. Edited by Giacomo Corazzol. 2 vols. Turin: Nino Aragno Editore, 2008.

Reuchlin, Johann. *De verbo mirifico, 1494. De arte cabalistica, 1517*. Facsimile reprint edition. Stuttgart-Bad Cannstatt: Frommann, 1964.

Severinus, Peter. *Idea medicinae philosophiae*. Basel, 1571.

Shackelford, Jole. *A Philosophical Path for Paracelsian Medicine: The Ideas, Intellectual Context, and Influence of Petrus Severinus*. Copenhagen: Museum Tusculanum Press, 2004.

Szulakow, Urszula. "John Dee and European Alchemy." Durham Thomas Harriot Seminar, Occasional Paper No. 21. Durham, NC: University of Durham, n.d.

Thorndike, Lynn. "Alchemy during the First Half of the Sixteenth Century." *Ambix* 2 (1938): 30–32.

———. *A History of Magic and Experimental Science.* 8 vols. New York: Columbia University Press, 1941.

Walton, Michael T. "Genesis and Chemistry in the Sixteenth Century." In *Reading the Book of Nature,* edited by Allen G. Debus and Michael T. Walton, 1–14. Kirksville, MO: Sixteenth Century Journal Publishers, 1998.

———. *Genesis and the Chemical Philosophy: True Christian Science in the Sixteenth and Seventeenth Centuries.* Brooklyn, NY: AMS Press, 2012.

———. "John Dee's *Monas Hieroglyphica;* Geometrical Cabala." *Ambix* 23 (July 1976): 116–23.

Walton, Michael T., and Phyllis J. Walton, "The Geometrical Kabbalah of John Dee and Johannes Kepler." In *Experiencing Nature,* edited by Paul H. Theerman and Karen Hunger Parshall, 43–59. Dordrecht: Kluwer Academic Publishers, 1997.

Wirszubski, Chaim. *Pico della Mirandola's Encounter with Jewish Mysticism.* Cambridge, MA: Harvard University Press, 1989.

Part Three

Evaluations and Perceptions

Chapter 9

Paracelsus on the Sidereal Powers

Revisiting the Historiographical Debate between Walter Pagel and Kurt Goldammer

Dane T. Daniel

The Pagel vs. Goldammer Dispute

The powers of the stars, and their relationship to humans, were clearly a principal concern of Theophrastus Bombastus von Hohenheim, or Paracelsus (1493–1541), the Swiss-German alchemist, medical practitioner, philosopher, and lay theologian. When reading of Paracelsus's belief that the human spirit consists of sidereal matter and that the human can draw knowledge and magical powers from raining stardust, one is tempted quickly to tie him to Renaissance Neoplatonism and natural magic rather than to the religious milieu of the early Reformation. A number of scholars have indeed seen in Paracelsus's discussion of the stars a direct tie to Marsilio Ficino's (1433–99) concept of the *spiritus mundi*, the subtle material substance that functions as a medium between the heavenly and sublunar world; such "spirit bodies" and their magical and medicinal powers can be attracted and employed.[1] Paracelsus did refer to Ficino on several occasions, even calling him the "best Italian physician," and he probably knew Ficino's *De vita* in German translation.[2] As we shall see, with regard to their respective views on magic and the sidereal bodies, there are striking parallels between Paracelsus on the one hand, and such Renaissance luminaries as Ficino and Giovanni Pico della Mirandola (1463–94) on the other. The characterization of Paracelsus as a Renaissance Neoplatonist is exceptionally problematic, however, and has evoked considerable debate.[3]

On this subject, the present chapter revisits an old quarrel regarding the sources of Paracelsus and focuses on Paracelsus's concept of the stars, which exist both in the world (macrocosm) and human being (microcosm). On one side of the debate, Walter Pagel strongly linked Paracelsus to Renaissance Neoplatonism and gnosticism, emphasizing commonalities with Ficino and Pico della Mirandola.[4] On the other, Kurt Goldammer, who edited dozens of Paracelsus's explicitly theological tracts, distanced Paracelsus from the Italian Renaissance and focused instead on the religious context, particularly that of the early Reformation in German-speaking areas.[5]

Opinions in the pertinent scholarship certainly tend to the extremes. (This is no surprise considering the exceptional difficulties in general that await the modern scholar who attempts to translate the early modern German while navigating Paracelsus's unique and often misunderstood terminology and concepts,

1. See Hirai, "Concepts of Seeds and Nature," esp. 274–76.

2. See Ficino, *Three Books on Life*, ed. Kaske and Clark, 12: "A German translation of *De vita* 1 and 2 was done by Johannes Adelphus Müling and published at Strassburg in 1505. Wilhelm Kahl and Dieter Benesch have identified seven more editions of the translation, *Das buoch des lebens*, between 1507 and 1537." See also the manuscripts that have parts of the work in German translation: Heidelberg University Library, *Codex palatinus germanicus* 730, fols. 1–35v, and *Codex palatinus germanicus* 452. See also Paracelsus's letter to Christoph Clauser (1527) in Paracelsus, *Sämtliche Werke I*, 4:71, in which he called Ficino the "Italorum medicorum optimus." The fourteen-volume first division of the modern edition of Paracelsus's collected works (treating medical, natural philosophical, and philosophical works) was edited by Karl Sudhoff and the seven-volume second division (theological writings) was edited by Wilhelm Matthiessen and Kurt Goldammer (with Matthiessen editing just the first volume and Goldammer the remaining editions).

3. Schütze argued that Ficino had no direct—if any substantial—influence on Paracelsus. See Schütze, "Zur Ficino Rezeption bei Paracelsus." Clearly, Renaissance Neoplatonism is a difficult category to delimit. Here, it encompasses, in general, such famous Italian thinkers as Ficino and Pico and their influential invocation and transformation of ancient theories concerning magic and sidereal powers.

4. Two representative books are Pagel, *Medizinische Weltbild des Paracelsus*; and Pagel, *Paracelsus: Introduction to Philosophical Medicine*.

5. For example, see Goldammer, *Paracelsus in neuen Horizonten*; and Goldammer, *Paracelsus: Natur und Offenbarung*. Regarding the theological tracts, see Paracelsus, *Sämtliche Werke II*.

ranging from *archeus* to *nectromantia,* and from *Limbus* to *chirurgia coelestis.*) Pagel, a pioneering scholar in the history of medicine and science—who boldly examined fresh intellectual contexts with a willingness to evaluate and incorporate such hitherto fringe topics as alchemy and natural magic and their sources and significance—at one point even wrote that "Paracelsus' whole life and work seems to be an attempt at implementing the ideal of Ficino's priest-physician," adding that Paracelsus derived his philosophical inspiration from "Ficino as the exponent of Neo-Platonism."[6] Pagel was not looking at the explicitly theological tracts by Paracelsus (of which dozens of volumes exist, half of which are still not available in printed form), and given what he could examine, he deemphasized how the religious texts and Reformation context influenced Paracelsus's natural philosophy. Some of this is attributable to Karl Sudhoff's decision to present the modern edition of Paracelsus's collected works in two divisions: (1) *Medizinische, naturwissenschaftliche, und philosophische Schriften*; and (2) *Theologische und Religionsphilosophische Schriften* (the second division of which became available only to a post-Pagel generation of scholars).[7] Another brilliant and trail blazing scholar, Allen G. Debus, echoed this sentiment in his book *The French Paracelsians,* wherein he added that the religious issues surrounding Paracelsus's thought should be relegated to a "sociological study," for they are extraneous to "intellectual history."[8] Debus was, of course, focusing on the explicitly chemical thought of Paracelsus and his followers, a monumental task in itself, and in his amazing contributions to *Paracelsica*—even without access to the *theologica* within Paracelsian studies—Debus gave some of the most lucid commentaries on Paracelsian cosmogony.[9]

Goldammer, in contrast (and with access to the entirety of Paracelsus's manuscripts), devoted nearly fifty years to the study of Paracelsus's biblical exegeses, their significance within his natural philosophy, and their Reformation context. With regard to the relationship between Paracelsus and Renaissance Neoplatonists, he often whittled the connection down to mostly just a common thought world and common sources, for example, the Augustin-

6. Pagel, *Paracelsus: Introduction to Philosophical Medicine,* 223. Andrew Weeks pointed to Pagel's thoughts on this subject in *Paracelsus: Speculative Theory,* 57.

7. See note 2 above. Sudhoff's work on the collected edition was restricted to the first division.

8. Debus, *French Paracelsians,* xv, 6. Weeks notes Debus's point in *Paracelsus: Speculative Theory,* 23.

9. It is well known that biblical cosmogony is an essential tenet of Paracelsianism, but it remains an underdeveloped topic in Paracelsus studies. Debus gave inspiring attention to the subject in several works, including *English Paracelsians.*

ian sources of Paracelsus's Neoplatonism. However, near the end of his life, Goldammer explored more deeply the relationship between the magic of the Italians and Paracelsus, a topic captured in his remarkable study *Der Göttliche Magier und die Magierin Natur: Religion, Naturmagie und die Anfänge der Naturwissenschaft vom Spätmittelalter bis zur Renaissance, mit Betrăgen zum Magie-Verständnis des Paracelsus.*[10] Goldammer's emphasis was on Paracelsus's uniqueness and the extraordinarily close link between Paracelsus's natural philosophy and such Christian topics as the contemporary debates surrounding the Eucharist, apocalypse, and nature of resurrected bodies. Somewhat akin to Goldammer in focus, Charles Webster, in his recent book on Paracelsus, wrote that Paracelsus had mere "generic affinities with figures such as Ficino, Pico, or Agrippa," but that "there is surprisingly little evidence of specific reliance on their writings in the work of Paracelsus."[11] Despite the "generic affinities," with regard to the general intellectual climate he added that "iconic figures such as . . . Johannes Reuchlin and Johannes Trithemius established Neoplatonism as the fashion of the moment and promoted the ideas of Ficino and Pico with enthusiasm. . . [This helped] to create a climate of opinion in Germany that encouraged the broadest eclecticism. . . and. . . generated a much greater tolerance of the arts associated with magic."[12]

Concerning the Pagel vs. Goldammer debate that still flavors the scholarship of *Paracelsica*, although I tend to side with Goldammer, one should take more of a *via media*. One should not be so dogmatic as not even to look for direct influences. The search for similarities and differences between Paracelsus and the likes of Ficino and Pico not only sheds light on the early modern intellectual milieu but also helps to differentiate the authentic from the spurious within the prodigious corpus of writings attributed to Paracelsus; pseudo-Paracelsus writings contain themes—such as emanation theories or talismans—that are antithetical to the authentic corpus as a whole. It is significant, however, that spurious tracts—for example, *Philosophia ad athenienses* and *De natura rerum*—were recognized by early moderns as authentic works by Paracelsus. Paracelsian ideas regarding "emanationism" or "preexistent matter," which do not conform to the authentic Paracelsian thinking, were held nevertheless to have been part of his *Weltbild*, even if the *Meister* himself

10. Goldammer, *Göttliche Magier und die Magierin Natur.*

11. Webster, *Paracelsus: Medicine, Magic and Mission*, 110.

12. Ibid., 47.

wished to present a Bible-based natural philosophy that focused on creation *ex nihilo*.[13] Clearly, comprehending Paracelsus's thought is very much complicated by the fact that the historical Paracelsus and the pseudo-Paracelsus were not differentiated in the sixteenth and seventeenth centuries. Paracelsianism as a tradition links to Ficino much more easily than to Paracelsus himself. Didier Kahn, for example, has pointed to some of the attempts in France to connect Paracelsianism with Ficino's Neoplatonism. He noted especially Adam von Bodenstein's (1528–77) cleverness in choosing the strongly alchemical *De vita longa* to introduce Paracelsus to the European intelligentsia. Like Ficino's similarly titled work, but even more so, *De vita longa* forwarded the medieval tradition of theoretical and practical thought on the prolongation of life, especially the medical alchemy of such notables as Roger Bacon (ca. 1214–94), pseudo(?)-Arnold of Villanova (ca. 1238–ca. 1310), John of Rupescissa (ca. 1300–ca. 1365), and pseudo-Llull. But significantly with regard to Neoplatonism, Bodenstein, in a conscious effort to echo somewhat Ficino's dedication to the Doge of Venice, dedicated *De vita longa* to an Italian prince. Thus Bodenstein played up Paracelsus's association with Ficinian Neoplatonism to make acceptable the doctrines associated with Paracelsus.[14] Regarding the search for Paracelsus's sources, however, Paracelsus does us few favors in our quest, for—with the exception of biblical passages—he generally was not one to name his sources, especially given his characteristic rant against those relying on ancient book knowledge rather than on the book of nature. Let us turn to Paracelsus's concept of the sidereal bodies, and evaluate several of the commonalities and dissimilarities between Paracelsus and Renaissance Neoplatonism. As this chapter will argue, with regard to the role of the stars in the cosmos and human beings, what perhaps most distinguishes Paracelsus from the Renaissance Neoplatonists is his overriding concern and pronounced search for a biblically based cosmogony and soteriology. His theology, mixed with alchemy and some concepts of Renaissance natural magic, created a unique *Weltbild*.[15]

13. This is one of the fundamental observations in Daniel, "Invisible Wombs."

14. Kahn, *Paracelsisme et alchimie en France*, 125.

15. For a discussion of the influence of medieval alchemical sources on Paracelsus's thought, especially his theology, see Daniel, "Medieval Alchemy and Paracelsus' Theology," 121–35.

Paracelsus on Magic and the Sidereal Component of the Macrocosm and Microcosm[16]

> The medium is the center; the center is the human. Thus, the heavenly powers [that is, natural sidereal agencies] may be brought through the human into the human, thus making possible [for the human] that which exists within a particular constellation. Thus, issuing forth from the human, into whom magic has brought these powers [*vires*], is precisely the same star as that which exists in itself, with the same secrets and *arcana*.[17]
>
> However, on account of the life that follows death, it is my advice that one live in the wisdom of the son rather than in nature. For the father regretted that he made the human, because the bodily and eternal aspects are meant to remain with each other, but they became separated [due to the Fall]. Consequently, a heathen is able to be a heathen, skilled in the light of nature, but unlearned in the Holy Spirit.[18]
>
> —Paracelsus, *Astronomia magna*

As illustrated by the above passages from Paracelsus's magnum opus, the *Astronomia magna*, Paracelsus not only engaged some of the conceptual framework of Renaissance natural magic, such as Pico della Mirandola's discussion of the human as the "center," but also incorporated into his philosophy his deep-seated religious beliefs, including the biblical teaching on the Fall and its costs.[19]

16. The first several pages of this section are derived from my dissertation: Daniel, "Paracelsus' 'Astronomia Magna,'" esp. 1–4 and 149–203.

17. "das medium ist der centrum, der centrum ist der mensch. also mag durch den menschen die himlische macht in den menschen gebracht werden, also das im selbigen constellation möglich ist. also wird aus dem selbigen menschen, in den die magica gebracht hat solche vires, gleich der selbig stern, wie er an im selbs ist, mit den selbigen secretis und arcanis"; Paracelsus, *Sämtliche Werke I*, 12:122 (my translation).

18. "mer ist aber mein rat, zu leben in der weisheit des sons dan in der natur von wegen des lebens, so dem tot nachfolget, dieweil den vatter gerauwen hat, das er den menschen gemacht hat. dan da sollen bei einander bleiben, dem leib das sein, dem ewigen das sein, das dan da gescheiden ist worden. dan ursach, ein heid mag ein heid sein, groß im liecht der natur vnd aber unbekant im heiligen geist"; Paracelsus, *Sämtliche Werke I*, 12:29 (my translation).

19. An example of Pico della Mirandola's discussion of this "center" can be found in the *Oration on the Dignity of Man*. In a manner similar to Pico, Paracelsus often wrote that the human is the "center," for the human is where the material, both elemental and sidereal, and divine realms intersect. See Pico della Mirandola, *De hominis dignitate*, in *Opera Omnia*, 1:313–31. "Nascenti homini omnifaria semina, & omnigenae uitae germina indidit pater. Quae quisquae excoluerit, illa adolescent, & fructus suos ferent in illo. Si vegetalia, planta fiet. Si sensualia, obbrutescet. Si rationalia, coeleste euadet animal. Si intellectualia, angelus erit et Dei filius. Et si nulla creaturarum sorte contentus, in unitatis centrum suae se receperit, unus cum Deo spiritus factus, in sol-

To underscore the present argument that it is unnecessary to take an extreme position in the Pagel vs. Goldammer debate, this chapter will show that Paracelsus did not consider Renaissance natural magic and a *literal rendering* of scripture to be incompatible, but rather sought to reconcile his alchemical and Neoplatonic heritage with God's divinely revealed word. Paracelsus held striking affinities with the Renaissance Neoplatonists even as he made a radical departure by incorporating his idiosyncratic, but literal, biblical exegesis. The radical reformer who never left the Catholic Church was clearly influenced by the Reformation emphasis on *sola scriptura* as he created his own theology and Bible-based *Weltanschauung*.

Turning to Paracelsus's specific ideas concerning magic and the sidereal components of the universe (macrocosm) and human (microcosm), it is fitting to explore pieces from his analysis of the star powers in his *Meisterstück*, the *Astronomia magna*, written in 1537 and 1538, and first published by several of his followers in the late sixteenth century. Book 1 is called "The Book of the Philosophy of the Heavenly Firmament." The first three chapters of this opening book became the most widely reproduced and translated section of the *Astronomia magna*. There, Paracelsus depicted the ontological basis of his picture of man and the cosmos. Indicative of his unique exegesis, he differentiated the natural creation by God the father (the mortal realm) and the new eternal creation by God the son (immortal corporeality). In chapters 4 through 11, he broached the topic of "natural astronomy," which he discussed in terms of its nine divisions (*membra*) and ten gifts (*dona*). Paracelsus's nine *membra* specify the kinds of *adept* art—for example, magic, astrology, nigromancy, and adept medicine—that a natural astronomer can study and master. The *dona* are different in that they are "ethereal arts" brought about by nature herself without human assistance. These include impressions (such as wisdom and prudence), generation (outgrowths from the four elements), and *inanimatum* (including gnomes and nymphs). Therefore, Paracelsus's "natural astronomy" is in fact his term for "natural magic." And the focus concerns the arts and operations involving the subtle

itaria patris caligine qui est super omnia constitutus, omnibus antestabit"; ibid., 1:315. There are a few important differences here. Paracelsus did not view individual potential as radically as Pico, who held—as did Paracelsus—that the "seeds" of all creation exist within the human, but who added that the human, through the exertion of the will, can develop and harvest any abilities or talents that one desires; although Paracelsus also emphasized free will, nevertheless, in contrast to Pico, he emphasized inborn talents that limit one's potential. Such propensities are planted by the stars at one's conception in accordance with the alignment of the stars.

matter of the universe that proceeds from the stars. (He often used the terms "magus" and "astronomer" interchangeably.) Plainly, book 1 of the *Astronomia magna* is a catalogue of the types of occult arts, but it also features his natural magic, which is a systematic natural science that Paracelsus utilized to explain many natural phenomena. Book 1 also reveals Paracelsus's epistemology: the mental capabilities of mortals are associated with both the sidereal component within humans and the instruction provided by the stars. Furthermore, it is a universal cosmography: Paracelsus abandoned the Aristotelian-Scholastic teaching that the terrestrial and celestial regions each have their own matter and physics. Rather, he described the intimate interaction between the firmament and the earth, and the processes and elements that the two realms have in common. With regard to different types of practitioners within natural philosophy, Paracelsus added that the philosopher studies the elemental while the astronomer studies the sidereal, both inside and outside the human.[20]

In the second chapter of book 1 of the *Astronomia magna,* Paracelsus focused on "the matter out of which the human was created, what the dust of the earth [*limus*] is, and the quality that this mass [*massa*] possesses."[21] He wrote that the first creation occurred *ex nihilo*; through the word, God made the corpus and its spirit and then fashioned all creatures from this corpus. In contrast to all other creations, made from nothing, only the human is made out of something. The human was made from a *Stoff,* namely, the *limus terrae,* and Paracelsus liberally interpreted the *limus terrae* of Genesis 2:7 to include all essences within the domain of both elemental and sidereal things. He explained that God took into his hand the mass from the four elements, both visible (earth and water) and invisible (firmament and chaos). Then, he transformed this substance so that his image, that is, the human, came out of it. The operations of both the sidereal body and the corporeal elements were thus, in effect, in this image, hence lending their virtues and essences to the human in whom they live. In this way, the marvels of God's creation were to be seen through the human, whether specific

20. "also sol nun verstanden werden, das erstlich ein leib ist, von dem der astronomus nichts redet; er ist elementisch, gehört in die irdisch der elementen philosophei. aber von dem natürlichen geist im leib hat der astronomus zu reden gewalt und macht; dan der leib ist der elementen, der geist, der dem elementischen leib geben ist, der ist im vom firmament vermälet und eingeleibt. und also teilt sich die philosophia in zwen teil, in das wesen des geistes und zum andern in das wesen des leibs, das ist in den corpus und spiritum"; Paracelsus, *Sämtliche Werke I,* 12:16 (my translation).

21. "Das ander capitel, aus was der mensch gemacht sei, was der limus sei und was eigenschaft die selbige massa gehabt hat"; ibid., 12:31.

to the mechanical arts or to the wisdom of the light of nature (that is, scientific knowledge). The gifts, or the sidereal talents and knowledge rained upon individuals, were also to become manifest. A third part of the human, the immortal soul, came about when God breathed into Adam.[22]

Evaluating the Divide

Of course, Pagel and Goldammer differed in their interpretations of the mature Paracelsus's cosmological components, and each saw Paracelsus in a different light. For his part, Pagel addressed the concept of the sidereal powers by charting an evolutionary path from Plato to the Stoics, to Plotinus, to Ficino, and finally to Paracelsus. Pagel noted that Ficino, after inheriting from the Stoics and ancient Neoplatonists the concept of the star-body within humans, wrote that spirit is "a very tenuous body, as if now it were soul and not body, and now body and not soul."[23] These sidereal components exist in demons and heroes, and those who perfect them can manifest certain qualities, such as courage. As Pagel wrote, "[t]he astral body was one of the means by which neo-Platonism preserved its basic dualistic attitude without sacrificing the idea of cosmic continuity, coherence and unity."[24] The invisible spiritual spark provides an object with its form and virtues, not its coarse body. An impulse of the spirit, or will, causes an instantaneous conversion in matter. Like any other object, the human is also an intermediate being on the scale of emanation between the divine and material. Thus, the divine spark of light exists in the human body. As we have seen, however, Paracelsus did not adhere to the Neoplatonic

22. "So wissent, das ein massa corporalis aus den vier elementen in die hant gottes gefaßt ist worden, sichtigs und unsichtigs, das ist von den corporalischen elementen, erden und wasser und von den unsichtbaren elementen firmamentem und chaos, und in die substanz transformiret, das ein biltnus gottes daruas worden ist, in welcher der gestirnte leib auch die elementischen corporalitet ire operationes volbringen sollen, das ist, das sie sollen im menschen wonen mit iren virtutibus und essentiis und im menschen die selbigen eröfnen und durch in in das werk bringen. als die kunst mechanica, als sapienta liminis naturae, auf das die wunderwerk gottes gesehen werden durch den menschen und was in den eußern elementen got gelegt und gegabet hat. dan nach dem rechten grunt zu reden, so ist der mensch alein darumb geschaffen, das er der natur arbeiter sei, das zu tun, das got in sie geben, gelegt und geschaffen hat"; ibid., 12:52–53 (my translation).

23. On this topic, see Ficino, *Three Books on Life*, bk. 3, chap. 3: "Ipse vero est corpus tenuissimum, quasi non corpus et quasi iam anima, item quasi non anima, item quasi non anima et quasi iam corpus." See also Pagel's interpretation that the sidereal is "not body and almost soul, and not soul and almost body" in *Religion and Neoplatonism in Renaissance Medicine*, 129. Pagel cited Marsilius Ficinus, *De Vita coelitus comparanda*, 3.3 (Venetiis: Aldus, 1516), fol. 153r.

24. Pagel, *Religion and Neoplatonism in Renaissance Medicine*, 128.

theory of emanationism. His exegesis of Genesis 2:7 provided the basis of his macrocosm-microcosm theory.

Nevertheless, in his synthesis of Neoplatonic and alchemical themes, Ficino anticipated Paracelsus with his own natural and astrological magic.[25] In a manner similar to Paracelsus's discussion of the *archeus*, Ficino discussed human "spirits" that convert food to life.[26] In addition, Ficino thought that the magus is one who can extract the living essence, the *spiritus*, from things; an ethereal spirit lives in all things, and it brings generation and motion. In Paracelsus's system, the firmament rules every body of the material world, providing internal motion and direction. Furthermore, when discussing "adept philosophy," Paracelsus explicitly discussed firmamental virtues: "In addition to what they receive from the elements, all earthly bodies carry a firmamental power and virtue."[27] Paracelsus wrote that a common philosopher and/or alchemist can describe the natural or elemental power in plants, but the adept philosopher, skilled in the sidereal, can describe the firmamental power. Thus, he can create sidereal curatives.[28]

Goldammer, on the other hand, distanced Paracelsus from some of the salient aspects of Neoplatonism. He pointed out that the emanationism so characteristic of Neoplatonism is completely absent from Paracelsus's thought, even in the *Philosophia de generationibus et fructibus quatuor elementorum*, where Paracelsus noted that things of the natural world—including the elemental spheres, three principles, and fruits or substances of the elements—appeared via the *separation* of the *yliaster*. Related to this, he pointed to Paracelsus's commitment to biblical cosmogony, that is, creation *ex nihilo*. He contended that in Paracelsus's authentic writings one finds almost nothing reminiscent of Neoplatonic-gnostic cosmogonical processes.[29] It is important to stress that in Paracelsus's thought, the analogy between macrocosm and microcosm came about when God formed Adam from the *limus terrae*, the primordial mass of Genesis 2:7. He explained the interrelatedness of the human and cosmos in this manner rather than evoking the concept of a series

25. When discussing the origin of magical powers, however, Ficino evoked the World-Soul while Paracelsus preferred "the stars." Both discussed the role of a "magnet" inside the human that attracts the sidereal.

26. See Hirai, "Concepts of Seeds and Nature," 274–76.

27. "so merkent... das alle die irdische corpora uber das, das sie von elementen haben, ein firmamentische kraft und tugent mit tragen"; Paracelsus, *Sämtliche Werke I*, 12:97 (my translation).

28. Ibid.

29. Note these themes throughout Goldammer, "Paracelsische Eschatologie: Zum Verständnis der Anthropologie und Kosmologie Hohensteims," in his *Paracelsus in neuen Horizonten*, 45–85.

of emanations from a shared source. It seems that the more a text attributed to Paracelsus contains discussions of preexistent matter and emanation theory, the less likely its authenticity. In fact, Karl Sudhoff, the dean of Paracelsus studies, insisted that two oft-cited texts, the *Philosophy to the Athenians* and *De natura rerum*—both of which, incidentally, Pagel relied upon heavily—are spurious. Both possess Neoplatonic cosmogonies that the authentic Paracelsus rejected.[30] Goldammer also highlighted Paracelsus's belief that the material processes of the world change, and an apocalypse is on the horizon that will culminate in the destruction of all elemental and sidereal things via fire. Fifteenth-century Neoplatonists, he argued, generally saw material processes as irreversible and were mostly uninterested in teleology and eschatology.[31]

One can easily extend Goldammer's points. First, the dichotomy between immortal spirit and tangible body simply does not exist in Paracelsus's thought. In Paracelsus's theology and philosophy, both the mortal spirit and mortal elemental flesh return to their origins at death, dust to the earth and the stardust to the stars. Note how he addressed the resurrection: "all that comes from the earth will not resurrect, only that which is from God will resurrect. Therefore, if we do not resurrect in Christ, we will resurrect, but not in Christ, and not in nature, but rather in the spirit of Hell."[32] Building on his unconventional reading of scripture, Paracelsus reasoned that the immortal soul requires a body in paradise (for Job says that he will see God with his eyes after his earthly walk[33]), and so he creatively merged a variety of scriptural passages to come up with his theory of an immortal resurrection body: God the Son creates the intangible immortal body, the resurrection body, to accompany the soul in the afterlife. Paracelsus's sacramental thought found its bases in his literal interpretations of such biblical passages as 1 Corinthians 15 and John 6:27. The latter verse reads, "Labour not for the meat which perisheth, but for that meat which endureth unto everlasting life, which the Son of man shall give unto you: for him hath God the Father

30. See the careful discussion throughout Sudhoff, *Versuch einer Kritik der echtheit der Paracelsischen Schriften*.

31. Ibid.

32. "alles was von der erden ist nit auferstehen wird, alein was von got ist, das wird auferstên. darumb so werden wir in Christo auferstehen, so wir aber nit in Christo seind, so werden wir auferstehen, aber nit in Christo, nit in der natur, sonder in dem geist der hellen"; Paracelsus, *Sämtliche Werke I*, 12:290 (my translation).

33. Job 19:26–27.

sealed."[34] Paracelsus took this to mean that Christ gives humans, via baptism, an eternal flesh; this coexists with the elemental body, mortal spirit, and soul. He also evoked Matthew 26:26 in this context: "And as they were eating, Jesus took bread, and blessed it, and brake it, and gave it to the disciples, and said, Take, eat; this is my body."[35] Paracelsus provided the remarkable explanation that Christ's breaking of the bread actually was the creative act that produced all of the individual resurrection bodies.[36]

Thus, in contrast to the common Neoplatonic and gnostic urge, Paracelsus clearly did not wish to free immortal spirit from body. He gave the immortal soul a resurrection body! And actually (and again) the spirit of humans, according to Paracelsus, is mortal—and even the sidereal bodies and all their functions (including *mortal* thought itself)—will be irrevocably destroyed. One can supplement further Goldammer's points with the observations that Paracelsus only referred to Plato and Ficino a few times in passing, that he never mentioned Plotinus—that he was completely uninterested in mathematics, music, or the geometric solids—and that he was antagonistic toward ancient texts.[37]

On the other hand, Pagel is correct that some correlations—terminologically and conceptually—between Renaissance Neoplatonism and Paracelsus's philosophy are often striking. I add that Ficino and Pico were in fact well known in the German-speaking realm, in part through the work of Johannes Reuchlin (1455–1522).[38] Also, Paracelsus's discussion of bestial humans, and more significantly his proof for the art of magic, are at times almost identical to passages in Pico's *Oration on the Dignity of Man*.[39] In addition, in book 1 of the *Astronomia magna*, Paracelsus began his elaboration of magic with a biblical example (lending immediate authority to the craft). In so doing, he followed in the footsteps of the Renaissance natural magicians.[40] Evoking a motif common to his Matthew commentaries, Paracelsus wrote,

34. John 6:27. All biblical quotes are from the King James translation.

35. See also Mark 14:22: "And as they did eat, Jesus took bread, and blessed, and brake it, and gave to them, and said, Take, eat: this is my body."

36. See Daniel, "Paracelsus' 'Declaratio' on the Lord's Supper.'"

37. Now, on the basis of this philosophy of the Lord's Supper, Paracelsus *does* erect an entire realm of eternal celestial influences that are analogous to the sidereal powers, but these function entirely in the realm of the immortal soul of humans rather than the mortal spirit, and they are available only to Christians. Paracelsus discussed this at length in book 2 of *Astronomia magna*.

38. See Müller-Jahncke, *Astrologisch-magische Theorie und Praxis*, 67–89.

39. See note 19 above.

40. See Brann, *Trithemius and Magical Theology*, 28.

> Concerning what magic and a magician is, you should understand the following: Magic is divided into six types. Its first part is an interpretation of unnatural signs, that is, recognizing that which God places in heaven supernaturally; however, they still appear to be natural like the others, just as did the oriental star of Bethlehem. And just as Christ walked on earth as a human among humans, so too did this star stand among the other stars. And just as Christ was recognized only by his own, so too are these stars understood by magicians alone. Thus, magicians are interpreters of such supernatural signs in heaven, just as apostles recognize Christ, and are placed before the word to interpret Christ's statement that, "there will be signs in the sun, moon, and stars, etc."[41] (Luke 21:25) Thus, the magicians are also interpreters of all prophecies and of the apocalyptic revelation.[42]

Although this does not demonstrate a direct tie to Ficino, nevertheless it was Ficino and his Medici patrons who venerated the cult of the three Magi and looked upon the Magi as biblically based support for the magical craft.[43]

Toward a Via Media

It is not unprecedented for scholars to point to Paracelsus's affinity with, and then departure from, the natural magic of the Italians. For example, Will-Erich Peuckert addressed the issue in 1941, highlighting similarities between Paracelsus and the Renaissance Neoplatonists, but pointing to what he saw as traditionally "German" elements within Paracelsus, especially (and problematically for the medical practitioner Paracelsus) a "German" preference for knowledge over praxis.[44] (Clearly, one of Pagel's significant contributions to *Paracelsica* was to emphasize the Renaissance context and thus rescue the field from some of the

41. Luke 21:25: "And there shall be signs in the sun, and in the moon, and in the stars; and upon the earth distress of nations, with perplexity; the sea and the waves roaring."

42. "Was magica sei und magus verstehet also: sie wird geteilt in sechs species. nun ist der anfang der magica ein auslegung auf die unnatürlichen zeichen, sie zu erkennen, wie sie got ubernatürlich in himel stellt und doch wie natürlich erscheine und erkent werden under andern, als der orientisch stern auf Bethlehem. und zu gleicher weis, wie Christus auf ertrich wie ein mensch gewantlet hat under andern menschen, also auch der stern under andern sternen gestanden. und wie Christus alein von den seinen erkant ist worden, also werden die sternen alein von den magis erkennet. also seind magi ausleger solcher ubernatürlicher zeichen im himel, wie die apostel erkenner Christi, und werden auf das wort gesezt auszulegen, da Christus spricht: es werden zeichen in sonn, mon und stern etc, also seind auch die magi ausleger aller propheten und der apocalypsischen offenbarung"; Paracelsus, *Sämtliche Werke I*, 12:83 (my translation).

43. See note 40 above.

44. Peuckert, *Theophrastus Paracelsus*, 370.

less palatable subjectivity of the German scholarship of the 1930s and 1940s.) However, the Pagel-Goldammer divide still exists in the field of Paracelsus studies, and a more nuanced approach is necessary.

Among those who are currently seeking a middle path in this debate is Hartmut Rudolph. For years, Rudolph sided mostly with Goldammer and carried on a long, but amiable, debate with Pagel regarding Paracelsus's sources. At this time, Rudolph explored the Reformation setting of Paracelsus's thought. While demonstrating that Paracelsus's theology and medicine are inextricably linked, Rudolph noted that there are many queries still to be answered about Paracelsus's biography that pertain to the question of his religious affiliation.[45] He asked whether Paracelsus's religious attitudes affected his choice of Strasbourg as a dwelling place or motivated his patrons to call him to Basel. He added that Paracelsus did indeed find the healing art and the Christian faith to be fused, and yet, it is unclear whether he can be linked to the Nicodemism of Cornelius Agrippa von Nettesheim (1486–1535) or the thought of Caspar von Schwenckfeld (1489–1561), and if so, when in his development that linkage is valid. He pointed to Paracelsus's teaching on the Eucharist to show that the linking of Paracelsus to Schwenckfeld is problematic and stressed Paracelsus's Catholicism, especially the Franciscan/Erasmian influence on his thought during his first Salzburg period (mid-1520s), even if he is rightly tentative relative to the outright endorsement of a particular designation.[46] More recently, however, Rudolph has begun to argue that Paracelsus's anthropology—which, incidentally, runs throughout his entire opus and not merely his theological works—corresponds much more closely to the ideas of the humanists Ficino and Pico della Mirandola than to those of Protestant reformers like Martin Bucer (1491–1551) and Ulrich Zwingli (1484–1531).[47] Furthermore, in stressing some of the Ficinian aspects of Paracelsus's thought, Rudolph paid particular attention to the idea—shared by Ficino and Paracelsus—that the stars incline rather than determine human behavior.[48] Indeed, Rudolph is aware that Paracelsus wrote that the stars stamp talents and propensities onto a person's mind and senses at the moment of conception. As we have seen, the stars also continually rain down *dona* (gifts) and the light of nature (which accounts for Paracelsus's epistemology). A person is *responsible* for

45. Rudolph, "Paracelsus' Laientheologie in traditionsgeschichtlicher Sicht."

46. Ibid., 85.

47. Rudolph, "Hohenheim's Anthropology," esp. 191–92.

48. Ibid.

harvesting the gifts, the innovations and knowledge bestowed by the stars; such is part of one's divine calling. Note also the following conclusion of Rudolph:

> Paracelsus views and describes sacramental doctrine, biblical soteriology and eschatology from the same horizon of expectation as that which had governed the anthropology of a Pico or a Ficino. It cannot be denied that he shares with these figures the concept of natural magic as set forth, for example, in the *Picatrix*, in *De dignitate hominis*, in *De vita libri tres*, where it constitutes the "pars practica scientiae naturalis"; with this concept, man, being the centre of the macrocosm, is enabled to "coelo maritare mundum": "nulla est virtus in coelo aut in terra seminaliter et separata, quam et actuare et unire magus non possit", Pico says (*Conclusiones*, 3, 10ff.). But it is also undeniable that Hohenheim's theological anthropology, his "philosophia de limbo aeterno," pursued the same goal as that which was held by the neoplatonists to be man's final and actual determining to speak with Plato's *Theaetetus*.[49]

Rudolph's writings capture a correct appreciation and synthesis of the arguments of both Pagel and Goldammer. As we have seen, there are too many elements in Paracelsus's thought that have Neoplatonic roots, even as he baptized such concepts in his unique brand of radical-reform Christianity. Significant in the case of Paracelsus—and what would become a salient current of the early modern intellectual milieu via the Paracelsian traditions—was the combination of Reformation and Neoplatonic motifs in the expression of his astronomy.

Works Cited

Brann, Noel. *Trithemius and Magical Theology: A Chapter in the Controversy over Occult Studies in Early Modern Europe.* Albany: State University of New York Press, 1999.

Daniel, Dane T. "Invisible Wombs: Rethinking Paracelsus's Concept of Body and Matter." *Ambix* 53, no. 2 (2006): 129–42.

———. "Medieval Alchemy and Paracelsus' Theology: Pseudo-Llull's *Testamentum* and Paracelsus' *Astronomia Magna." Nova Acta Paracelsica*, n. s. 22, 23 (2008–2009): 121–35.

———. "Paracelsus' 'Astronomia Magna' (1537/38): Bible-Based Science and the Religious Roots of the Scientific Revolution." PhD diss., Indiana University, 2003.

———. "Paracelsus' 'Declaratio' on the Lord's Supper: A Summary with Remarks on the Term 'Limbus.'" *Nova Acta Paracelsica* 16 (2002): 141–62.

49. Ibid., 205. Rudolph cited Pico della Mirandola, *De dignitate hominis*, ed. Garin, 17.

Debus, Allen G. *The English Paracelsians*. New York: Franklin Watts, 1966.

———. *The French Paracelsians: The Chemical Challenge to Medical and Scientific Tradition in Early Modern France*. Cambridge: Cambridge University Press, 1991.

Ficino, Marsilio. *Three Books on Life: A Critical Edition and Translation*. Edited and translated by Carol Kaske and John Clark. Binghamton, NY: Center for Medieval and Early Renaissance Studies, 1989.

Goldammer, Kurt. *Der Göttliche Magier und die Magierin Natur: Religion, Naturmagie und die Anfänge der Naturwissenschaft vom Spätmittelalter bis zur Renaissance, mit Beträgen zum Magie-Verständnis des Paracelsus*. Stuttgart: F. Steiner, 1991.

———. *Paracelsus in neuen Horizonten: Gesammelte Aufsätze*. Vienna: Verband der wissenschaftlichen Gesellschaften Österreichs, Verlag, 1986.

———. *Paracelsus: Natur und Offenbarung*. Hanover-Kirchrode: Oppermann, 1953.

Hirai, Hiroshi. "Concepts of Seeds and Nature in Marsilio Ficino." In *Marsilio Ficino: His Theology, His Philosophy, His Legacy*, edited by Michael J. B. Allen and Valery Rees, 257–84. Leiden: Brill, 2002.

Kahn, Didier. "Paracelsisme et alchimie en France à la fin de la Renaissance (1567–1625)." PhD diss., University of Paris 4, 1998.

Müller-Jahncke, Wolf-Dieter. *Astrologisch-magische Theorie und Praxis in der Heilkunde der frühen Neuzeit*. Stuttgart: F. Steiner, 1985.

Pagel, Walter. *Das medizinische Weltbild des Paracelsus: Seine Zusammenhänge mit Neuplatonismus und Gnosis*. Wiesbaden: F. Steiner, 1962.

———. *Paracelsus: An Introduction to Philosophical Medicine in the Era of the Renaissance*. Basel and New York: Karger, 1958.

———. *Religion and Neoplatonism in Renaissance Medicine*. Edited by Marianne Winder. London: Variorum Reprints, 1985.

Paracelsus. *Sämtliche Werke I: Medizinische, naturwissenschaftliche und philosophische Schriften*. Edited by Karl Sudhoff. 14 vols. Munich: R. Oldenbourg, 1922–33.

———. *Sämtliche Werke II: Theologische und Religionsphilosophische Schriften*. Edited by Wilhelm Matthiessen. Vol. 1, *Philosophia Magna*. Munich: Otto Wilhelm Barth, 1923.

———. *Sämtliche Werke II: Theologische und Religionsphilosophische Schriften*. Edited by Kurt Goldammer. 6 vols. Wiesbaden (Stuttgart): F. Steiner, 1955–86.

Peuckert, Will-Erich. *Theophrastus Paracelsus*. Stuttgart: W. Kohlhammer, 1941.

Pico della Mirandola, Giovanni. *Opera Omnia*. Edited by Eugenio Garin. 2 vols. Turin: Bottega d'Erasmo, 1971.

Rudolph, Hartmut. "Hohenheim's Anthropology in the Light of His Writings on the Eucharist." In *Paracelsus: The Man and His Reputation, His Ideas and Their Transformation*, edited by Ole Peter Grell, 187–206. Leiden: Brill, 1998.

———. "Paracelsus' Laientheologie in traditionsgeschichtlicher Sicht und in ihrer Zuordnung zu Reformation und katholischer Reform." In *Resultate und Desiderata der Paracelsus-Forschung*, edited by Peter Dilg and Hartmut Rudolph, 79–98. Stuttgart: F. Steiner, 1993.

Schütze, Ingo. "Zur Ficino Rezeption bei Paracelsus." In *Parega Paracelsica: Paracelsus in Vergangenheit und Gegenwart*, edited by Joachim Telle, 39–44. Stuttgart: F. Steiner, 1991.

Sudhoff, Karl. *Versuch einer Kritik der echtheit der Paracelsischen Schriften*. Berlin: G. Riemer, 1894–99.

Webster, Charles. *Paracelsus: Medicine, Magic and Mission at the End of Time*. New Haven, CT: Yale University Press, 2008.

Weeks, Andrew. *Paracelsus: Speculative Theory and the Crisis of the Early Reformation*. Albany: State University of New York Press, 1997.

Chapter 10

John Dee at 400

Still an Enigma

Nicholas H. Clulee

The year 2009 marked the four hundredth anniversary of the death of John Dee (1527–1609),[1] yet scholarly studies of this enigmatic figure's life and work continue to seek a fuller understanding of his place in Renaissance history. In fact, something of a John Dee "industry" has blossomed in the last few decades. There have been three John Dee colloquia spurred first by a two-day meeting in London in 1995 and, most recently, the 2009 Quartercentenary Conference at St. John's College, Cambridge, both of which generated published proceedings.[2] Besides a number of scholarly articles, there have appeared Benjamin Woolley's popular biography, Michael Wilding's account of the adventures of Dee and Edward Kelley (1581–97), James Fenton's edition of Dee's diaries, and a host of occultist editions and studies of Dee's magic.[3] Dee has also figured as a subject or character in modern literature and film.[4] Most recently, Robert Barone has

1. It also marked the death of Allen Debus, to whom the present study is dedicated. See also note 27 below.

2. Clucas, *John Dee*; and Rampling, "John Dee and the Sciences," respectively.

3. Woolley, *Queen's Conjurer*; Wilding, *Raising Spirits*; and Dee, *Diaries of John Dee*. Occultist editions and studies related to Dee are too extensive to list here. Searching for "john dee" in the online catalogues of the British Library and the Library of Congress will yield most of what is available.

4. Szönyi and Wymer, "John Dee as a Cultural Hero"; and Wymer, "Dr. Dee and Derek Jarman."

published a history of Dee that traces Dee's reputation from negative to warranting serious attention.[5] There have also been six major scholarly monographs that represent substantial and creative contributions to understanding Dee and his relation both to the social and political worlds in which he sought patronage and to sixteenth-century currents in scholarship, the sciences and mathematics, occultism, and religion. Yet, despite my monograph (1988) and the rich contributions of William H. Sherman (1995), Deborah E. Harkness (1999), Håkan Håkansson (2001), György Szönyi (2004), and Glyn Parry (2011), there is still no consensus on a comprehensive understanding of Dee.[6] He remains an enigma.

What follows, after an overview of earlier studies of Dee, is a review of these major monographs. The aim is not to provide a thorough survey of their contributions but to highlight the continued problematic of Dee. One fundamental issue that underlies Dee's modern historiography is the diversity and apparent contradictions among his many writings and activities. Was Dee of one or many minds, dividing a Dee embodying a continuous coherent vision throughout his career and a Dee who dynamically rethought positions as his career evolved? This dichotomy is not necessarily absolute—there may be a spectrum between the poles—but it continues to map the terrain of Dee scholarship. How those studying Dee have presented him is entwined with two other divisions: one involving historical methodology, the other involving the nature of the Renaissance as an intellectual and cultural phenomenon.

Modern scholarship on Dee started in the past century with Charlotte Fell-Smith's *John Dee (1527–1608)*, prompted by the previous centenary of Dee's death and exemplifying a central challenge of Dee.[7] Claiming to replace centuries of misjudgment and slander by presenting the "facts" of Dee's life "calmly and impartially in the light of reason and science," Fell-Smith drew upon a variety of Dee's autobiographical material published in the nineteenth century as well as writing published in his lifetime and in the seventeenth century.[8] The result is a summary of Dee's biographical record with some romantic effusions but little probing inquiry into controversial episodes—such as Dee's arrest during Mary's

5. Barone, *A Reputation History of John Dee.*

6. Clulee, *John Dee's Natural Philosophy*; Sherman, *John Dee*; Harkness, *John Dee's Conversations with Angels*; Håkansson, *Seeing the Word*; Szönyi, *John Dee's Occultism*; and Parry, *Arch-Conjuror of England.*

7. Fell-Smith, *John Dee.*

8. Ibid., 1–5.

reign—or deep intellectual analysis of his writings to support claims for his "solid achievements in science" and "thought too advanced for his own age."[9] Most notably, the emphasis of Fell-Smith's account is skewed by the character of Dee's *Nachlass*: the large bulk of the records of his scrying activities—known variously as the "spiritual diaries" or "conversations with angels"—serves as a "great attractor" drawing attention to a limited part of Dee's career at the expense of his other activities. Over half of Fell-Smith's account is devoted to the single decade from 1580 to 1590, out of a complete life of eight decades.

The challenge of this imbalance in the records for the study of Dee is how to recapture sufficiently the other periods of his career and to integrate the "actions with spirits" with his other intellectual production. Are these "actions," in Meric Casaubon's seventeenth-century expression of the continuity of Dee's mind, that "Dr. Dee, of himself, long before any Apparition, was a Cabalistical man, up to his ears, as I may say: as may appear to any man by his *Monas Hieroglyphica*," from which he could "extract no sense or reason"?[10] Or are they a "new phase," as Fell-Smith would have it and a departure from his "sound" pursuits?[11] If so, how do we explain and assess this departure? Fell-Smith does not see it as consistent with "solid achievements in science" and suggests that Dee was duped and manipulated under the inordinate sway of Edward Kelley, his principal scryer.[12]

The inclination to give most attention to the 1580s in Dee's career has characterized most of the subsequent popular biographies. The sensationalism certainly sells, and these activities have resonated with currents of interest in the occult and the psychic. Gertrude M. Hort's brief biography basically follows Fell-Smith, but more positively recommends Dee's contribution to modern psychical research.[13] Resonating with a different cultural milieu, Richard Deacon presents Dee as a founder of extrasensory perception and telepathy with kinship to the psychedelic interests of the 1960s.[14] His solution to interpreting Dee's scrying activities is to resurrect Robert Hooke's seventeenth-century contention, dismissed by Fell-Smith and Hort, that the angelic revelations were a form of cryptography, so that, for Deacon, Dee was an Elizabethan 007 engaged in

9. Ibid., 2.

10. Dee, *True and Faithful Relation*.

11. Fell-Smith, *John Dee*, 60–61.

12. Ibid., 124.

13. Hort, *Dr. John Dee*. This is listed as part of Rider's Mystics and Occultists series.

14. Deacon, *John Dee*.

Figure 10.1: Title page from John Dee, *Monas hieroglyphica,* 1564. Courtesy of Beinecke Rare Book and Manuscript Library, Yale University.

espionage for England as an early modern James Bond. Unfortunately, Deacon's interpretive elaborations on the prosaic record of Dee's career are largely fantasy and speculation, lacking documentation and serious analysis of Dee's writings.

Benjamin Woolley's *The Queen's Conjurer: The Science and Magic of Dr. John Dee, Adviser to Queen Elizabeth I* (2001) draws on the substantial body of scholarship since Fell-Smith and is better grounded and documented than Deacon.[15] Although the title, by sandwiching Dee between "Queen's Conjurer" and "Adviser to Elizabeth I," might portend some sensationalism, Woolley gives a considered and balanced account of Dee that is engaging and readable. There are some brief sidetracks on science. Dee's birth chart provides the occasion to explain the rudiments of astrology, and there are brief synopses of his *Propaedeumata aphoristica* (1558) and *Monas hieroglyphica* (1564), but no special treatment of the "Mathematicall Praeface" to Euclid (1570). The emphasis is on a narrative of Dee's personal and public life more than on his role in contemporary science. The bulk of Woolley's account is devoted to 1582 to 1589, the period of Dee's conversations with angels through the scryer Edward Kelley, which took Dee and Kelley on an adventurous journey to Poland and the Prague of Rudolf II and culminated in the famous episode of their sharing wives at the command of the angels. If Woolley has a thesis, it is only implied. Dee's turn to the spirits results from his growing despair with human sources of knowledge compounded by financial difficulties and a lack of support at court from William Cecil (1520–98). On the perplexing relationship with Kelley, Woolley does not think Dee was merely a dupe. He concludes that Dee's hope in Kelley was that "they would navigate a northwest passage to universal truth before, with the onset of the modern, mechanistic age his own work helped to inaugurate, unity would become irrecoverable, and the magic would go out of the world forever."[16] Woolley's basic touchstone remains the older view of Frances Yates and Peter French (see below) of Dee as a key figure in the English Renaissance. Unfortunately, Dee's Copernicanism, his associations at court with pro-Protestant activists (the so-called Sidney group, see below), and with the group of men known as the School of Night animated by Sir Walter Raleigh (ca. 1554–1618), all require a good deal of speculation and leaps from slim evidence. Deacon's ghost also haunts

15. Woolley, *Queen's Conjurer*.

16. Ibid., 298.

Woolley's suggestions of clandestine involvements by Dee, Kelley, and others in covert espionage activities.

Modern scholarship has taken a variety of nuanced perspectives on Dee's actions with spirits and their relationship to his other writings and intellectual formations,[17] but in the mid-1900s their overpowering bulk and antithesis to conventional ideas of science posed challenges for anyone who wanted to portray Dee as a scientist or significant figure in a progressive view of sixteenth-century culture. The first scholars to present Dee as important for early modern science isolated bits of his work for attention because of their relation to scientific progress.[18] Francis R. Johnson's *Astronomical Thought in Renaissance England* and Eva G. R. Taylor's studies of Tudor geography and mathematical practitioners are selective in this way.[19] The implications were that Dee's scientific thought was or could be disconnected from the angel conversations and his major early publications, the *Propaedeumata aphoristica* and the *Monas hieroglyphica*, both of which had occult dimensions.

Ian R. F. Calder attempted to reunify these different minds in his dissertation "John Dee Studied as an English Neoplatonist," the first truly scholarly treatment of Dee.[20] He was the first to study thoroughly and to attempt to explicate all of Dee's surviving writings, that is, manuscript remains in addition to the published works. Although biography is only a secondary concern to Calder, he unearthed archival records with much new biographical information in addition to Dee's own autobiographical records. Calder presents a unified Dee, all of whose activities and writings are organically related as expressions of a "scientifically oriented 'neoplatonism'" that was an important feature of Renaissance philosophy and that "made important contributions to the development of scientific theory and practice."[21] In addition to interpreting the Renaissance as reviving Neoplatonism, and with it a mathematically oriented approach to natural science, Calder's methodology was shaped more by his philosophical interests than by history. He opened by laying out his characterization of this Renaissance "scientifically oriented 'neoplatonism'" built from a variety of recognized Renaissance

17. Clucas, "John Dee's Angelic Conversations."

18. Fell-Smith, *John Dee*.

19. Johnson, *Astronomical Thought in Renaissance England*; Taylor, *Tudor Geography*; Taylor, *Mathematical Practitioners*. Both Johnson and Taylor have also written articles on this topic.

20. Calder, "John Dee Studied as an English Neoplatonist."

21. Ibid., 1:8–18, 865.

authors.[22] Dee's activities and writings were then presented and interpreted as illustrations and expressions of this intellectual formation. While he related all of Dee's writings to this, there was little sense of dynamic development over time and progression from one writing to the next. This "Neo-platonic mathematicism" could accommodate scientific, metaphysical, and mystical elements, bridging Dee's various interests but also grounding his importance in progress in a more modern science.[23] Although he acknowledged passing hints of the "angelic conversations" in the *Propaedeumata aphoristica* and the *Monas hieroglyphica,* he emphasized in these, as well as the "Mathematicall Praeface" to Euclid's *Elements,* how they presage progressive development in science: a rationally mechanical and quantitative causality in the *Propaedeumata* and a geometrical alchemical atomism in the *Monas* that look forward to seventeenth-century mechanism and corpuscularism. Acceptance of a spirit world was part of this intellectual formation, thus the "angelic conversations" are not discordant, but they were presented neither as a logical progression from his earlier works nor as integral to his contributions to science or natural knowledge. Calder saw them as the expression of Dee's growing apocalyptic sense of impending change influenced by the prophetic tradition and astrological theories of cycles of historical ages.[24]

Calder established two key frameworks for the unitary view of Dee: a unified Renaissance intellectual tradition, in this case Neoplatonism, in which Dee fit; and a privileging of archetypal authors and works, such as Marcilio Ficino (1433–99), Giovanni Pico della Mirandola (1463–94), Cornelius Agrippa (1486–1535), etc., against which Dee's writings are read as reflections and instances. Frances Yates, Calder's mentor, absorbed this approach to Dee into her conception of a unified Hermetic tradition in the Renaissance.[25] Relying largely on Calder's research and readings of Dee's texts and his positive assessment of Dee's contribution to science, Yates adopted Dee as a prime example of how the impulse embodied in Hermetic magic to employ the secrets of nature to operate on the world contributed to scientific progress. Peter French developed

22. Ibid., chap. 1.

23. Ibid., 1:865.

24. Calder, "John Dee Studied as an English Neoplatonist," 1:743, 781.

25. Yates, *Giordano Bruno and the Hermetic Tradition*; Yates, "Hermetic Tradition in Renaissance Science"; and Yates, *Occult Philosophy in the Elizabethan Age.*

Yates's interpretation of Dee in *John Dee: The World of an Elizabethan Magus.*[26] Following chapters on "The Development of an English Magus" and "John Dee and the Hermetic Philosophy," which uncritically flesh out Yates's idea of Dee, French covered all of Dee's philosophical writings, from the *Propaedeumata*, the *Monas*, and the "Mathematicall Praeface" to the angelic conversations in a single chapter on "Magic, Science and Religion," suggesting a coherently unified intellectual formation without historical development. French did broaden earlier emphases on Dee's science by presenting Dee's importance in the broader context of Elizabethan Renaissance culture. So, in addition to a chapter on Dee and applied science, he covered Dee's influence on aesthetic culture through the Sidney group as well as through his antiquarianism and its relation to Elizabethan imperial ideas. Nonetheless, all aspects of Dee's activities and important contributions are presented as manifestations of a unified Hermetic inspiration.

From Calder to Yates we see the articulation of an encompassing, even totalizing, tradition defining a mental world, and an approach to and interpretation of nature rooted in the idea of the efficacy of magical, manipulative powers to reveal the secrets of the cosmos. Also, from Calder through Yates to French, the interpretation of Dee degenerated from attentive readings of his texts to searching those texts for instances of supposed archetypes. A central theme of this approach to Dee is occultism: that central to Renaissance natural philosophy and Dee is a conviction that occult virtues and spiritual influences and forces are not only central explanatory factors in understanding nature but also vehicles for the integration and elevation of the human spirit within the cosmos.

In 1988, I challenged this approach to and view of Dee in *John Dee's Natural Philosophy*.[27] I made the methodological decision to take Dee's writings

26. French, *John Dee*.

27. See note 6 above. I came to this study as a student of Allen Debus, although I must admit that Dee's interests only peripherally intersect with Allen's devotion to alchemy and the "Chemical Philosophy." I was first stimulated to consider the history of science by Burtt's *Metaphysical Foundations of Modern Physical Science* and was particularly interested in the role of mathematical presuppositions. I had no knowledge of alchemy, but as a student at the University of Chicago, of necessity, studying the history of science meant working with Allen Debus. In his seminar on Renaissance science, I remember his sending us to the rare book room to read Dee's "Mathematicall Praeface" in the original edition. Besides being fascinated by the text, I was awed and seduced by the physical contact with the binding, paper, typography, touch, and smell of that book. Allen also had us read Dee's *Monas hieroglyphica*, which seemed to defy understanding and should have repelled anyone interested in understanding the Scientific Revolution as then embodied in the grand narrative from Copernicus to Newton. Nonetheless, when casting about for a dissertation subject, I thought of Dee, and Allen was supportive and helped me over a number of obstacles and discouragements. For better or worse, I've been entangled with Dee ever since. I think this is a telling tribute to Allen's comprehensive interest both in

as particular moments when his thinking crystallized. Close readings of the texts were historically contextualized in terms of Dee's biography and social situation as well as of indications of the specific intellectual problems he was addressing and the influences that could be documented from sources he demonstrably read, either from references in his texts or datable books he owned and preferably annotated. Although my work predated Roberts and Watson's now fundamental edition of Dee's library catalogues, I corresponded with them extensively and consulted many of Dee's copies of books and manuscripts.[28] The other side of this methodology was a refusal to read Dee's earlier writings in terms of later ones or in terms of some broader tradition. The result was a dynamic view of Dee over time, confronting new sources, broaching new issues and problems, modifying ideas and formulations, and adapting to changing social circumstances as he sought to support himself and his family through patronage. The continuity that emerged was a unity of intention to discover the best path to natural knowledge and to understand the natural world by uncovering the architecture of its divine creation.

Dee's different writings, however, despite this continuity of purpose and the persistence of some themes, revealed some ruptures and significant shifts in emphasis and direction. These shifts reflected significantly different ideas on how to achieve his goals, from the mathematical-physical causality of the *Propaedeumata aphoristica* in the late 1550s to the direct revelations of the "angelic conversations" in the 1580s. His readings included texts associated with the Renaissance Neoplatonic and Hermetic traditions, but there is little indication that he identified himself with that tradition. His sources were highly eclectic, and in many cases medieval, rather than classical or Renaissance, sources were more influential. My work was also situated in the specific science history context of the debate about the relationship of the occult sciences to the emergence of a more modern science in the early modern period. My answer here was that it depended on the particular methodological values of the occult science in question: the astrological physics of the *Propaedeumata* looks forward, whereas the angelic conversations represent an abandonment of naturally acquired human knowledge.

From Johnson and Taylor through the Warburg school and my study, Dee's relation to science was central: it was what rescued him from obscurity and made

Renaissance science and in mentoring students.

28. Roberts and Watson, *John Dee's Library Catalogue.*

him an important case study in the relation of the occult sciences to the Scientific Revolution that was a major historiographic theme in the history of science from the 1960s through the 1980s, a theme to which Allen Debus made important contributions. Dee, however, was more than a natural philosopher. He was a polymath or even omnimath who, beyond natural philosophy and mathematics, wrote extensively on geography, exploration, and British imperial claims, as well as on economic and mercantile policy, the British Navy, and the strengthening of the British state. And then there are the extensive "angel conversations." Here are more minds of John Dee that challenge integration into a unified vision.

Graham Yewbrey addressed Dee's political writings—many of them in manuscripts that had previously received little or no attention—in a 1981 dissertation. There, he related them to a larger unity in Dee's thought rooted in a Yatesean framework of a broader Christian-Neoplatonic-Hermetic-Cabalist tradition.[29] Yewbrey presented Dee as primarily a political philosopher advocating a "cosmopolitics" that would assure the redemption of humanity through the establishment of a pan-European theocratic state guided by Dee's magical religious philosophy.[30] Dee saw his mission as the prophet of this religiously focused political reform as the world entered the final, apocalyptic stages of its six-thousand-year history. This "cosmopolitics" provides the key to identifying a fundamental coherence in Dee's thought running through the entire span of his career and all of his writings. In the absence of any single, explicit, written statement of this philosophy, however, Yewbrey ranged over Dee's entire body of writings, finding hints, identifying latent ideas, and reading between the lines to construct a rather disembodied and speculative notion of this "cosmopolitics," which he then finds implicit in the specifically political texts. His key text was the *Brytannicae reipublicae synopsis* of 1570, a bracketed Ramist outline on a large manuscript roll. With, for Yewbrey, the most important section on "Wisedome (*Per me, Reges regnant*)" destroyed by fire, Yewbrey fleshed this section out from the esoteric lore of Dee's other writings, with results that are, in William H. Sherman's characterization, "somewhat overblown."[31] While presenting a single Dee, Yewbrey broke ground in his close and meticulous reading of some of Dee's unpublished writings that had been marginalized,

29. Yewbrey, "John Dee and the 'Sidney Group.'"

30. Ibid., 3–4.

31. Ibid., 42–55, 200–27; and Sherman, *John Dee*, 141.

and was attentive to the social dimensions of Dee's activities by relating his political writings to the interests of the Sidney group at court.

The enigma of Dee presents itself starkly when we turn from Yewbrey to William H. Sherman's *John Dee: The Politics of Reading and Writing in the English Renaissance* of 1995. In part 3, on the politics of Dee's writings, Sherman focused on the *Brytannicae reipublicae synopsis*, the *General and Rare Memorials*, the *Of Famous and Rich Discoveries*, the *Brytanici imperii limites*, and the *THALATTOKRATIA BRETTANIKI*. With the exception of the recently discovered *Brytanici imperii limites*, these are the same texts on which Yewbrey's attention was centered. In contrast to studies of Dee's natural writings and Yewbrey's reading of the political writings, Sherman found no occult philosophy, no theocracy, no apocalyptic prophet, and nothing "cosmic" in Dee's politics. Sherman's Dee is a practical if enthusiastic author of policy papers much in the mainstream, offering his expertise and advocacy on contemporary issues.

How can the same texts yield such disparate views of the same person? Beyond expressions of what may be seen as a growing academic imperative to build a young career on the ashes, rather than the shoulders, of previous scholars, there are two clear roots of these two Dees. One is methodological. Sherman insisted that "Dee's writings must be read on their own terms—which is emphatically not to say divorced from the conventional terms of his period—and not made to conform to a grand, pre-fabricated version of his intentions."[32] This approach is not markedly different from the one I adopted toward Dee's natural philosophy texts and strikes me as much preferable to Yewbrey's. The other root lies in the radically different frameworks of the two approaches. Yewbrey pursued coherence and embedded Dee's political writings within a Yatesean frame. Sherman, rejecting Yates's "myth of the magus," considered the political texts divorced from not only Dee's non-political writings but also from any references in the political writings that might resonate with the occult strands of Dee's other writings and activities. Not that Sherman denied that there is a dimension of magic in some of Dee's writings. The myth that he rejected is Yates's construction of Dee as a philosopher-magician that "essentializes Dee by isolating him from his social and spatial circumstances" and overlooks other dimensions of Dee's activities and writings.[33]

32. Sherman, *John Dee*, 127.
33. Ibid., 12–13, 19–20.

In the course of Sherman's thorough treatment of all of the political texts in terms of their content, of their relation to other contemporary commonwealth advocacy, and of their positions on the contemporary issues confronting the court, Dee emerges as an accomplished textual and historical scholar, marshalling his learning to contribute to the policy debates at court regarding England's maritime enterprises and imperial claims.[34] Rather than seeing these writings in the context of an overarching "philosophy," Sherman treated them in their own right and within the political framework to which they were addressed.

Sherman's other important initiative entails contextualizing these writings in terms of Dee's library and reading practices as revealed in his marginalia. Sherman was the first scholar to benefit from Roberts and Watson's edition of Dee's library catalog and other book lists, which identified many of the surviving copies of Dee's books. Dee's library and marginalia are the subject of "the politics of reading." This yields a new Dee and places him more in the tradition of Renaissance humanist textual scholarship than in any Renaissance philosophical tradition, and serves as the foundation for the historical scholarship and the largely Ciceronian and Stoic cast of his political arguments. Rejecting Dee the "philosopher" and the "myth of the magus," Sherman saw Dee's socio-professional role as that of a scholarly mediator between "bodies of knowledge" and a "body of political readers," an "intelligencer" who used his library, reading, and scholarship to serve as a "retailer of special (often secret) knowledge" among "a wide range of courtly and commercial contacts."[35]

Despite Sherman's demur that he does "not claim to present 'the whole Dee,'" he explicitly challenged previous renditions of Dee without providing an alternative reading of the non-political writings. Sherman's Dee seems isolated, as if the other parts or minds of Dee do not exist.[36] Synthesizing this Dee with the Dee of the natural-philosophical texts is still an open legacy. While Sherman reclaimed some of Dee's manuscript writings, he did not treat them all. He touched only briefly on Dee's calendar reform and only for an example of the process of policy proposal and deliberation at court. He left the "angel

34. MacMillan, in several recent studies of Dee's *Brytanici imperii limites*, has emphasized Dee's important role in English overseas enterprises, in addition to his erudition in marshalling arguments from history, geography, and law in order to ground Elizabeth's title to new lands. He does not, however, address any overall interpretation of Dee. See Dee, *Limits of the British Empire*; MacMillan, "Discourse on History"; and MacMillan, "John Dee's 'Brytanici Imperii Limites.'"

35. Sherman, *John Dee*, xiii, 129.

36. Ibid., xiii. Lehrich, *Occult Mind*, 58–60, is clearly aware of this problem.

conversations" entirely to the side.[37] As we shall see, Sherman's rendition of Dee the courtier as successful and sought after and appreciated is rather rosy; it sidestepped the matter of growing financial straits that forced Dee to seek his fortunes elsewhere.

Contra Sherman's attempt to exorcize the "myth of the magus," the other major group of Dee's manuscript writings have militated against banishing Dee's occultist and magical interests, one might even say obsessions. These texts, some of which were published by Meric Casaubon in the seventeenth century, dealt with the "actions with spirits," or "spiritual diaries," or "conversations with angels." The activities embodied in this material dominated Woolley's biography and Wilding's account of Dee and Kelley's adventures; substantial excerpts from this material have been incorporated into Fenton's edition of Dee's diaries; and this part of Dee's legacy has become the object of an extensive occultist industry.[38] The earliest "spiritual diaries," those from December 1581 through May 1583, were not part of the manuscript that Casaubon published and have been edited and studied by Christopher Whitby, without, however, a consideration of their larger significance in Dee's career and natural philosophy.[39] Trying to understand Dee's conversations with angels, the puzzling *Monas hieroglyphica,* and more broadly his occultism and magic, is in varying ways the problem addressed in a trio of closely spaced scholarly works: Deborah E. Harkness's *John Dee's Conversations with Angels* (1999), Håkan Håkansson's *Seeing the Word: John Dee and Renaissance Occultism* (2001), and György Szönyi's *John Dee's Occultism: Magical Exaltation through Powerful Signs* (2004).[40] These studies present three somewhat different Dees, all very different from Sherman's.

While varying in their focus and emphasis—the conversations with angels for Harkness, the *Monas hieroglyphica* and its reflections in the conversations for Håkansson, and the conversations primarily for Szönyi—these three works reflect remarkable consensus about the central common themes of Dee's occultism that provide a coherence to his intellectual career. Prominent among these themes is Dee's quest to recover the divine language, the *lingua adamica,* the Word of God, or *logos.* For Dee, this language would give direct access to the

37. On Dee's calendar work, see Poole, "John Dee and the English Calendar"; and Poole, *Time's Alteration.*

38. Recall note 3 above.

39. Whitby, *John Dee's Actions with Spirits.*

40. See note 6 above.

divine realm by communicating God's pure redemptive message for mankind. The divine truths about the nature of the created world would then serve as a foundation for a unified comprehensive knowledge of all things. This restoration of the human connection with divinity would lead to the attainment of redemption and salvation through the reformation of man or, for Szönyi, his *exaltatio* (divinization). These divine truths about nature would also reveal the true alchemy, giving access to the *medicina dei* that would be a cure for the fallen, decayed state of nature and mankind. Through this special divine knowledge, Dee also became the purveyor of apocalyptic prophecies of the eschaton and the need for a religious-political reformation transforming all of Europe into a universal religious and political unity.

These themes emerge fully in the angelic conversations, but all three authors see them as expressions in various ways of a unified and coherent philosophical enterprise. So we return to coherent Dees, but these pictures of coherence have different characters arising from the different frameworks of each author. For Harkness, as a historian of science, the problematic is to present these activities as the culmination and "pinnacle" of his "Christian natural philosophy,"[41] whereas they are usually dismissed and ignored, or seen as a turn away from natural philosophy and science to religion, or interpreted as a manifestation of Dee just as going off the deep end, as it were. Thus, with "angelology as natural philosophy," the angel conversations are a legitimate part of the history of science. Pervading Dee's endeavors is the context of a late sixteenth-century humanist pessimism about the growing decay of nature and despair over the limits of human knowledge that frustrates his intellectual project to construct a universal science. Dee's natural philosophy writings are presented as a continuous and progressive search for the means to overcome and repair this decay, with the conversations growing out of and transcending the earlier works. The metaphysics of light that underlies Dee's optical theories of celestial influence in the *Propaedeumata aphoristica* is seen as including a spiritual dimension pointing to the crystallomancy of the angel conversations.[42] Likewise, the influence in the *Monas hieroglyphica* of the alchemical cabala of the *Voarchadumia* (1550) by the sixteenth-century Italian priest and alchemist Giovanni Agostino Pantheus also has spiritual dimensions and presages the "true cabala of nature" revealed by the

41. Harkness, *John Dee's Conversations with Angels*, 224–25.

42. Ibid., 74–77, 118–19.

angels for understanding the decayed book of nature.[43] Neither of these readings, however, is unproblematic. In the most recent study of the *Propaedeumata*, Steven Vanden Broecke argues for a much more restrictive reading than Harkness.[44] Likewise, Hilde Norrgrén, in the most thorough study of Dee's annotations to Pantheus to date, does not find anything spiritual in Pantheus's alchemy.[45]

Both Håkansson and Szönyi have disciplinary backgrounds different from the history of science—cultural studies and English and American studies, respectively—and both explicitly seek to free Dee from a science history perspective and from a progressive narrative of the Scientific Revolution. Both acknowledge that Dee was not static but had changing interests, appropriated new sources, and exhibited "variety" and "inconsistencies" in intellectual development.[46] Szönyi even takes issue with Harkness's consideration of Dee's earlier writings as "the genesis of the angel conversations."[47] Yet both, in their effort to recover and understand Dee's "otherness," present a coherent Dee, Håkansson by relating Dee's endeavors to Renaissance practices of symbolic exegesis, Szönyi by embedding Dee's magic in a coherent western esoteric tradition. Dee's engagement with "symbolic exegesis," for Håkansson, brings into relief the effort to grasp God's word as the "central motif" in his career as a natural philosopher.[48] For Szönyi, Dee's magic is not the culmination of natural philosophy and science, but their abandonment for the more elevated wisdom and understanding embodied in occultism as a separate discipline leading to *exaltatio*. Szönyi admittedly returns to a Yatesean "master narrative," while Håkansson's focus on exegetical practice emphasizes the heterogeneous, syncretic, and idiosyncratic nature of ideas about language and of Renaissance occultism, and of Dee's place within this context. In common with Szönyi, Håkansson weaves around Dee's effort to "See the Word" a comprehensive treatment of Dee's magic and occultism and their relation to ancient, medieval, and Renaissance ideas on these subjects.

These three approaches vary in their methodology and contextualization of Dee's occult activities. While all are post–Roberts and Watson, Harkness is most rigorous in tying Dee to demonstrably used sources and a sense of his-

43. Ibid., 88–89.

44. Broecke, *Limits of Influence*, 174n115.

45. Norrgrén, "Interpretation and the Hieroglyphic Monad," 217–45.

46. Håkansson, *Seeing the Word*, 332; and Szönyi, *John Dee's Occultism*, 181.

47. Håkansson, *Seeing the Word*, 60; and Szönyi, *John Dee's Occultism*, 181.

48. Håkansson, *Seeing the Word*, 332.

torical chronology. She thoroughly illuminates the practices of the angelic conversations and relates them to contemporary intellectual, cultural, and religious strands. Håkansson and Szönyi are both partisans of the "linguistic turn," engaging in freer interpretations, associations, and reconstructions, and are less systematic in referencing Dee's annotated books. Harkness does not embed Dee in a Yatesean tradition, but emphasizes associations with the Yatesean Renaissance canon of Ficino, Pico, Johannes Reuchlin (1455–1522), and Agrippa over any medieval influences, despite Stephen Clucas's work on the medieval models for Dee's magic.[49] While Szönyi recognizes a component of medieval ceremonial magic, he has an admittedly Yatesean take on the uniqueness of a Renaissance occultist view of man's position in the cosmos. Håkansson is very syncretic in finding associations with a broad range of ancient, biblical, early Christian, medieval, and Renaissance texts and ideas without positing a single coherent tradition.[50] One area where Szönyi adds a new perspective is his exploration, through his own eyes as an eastern European, of the attractions of eastern Europe for Dee, and the allure Dee and Kelley had for eastern Europeans, including Dee's continuing influence there into the seventeenth and eighteenth centuries. All three are also highly intellectualized approaches that sideline Dee's biographical realities, social context, and political and patronage activities. They also give little attention to the dynamic interaction of Dee and Kelley, which is by silence largely considered unproblematic, or to the changing social/political circumstances that lay behind the angelic conversations.

Despite the extensive work on Dee's occultism and the angelic conversations represented by these three studies, we are still left with the issue of the relation of these and Dee's larger natural philosophy to his actual career and pursuit of patronage, his political writings, his practical mathematics, and aspects of his science and natural philosophy that do not seem closely related to occultism. We still seem to have no clear reconciliation between the various Dees of Yates/French, Yewbrey, Clulee, Sherman, and Harkness/Håkansson/Szönyi. Is there one Dee of one mind or many Dees of many minds? Christopher Lehrich has

49. Clucas, "John Dee's Angelic Conversations and the *Ars Notoria*." Although only published in 2006, Clucas originally presented this paper at the 1995 conference where Harkness also presented.

50. Håkansson, more than most others, has taken cognizance of Clucas's work showing the extent to which Dee's angelic magic has extensive parallels with medieval theurgy and the Pseudo-Solomonic *ars notoria*. See Clucas, "John Dee's Angelic Conversations and the *Ars Notoria*," in his collection *Magic, Memory, and Natural Philosophy*, 1:231–73; and Clucas, "'Non est legendum sed inspicendum solum'" in ibid., 3:109–32.

suggested, apparently without awareness of Yewbrey's dissertation, that Dee's reference to "cosmopolitics" in the *Monas hieroglyphica* could link the political writings with occultism, but he does not develop this in detail.[51]

Into this mix is now added a new Dee in Glyn Parry's *The Arch-Conjuror of England: John Dee*, which is the most seriously scholarly biography of Dee to date.[52] Parry is a historian of Tudor Britain with interests in religious history, science, magic, witchcraft, and maritime history, but he is not a specialist exclusively in the history of science or cultural history. He brings to the study of Dee a solid command of Tudor politics and the workings of the court and ecclesiastical institutions based on a thorough command of archival sources more extensive than any previous student of Dee. With these tools, he has uncovered many overlooked or unknown aspects of Dee's career. Tudor history and politics, and the workings of the court—rather than the history of science or intellectual/cultural history—thus serve in his study to contextualize Dee's career. There is no systematic exposition of Dee's writings and ideas; they are only introduced as they intersect with episodes in the biography. Parry's work is therefore very much a political biography with the center of focus not so much Dee himself as his interactions and fortunes within his political context, especially the Elizabethan and eastern European courts. Contra Sherman, Parry argues that the key to these courts' interest in Dee was his occult philosophy and practice. Parry's "occult Dee" is the conjuror, the alchemist, and the prophet of an astrologically imminent new age of religious reform. Parry moves Dee to the mainstream of the Elizabethan world by evoking the extent to which Elizabeth and members of her court were immersed in occult philosophy, seeking sources of wealth through alchemy and insight into the future through divination. Dee appears in this context less as an active agent than as subject to and constantly responding to larger forces, particularly the political machinations at court, with many of his writings being responses to the needs of the moment. Since Dee offered esoteric wisdom rather than the practical benefits that had political currency, his uneven fortune with patronage was at the mercy of competition both from other occultists vying for favor and from the shifting power relations among factions at court. In a number of cases, Dee's temporary success at court was a function

51. Lehrich, *Occult Mind*, 50, 55–56.

52. Parry, *Arch-Conjuror*. Parry prepared for this in a number of earlier studies: Parry, "John Dee and the Elizabethan British Empire in Its European Context"; Parry, "English Magicians and the Crown of Poland"; and Parry, "John Dee's Occult Philosophy."

of his being used by one faction or another to further their agenda, only to be dropped once his usefulness passed.

What emerges is a unified Dee grounded in popular magic and the religious magic of the pre-reformed Catholic Church rather than a unity based on philosophical continuities among his writings or between his ideas and a larger Renaissance tradition. This was the basis for a lifelong practice of conjuring to evoke special knowledge that he attempted to parlay into support at court in England and in eastern Europe. Three key events in the early 1550s, but rooted in Dee's earlier experiences, were to shape Dee's life. First, the collapse of his father's fortunes left him impecunious and drove his ambitious search for patronage. Second, his ordination as a Catholic priest by Bishop Bonner and his collaboration in the interrogation of Protestants gave him some security during Mary's reign but tainted his reputation during Elizabeth's. More than a whim or expedient, his receptivity to Catholicism built on his preference for Catholic conservative mentors at St. John's College, Cambridge, and his choice of the University of Louvain for foreign study. Third, he was arrested for providing some occult service to Princess Elizabeth in 1555. This likely entailed astrology, but the charges also came to include "conjuring," suggesting to Parry that he was also summoning spirits at this point, a practice that might have had roots in popular magic, popular Catholicism, and his time at Cambridge. The records and specific content of this "conjuring" are scant and elusive, so I am less sure than Parry that Dee was doing spirit magic rather than some other form of divination prior to the explicit records that emerge in the 1570s. Dee's comment of 1561 or later on a text concerning the revelation of secrets through angels—"May God once grant this to us"—suggests that he had not had the experience of such revelations as early as the 1550s.[53] Dee was cleared of these charges, but the damage had been done, providing the touchstone for Dee's persistent reputation as a conjuror, with the ambivalent result that, while he was sought after for this magic, he carried a black reputation. A key feature of many of Dee's consultations by Elizabeth and others at court was his performance of this divinatory magic. This unites the occult and political/courtly dimensions of Dee's activities, because Parry argues that the political writings contain two messages: first, pragmatic policy advocacy accessible to the general public and courtiers and, second, allusions to an esoteric ideology accessible to those in the government

53. Gilly, "Tra Paracelso, Pelagio e Ganello," 1:277 (Italian), 288 (English).

and at court who could decipher it. Parry argues that what is cutting edge in the political writings is not Dee's claims for an Atlantic empire but his advocacy of claims of European empire addressing the context of the crisis of the revolt of the Netherlands. The tide of Dee's support at court waned along with the currency of these issues in the late 1580s. This was followed by two debacles: the scuttling by bishops of his calendar reform proposal, and his association with Albert Laski of Poland, who fell under the increasing displeasure of the political elite. These two developments intersect with the beginning of the documented practice of angel summoning via scryers in 1579. All the negatives were enhanced with the arrival of Edward Kelley as Dee's chief scryer, whose "weak grasp of political realities meant that at crucial moments the angels' commands he transmitted badly mislead Dee in dealing with the powerful."[54] The decline of his fortunes in England and the angels' promises about Laski led to Dee's seeking his fortunes in eastern Europe, where he had no better success. Thus, there is no radical break between what seems to be the height of Dee's success at court and political advocacy in the 1570s and his absorption in the angelic conversations in the 1580s. Parry's treatment of the conversations gives most attention to the chronological progression and dynamics of the Dee-Kelley relationship in the context of their fortunes at the courts in Poland and Bohemia in which Kelley's "weak grasp of political realities" continued to serve Dee poorly.

The consensus of studies over the last decade or so seems to be that Dee was fundamentally of one mind throughout his career, although Parry acknowledges elements of development and elaboration in his basic ideas over time. But none of these studies seems to present quite the same Dee, and we still cannot be sure which of these Dees, if any, is the right one. Can they be reconciled or synthesized? Juxtaposing Parry's revisionist portrait with previous studies illustrates the dimensions of this problem.

Like Yates, Parry's Dee is one, if not *the*, "characteristic philosopher of the Elizabethan age" and thoroughly mainstream with the central role that Parry posits for the occult in the Elizabethan court.[55] Parry, however, eschews characterizing Dee's occult philosophy in terms of any overarching, coherent intellectual formation, whether Yates's Hermetic tradition in its various guises or following those who have taken her lead, including French, Yewbrey, and Szönyi.

54. Parry, *Arch-Conjuror*, 147.

55. Yates, *Occult Philosophy in the Elizabethan Age*, 75.

What constitutes "occult philosophy" for Parry is more the acceptance of occult practices—alchemy, astrology, divination including spirit summoning—for their practical benefits than the expression of any specific intellectual formation. Further, Parry's definitive establishment of Dee's Catholic allegiances should put to rest the inclinations, explicit or implicit, of many previous studies to ally Dee with Protestantism. By focusing on Dee's political biography, however, Parry does not systematically address Dee's writings and thinking and, so, does not help to resolve the disparities among earlier studies concerning the motives, aims, and meanings for him that his writings expressed.

While Parry does not embrace the exalted concept of the magus, or its myth, that attracted Yates, French, Yewbrey, and Szönyi, he does give his own endorsement to Dee as a magus and presents his political advocacy and consultation by the Elizabethan elite as much more than Sherman's humanist scholarship and Ciceroneanism. There is a "secret" dimension to Dee's political writings that entails the previously unrecognized element of justification for English imperial dominion in northern Europe, not just the Atlantic, in conjunction with a sense of his prophetic mission proclaiming a new astrological age. In this, however, he eschews the "overblown" superstructure of a cosmopolitical philosophy erected by Yewbrey.

Parry also renders problematic what scholars do with the bulk of the records of the angelic conversations that issued from Kelley's scrying. Parry makes quite clear what he thinks was going on when he says that "Kelley could concoct for the angels, out of the English Bible and magical texts, the most sonorous, majestic, apocalyptic language, sufficient to overawe Dee's occasional doubts. The fact that sometimes Kelley became psychologically disturbed enough to believe he actually saw angels also helped to convince Dee."[56] The implication that Dee was being duped and manipulated drains, or at least renders problematic, the content of the revelations of much of their significance as reflections of any vision intrinsic or exclusive to Dee, as they have been employed by Harkness, Szönyi, and Håkansson. Dee's acceptance of the revelations as meaningful and as coming from angels may render them evidence for things that resonated with his dreams and beliefs, but the dependence of their fluctuating content on Kelley's erratic inspiration creates challenges for their interpretation. Kelley tends to lurk in the background as Dee's scryer, but he may have been less Dee's spiritual assistant

56. Parry, *Arch-Conjuror*, 147.

than an equal partner with his own interests and purposes. Future consideration will need to give due attention to Kelley and his role, which will be supported by the recent contributions of Michael Wilding.[57]

Another issue is the relation of Dee's work to science, an issue that has practically disappeared from comprehensive treatments of Dee in the last decade or so. Particular specialized studies suggest that Dee was consequentially engaged with current developments in natural knowledge, mathematics, and science. Integrating this work with a Dee of occult philosophy and Dee, the practical magus at princely courts, is thus a challenge. Even at four hundred then, Dee is still an enigma that calls for further study.

Works Cited

Barone, Robert W. *A Reputation History of John Dee, 1527–1609: The Life of an Elizabethan Intellectual.* Lewiston, ME: Edwin Mellon Press, 2009.

Broecke, Steven Vanden. *The Limits of Influence: Pico, Louvain, and the Crisis of Renaissance Astrology.* Leiden: Brill, 2003.

Calder, Ian R. F. "John Dee Studied as an English Neoplatonist." 2 vols. PhD diss., Warburg Institute, London University, 1952.

Clucas, Stephen, ed. *John Dee: Interdisciplinary Studies in English Renaissance Thought.* International Archives of the History of Ideas. Dordrecht: Springer, 2006.

———. "John Dee's Angelic Conversations and the *Ars Notoria*: Renaissance Magic and Medieval Theurgy." In Clucas, *John Dee*, 231–73, and in Clucas, *Magic, Memory and Natural Philosophy in the Sixteenth and Seventeenth Centuries,* 1:231–73.

———. *Magic, Memory and Natural Philosophy in the Sixteenth and Seventeenth Centuries.* Burlington, VT: Ashgate, 2011.

Clulee, Nicholas H. *John Dee's Natural Philosophy: Between Science and Religion.* London: Routledge, 1988.

Deacon, Richard. *John Dee: Scientists, Geographer, Astrologer and Secret Agent to Elizabeth I.* London: Frederick Muller, 1968.

57. Wilding, "Biography of Edward Kelly." Wilding, *Raising Spirits,* is largely a transcription of the angelic conversations tied together with narrative context and commentary by Wilding.

Dee, John. *The Diaries of John Dee*. Edited by Edward Fenton. Oxfordshire: Day Books, 1998.

———. *The Limits of the British Empire*. Edited by Kenneth MacMillan and Jennifer Abeles. Westport, CT: Praeger, 2004.

———. *A True and Faithful Relation of What Passed for Many Years between Dr. John Dee and Some Spirits*. Edited by Meric Casaubon. London: D. Maxwell for T. Gartnwait, 1659.

Fell-Smith, Charlotte. *John Dee (1527–1608)*. London: Constable and Co., 1909.

French, Peter J. *John Dee: The World of an Elizabethan Magus*. London: Routledge and Kegan Paul, 1972.

Gilly, Carlos. "Tra Paracelso, Pelagio e Ganello: L'ermetismo in John Dee." [Between Paracelsus, Pelagius and Ganellus: Hermeticism in John Dee.] In *Magia, alchimia, scienza dal '400 al '700/Magic, Alchemy and Science 15th–18th Centuries*, 2 vols., by Carlos Gilly and Cis van Heertun. Firenza: Centro Di, 2002, 1:275–85 [Italian], 1:286–94 [English].

Håkansson, Håkan. *Seeing the Word: John Dee and Renaissance Occultism*. Lund: Lund Universitet, 2001.

Harkness, Deborah E. *John Dee's Conversations with Angels: Cabala, Alchemy, and the End of Nature*. Cambridge: Cambridge University Press, 1999.

Hort, Gertrude M. *Dr. John Dee, Elizabethan Mystic and Astrologer*. London: William Rider and Son, 1922.

Johnson, Francis R. *Astronomical Thought in Renaissance England*. Baltimore: Johns Hopkins University Press, 1937.

Lehrich, Christopher I. *The Occult Mind: Magic in Theory and Practice*. Ithaca, NY: Cornell University Press, 2007.

MacMillan, Kenneth. "Discourse on History, Geography, and Law: John Dee and the Limits of the British Empire, 1576–80." *Canadian Journal of History* 36 (2001): 1–25.

MacMillan, Kenneth. "John Dee's 'Brytanici Imperii Limites.'" *Huntington Library Quarterly* 64 (2001): 151–60.

Norrgrén, Hilde. "Interpretation and the Hieroglyphic Monad: John Dee's Reading of Pantheus's *Voarchadumia*." *Ambix* 52 (2005): 217–45.

Parry, Glyn. "English Magicians and the Crown of Poland: John Dee, Edward Kelly and Albrecht Laski, 1583–1585." *New Zealand Slavonic Journal* 42 (2008): 79–100.

———. "John Dee and the Elizabethan British Empire in Its European Context." *Historical Journal* 49, no. 3 (2006): 643–75.

———. "John Dee's Occult Philosophy and Late Tudor Political Culture." Paper presented at John Dee Quatercentenary Conference. St. John's College, Cambridge, 2009.

———. *The Arch-Conjuror of England: John Dee*. London: Yale University Press, 2011.

Poole, Robert. "John Dee and the English Calendar: Science, Religion and Empire." In *Time's Alteration: Calendar Reform in Early Modern England*. London: UCL Press, 1998. Online at http://www.hermetic.ch/cal_stud/dee.html.

Rampling, Jennifer. "John Dee and the Sciences: Early Modern Networks of Knowledge." *Studies in the History and Philosophy of Science* 43 (2012): 432–549.

Roberts, Julian, and Andrew G. Watson, eds. *John Dee's Library Catalogue*. London: Bibliographical Society, 1990.

Sherman, William H. *John Dee: The Politics of Reading and Writing in the English Renaissance*. Amherst: University of Massachusetts Press, 1995.

Szönyi, György E. *John Dee's Occultism: Magical Exaltation through Powerful Signs*. Albany: State University of New York Press, 2004.

Szönyi, György E., and Rowland Wymer. "John Dee as a Cultural Hero." *European Journal of English Studies* 15, no. 3 (2011): 189–209.

Taylor, Eva G. R. *Mathematical Practitioners of Tudor and Stuart England*. Cambridge: Cambridge University Press, 1954.

———. *Tudor Geography, 1485–1583*. London: Methuen, 1930.

Whitby, Christopher. *John Dee's Actions with Spirits, 22 December 1581 to 23 May 1583*. New York: Garland Publishing, 1988.

Wilding, Michael. "A Biography of Edward Kelly, the English Alchemist and Associate of Dr. John Dee." In *Mystical Metal of Gold: Essays on Alchemy and Renaissance Culture*, edited by Stanton J. Linden, 35–89. New York: AMS Press, 2007.

———. *Raising Spirits, Making Gold, and Swapping Wives: The True Adventures of Dr. John Dee and Sir Edward Kelly*. Beeston, UK: Shoestring Press, 1999.

Woolley, Benjamin. *The Queen's Conjurer: The Science and Magic of Dr. John Dee, Adviser to Queen Elizabeth I*. New York: Henry Holt, 2001.

Wymer, Rowland. "Dr. Dee and Derek Jarman: The Art of Mirrors." Paper presented at John Dee Quatercentenary Conference. St. John's College, Cambridge, 2009.

Yates, Frances A. "The Hermetic Tradition in Renaissance Science." In *Art, Science, and History in the Renaissance*, edited by Charles S. Singleton, 255–74. Baltimore: Johns Hopkins University Press, 1968.

Yates, Frances A. *Giordano Bruno and the Hermetic Tradition*. Chicago: University of Chicago Press, 1964.

———. *The Occult Philosophy in the Elizabethan Age*. London: Routledge and Kegan Paul, 1969.

Yewbrey, Graham. "John Dee and the 'Sidney Group': Cosmopolitics and Protestant 'Activism' in the 1570s." PhD diss., University of Hull, 1981.

Chapter 11
On the Imagery of Nature in the Late Medieval and Early Modern Periods*

Heinz Schott

*Dedicated to the memory of Michael Walton.***

The misogynic legacy is well known in the medical and cultural historiography. Women, deemed physically and morally weaker and less healthy than men, were characterized in the terms of ancient Greek humoral pathology as being cold and wet, having less life energy, and associated with the moon and the dark earth. Men, on the other hand, were understood as being warm and dry, relatively powerful, and associated with the sun. While this assessment was differently

*Allen G. Debus, one of the greatest adepts of Paracelsianism in the historiography of science and medicine, continued and completed the "vast studies" of Walter Pagel, whom he considered "the doyen of Paracelsian scholars" and who "served as the solid bedrock" for Debus's interpretations of the history of Renaissance and early modern chymistry. (See Debus, *Chemical Philosophy,* xvi.) Walter Pagel's analyses were also quite fundamental for my own understanding of Paracelsus and Jan Baptist van Helmont, but the present contribution is embedded in a perspective that goes beyond the early modern history of Paracelsian alchemy and natural magic. This chapter stresses the learned imagery of the (female) Nature and its sources in the early modern age as a whole. See also note 2 below.

**Mike Walton kindly invited me to contribute to this volume on the intellectual legacy of Allen Debus. Without Mike's patient and friendly insistence I would not have accomplished it. In 2012, he chaired the session at the SCSC Annual Meeting in Cincinnati, where both of us spoke. My paper was, in some sense, an abstract of the present chapter. It was our last meeting.

articulated throughout the modern era,[1] it is remarkable that historiographically the other side of the coin was often neglected, namely, the explicit gynephilic attitude of distinguished Renaissance and early modern scholars in regard to the personification of nature as a cosmological female figure. This chapter explores that alternate attitude.[2]

Renaissance scholars intensively discussed two important traits of Nature, or what they termed "Natura" in Latin: first, Natura as a sort of Holy Scripture, which had to be studied and deciphered in order to learn "her" secret language and, second, Natura as a divine female figure, a goddess, who would only communicate her secrets when she was gently and reverently handled and not violently unveiled. The traits were essentially inseparable. It was the endeavor of natural magic (or *magia naturalis*) to find out the secrets of nature.[3] Those secrets, however, were not only to be admired; Nature had also to be analyzed and scientifically investigated as well as imitated and completed. In the early modern period, the newly established scientific academies committed themselves to such an approach. For example, the Italian physician Giambattista della Porta (1535?–1615), who worked in Naples and published his groundbreaking work *Magia naturalis* in 1558, founded one of the first natural scientific academies in Europe, namely, the Academia Secretorum Naturae (or the Accademia dei Segreti) in 1560.[4] Its sole mission was the exploration of Nature, and members were accepted only when they "could present a hitherto unknown secret in the field of medicine or the mechanical arts."[5] So-called encyclopedias of secrets were thus composed, and natural research was explicitly seen as a hunt.[6] The final task, however, was the rational explanation of those natural secrets, a task that the Italian physician and polymath Girolamo Cardano (1501–76) tried to make more acceptable through his concept of subtlety (*subtilitas*).

1. These differences are reflected in the contrast between the early modern witch hunt and the views expressed by the renowned twentieth-century German psychiatrist Paul Möbius, who proclaimed the "physiological mental deficiency of the woman." See Möbius, *Ueber den physiologischen*, title page.

2. These ideas are explored further in my recent book *Magie der Natur*. The present chapter, however, focuses on the imagery of Nature and the corresponding gynephilic ideas as they were propagated by Agrippa von Nettesheim. It does not consider, as does my book, Paracelsus's appreciation of Nature as a female magician (*maga*) nor does it treat the Paracelsians who also adopted this idea.

3. Compare Eamon, *Science and the Secrets of Nature.*

4. Ibid., 194–233.

5. "wer ein bisher unbekanntes Geheimnis auf dem Gebiet der Medizin oder der mechanischen Künste präsentieren konnte"; Gronemeyer, *Optische Magie*, 87 (my translation).

6. Ibid., 273–85.

In the occidental tradition, there was a certain ambivalence about making natural science a central theme, especially in the early modern period. On the one hand, it was not the custom to promote research activities freely; on the other hand, it was held that when they were carried out, such activities should be pursued humbly. Insofar as God originally chose to hide Nature's secrets, natural science and research (*curiositas*) seemed to be a criminal act, in which scientists violated Nature's integrity by forcible experimentation. The biblical narrative of God's interdiction to eat the fruits of the tree of knowledge corresponded to the Greek legend reported by Plutarch that it was prohibited to wrest the veil from the statue of the Goddess Isis in Sais. Moreover, the church father Lactantius (ca. 240–ca. 320) had stressed that God had created Adam in a final step so that Adam, the prototypic man, would not acquire knowledge of the act of creation. "In confirmation of this, the popular image of the goddess Natura implied that nature covers herself with a veil in order to hide her secrets from mortals."[7] Thus, mortals should be excluded from the secrets of divine Nature.

It is remarkable that the historiography of religion—and of culture more generally—ignores the concept of Natura. Even handbooks like that edited by Mircea Eliade and Ioan Culianu fail to mention it.[8] And whereas the images of goddesses—and of godlike women such as Sophia and Mary—were repeatedly discussed in Jewish studies and theology, their importance for the history of science and medicine per se were overlooked. It is thus a crucial question whether Natura was already perceived in the early kabbalah of the eleventh and twelfth centuries. Such ideas contributed to later natural philosophy and science, especially in regard to Mary and the Shekhina.[9] This chapter will explore some of the early modern iconography and emblematics that personified "veiled nature" in feminine form.

Natura as Nourishing Mother

The German teacher of Romance languages, Ernst Robert Curtius, addressed the "Goddess Natura" in a special chapter of his book *Europäische Literatur und*

7. Eamon, *Science and the Secrets of Nature*, 59.
8. See Eliade and Culianu, *Handbuch der Religionen.*
9. Schäfer, *Mirror of His Beauty.*

lateinisches Mittelalter.[10] There he offered rich sources for the occidental conceptions of Nature and of their imagery and iconography. Although Curtius did not specifically consider medicine and magic, his references nevertheless became essential for an understanding of Nature as a (female) magic artist. His classification of Natura in the sense of a "historische Topik" (historical topic), as he put it, focused on the topos of "Natura mater generationis" (Nature, generating mother) that originated in late antiquity. Although he distinguished fourteen categories,[11] three are of particular importance for the present study: "Natura artifex mundi" (Nature, world creator), "Natura Dei serva" (Nature, servant of God), and "Natura altrix hominum" (Nature, provider for men). As a "servant of God," Nature was equally appreciated as a "world artist" or "world creator" who shaped and produced everything.[12] The famous savant at the School of Chartres Alanus ab Insulis (ca. 1116/17–1202/3, also known as Alain de Lille) broached the issue of "the creation of the perfect man" at the end of the twelfth century in his *Anticlaudian.*[13] According to Curtius, Alanus held that "Christ's act of redemption does not seem to have given relief; relief can only be achieved by the creation of a new man; with him the golden age will return."[14] (As the statement is heretical, Curtius wryly noted that today Alanus, "the *doctor universalis,* would have failed to secure a clerical printing-license," given such views.[15]) Similarly, the anthroposophist Wilhelm Rath, an expert on the School of Chartres, underscored the obvious "secret" that there was "such an inner link between the Goddess Natura and the virgin mother of the gospels," although this was never explicit in the literature of the school.[16]

10. Curtius, *Europäische Literatur und lateinisches Mittelalter,* 116–37.

11. His fourteen categories were (1) Natura artifex mundi, (2) Natura parens omnium rerum, (3) Natura domina omnium rerum, (4) Natura plasmatrix terrae et locorum, (5) Natura dotatrix hominum, (6) Natura formatrix hominum, (7) Natura domitrix feritatis et mater pietatis, (8) Natura discretis veteris tumultus, (9) Natura deos Iovi famulos tradens, (10) Natura plangens, (11) Natura de Phoenicis avis immortalitate laborans, (12) Natura Pronuba, (13) Natura Dei serva, and (14) Natura altrix hominum; ibid., 182–84; and Curtius, "Zur Literaturästhetik des Mittelalters II."

12. Modersohn, *Natura als Göttin im Mittelalter,* 17.

13. "die Schöpfung des vollkommenen Menschen"; Curtius, *Europäische Literatur und lateinisches Mittelalter,* 131 (my translation).

14. "[D]ie Erlösungstat Christi scheint nicht geholfen zu haben; helfen kann nur die Schaffung. eines neuen Menschen; mit ihm kommt das goldene Zeitalter wieder"; ibid., 131 (my translation).

15. "Die kirchliche Druck-Erlaubnis würde der *doctor universalis* heute nicht erhalten"; ibid. (my translation).

16. "diese innere Beziehung der Göttin Natura zu der jungfräulichen Mutter der Evangelien"; Rath, "Einleitung," 76 (my translation).

Clerical dogma simply would not have allowed such an association. *Natura altrix hominum*—that is, Nature as a provider for man—implied "provider" in the sense both of a physical and a mental provider or *alma mater*. She was thus viewed simultaneously as an educator, a teacher, and a leader within the scope of magic and alchemy.

Based on Curtius's findings, this chapter will analyze some personifications of Natura as illuminating medicine and natural science in early modern natural magic. For example, the French sculptor Jean Baptiste Boudard (1710–68) edited an *Iconologie* in three volumes that included 630 engravings with legends that represents a remarkable repository of images. Boudard presented the personifications of diverse natural qualities, of arts, of sciences, and of human virtues, mostly in female figures. For instance, Nature appeared as a statue of the many-breasted Isis with birds sitting on her extended arms and leaping beasts decorating the pedestal (see fig. 11.1). For the Egyptians, the goddess's veiled head implied "that the most perfect secrets of nature are reserved for the Creator."[17] Since Nature's role as provider was reminiscent of Mary, the Mother of God, an iconographically blended female figure evolved.

Undoubtedly, Mary represented an ideal of the "wet nurse of God." In the words of the thirteenth-century German poet Konrad von Würzburg (d. 1287),

> The godly spirit from above
> chose you to be his bride
> and wanted you especially
> to ignite and inflame you
> as a wet nurse of God.[18]

Wilhelm Grimm commented that the milk of this wet nurse was considered an effective remedy. An old legend "that she would appear to the sick and feed them refreshing drops of milk from her breast or let them drink out of it, whereof they were soon healed, seems also to refer to the idea of the All-Mother Artemis."[19]

17. "que les plus parfaits secrets de la nature sont réservés au Créateur"; Boudard, *Iconologie tirée de divers auteurs*, 3:1 (my translation).

18. "do hete dich zů brůt erwelt / der frone geist her under / und wolde dich besunder / als ein gotheitsammen / entzünden und enflammen"; Würzburg, *Die goldene Schmiede*, 37–38, lines 290–94 (my translation).

19. "daß sie Frommen, die krank darnieder lagen, erschienen, und aus ihrer Brust labende Milchtropfen zugesprützt, oder sie daran hat trinken lassen, wovon sie alsbald genesen, scheint sich gleichfalls auf die Idee der All-Mutter, Artemis, zu beziehen"; ibid., 15 (my translation).

ICONOLOGIE 1

NATURE.

LA nature étant l'aſſemblage, & la perpétuation de tous les êtres créés, ſe repréſente par une jeune femme, dont la partie inferieure eſt priſe dans une gaine, qui eſt ornée de differentes ſortes d'animaux terreſtres, & ſur ſes bras qui ſont étendus ſont divers ſortes d'oiſeaux; elle a pluſieures mamelles pleines de lait. Sa tête couverte d'un voile ſignifie, ſelon l'opinion des Egyptiens, que les plus parfaits ſecrets de la nature ſont réſervés au Créateur.

NATURA.

LA natura, ch'è il compoſto, e la perpetuazione di tutte le coſe create, viene rappreſentata ſotto l'immagine di giovane donna, nella parte inferiore rinſerrata dentro una ſpecie di termine ornato nei lati di differenti ſorte d'animali terreſtri. Reca ella ſulle braccia ſteſe diverſe ſorte d'uccelli, e le ſi veggono varie poppe ſul petto gravi di latte. Ha il capo velato, per moſtrare, ſecondo il parer degli Egizj, che i più importanti di lei ſecreti ſono riſervati al ſolo Creatore.

Tom. III. A

Figure 11.1: "Nature," from Jean Baptiste Boudard, *Iconologie tirée de divers auteurs: Ouvrage utile aux gens de lettres, aux poëtes, aux artistes, et généralement à tous les amateurs des Beaux arts* (Parme: Sebstverl.; [Drucker:] Carmignani, 1759), vol. 3, fig. 1 © Herzog August Bibliothek Wolfenbüttel: Sign. M: Uk 4° 23.

Figure 11.2: Emblem XVIII, from Guillaume de La Perrière, *Le Théâtre des bons engins* . . . (Paris: Janot, 1539). © Herzog August Bibliothek Wolfenbüttel: Sign. M: Lm 2057 (PURL: http://diglib.hab.de/drucke/lm-2057/start.htm?image=00050).

Emblems of Nature

Emblems of Nature particularly flourished in sixteenth-century texts, wherein the image of a woman appeared in various figures, clothes, and poses. The Italian humanist Andrea Alciato, for example, introduced the classical form in his *Emblematum liber* published in 1531: image (*ikon, pictura, imago,* or *symbolon*) following the design of the hieroglyphics; lemma (that is, title); motto (*inscriptio* in the sense of *impresa*); and epigram (subscription).[20] Baroque emblematics—or *Sinnbildkunst*—developed as a consequence of this emblem book, with Nature often depicted as Venus.[21]

The French humanist Guillaume de La Perrière (ca. 1500–ca. 1565), for example, presented a nature emblem in the first edition of *Le Théâtre des bons engins* published in 1539 (fig. 11.2). It displayed a naked woman with wavy hair standing in a mountainous landscape and holding a large key in her right hand. Her left forefinger points to her mouth and her right foot rests on a turtle. The 1545 edition shows quite a different composition of the picture, however. There, below the title "Ce qu'est requis en la femme prudente" (What is required in the prudent woman), Venus sits partially clothed on a cushion beneath a baldachin. Key, turtle, and finger on the lips remain arranged as in the earlier representation of 1539 (fig. 11.3). The explanatory text provided the symbolic interpretation of the various symbols employed in both illustrations. The goddess represents the proper attitude of an honorable woman, while the turtle reflects ties to location, the forefinger indicates the avoidance of loquaciousness, and the key signifies the sagacious administration of the goods of the husband.

A similar picture can be found in Andrea Alciato's emblem book, *Mulierum famam non formam vulgatam esse oportere* (fig. 11.4). Enveloped only by a veil, Venus holds an apple instead of a key. Her adumbrated vulva and voluptuous facial expression characterize her as Venus at the same time that she suggests to the observer Eve beside the tree of life in paradise. Although Adam and the snake are both absent, the sun shines above, symbolizing God. Fostering Venus is thus a fostering mother.

The upgrading of Natura to a divine presence that shows man the guidelines of his own evolution was mainly influenced by Alanus ab Insulis's twelfth-

20. Wüseke, *Freimaurerische Bezüge,* 6.
21. Ibid., 9.

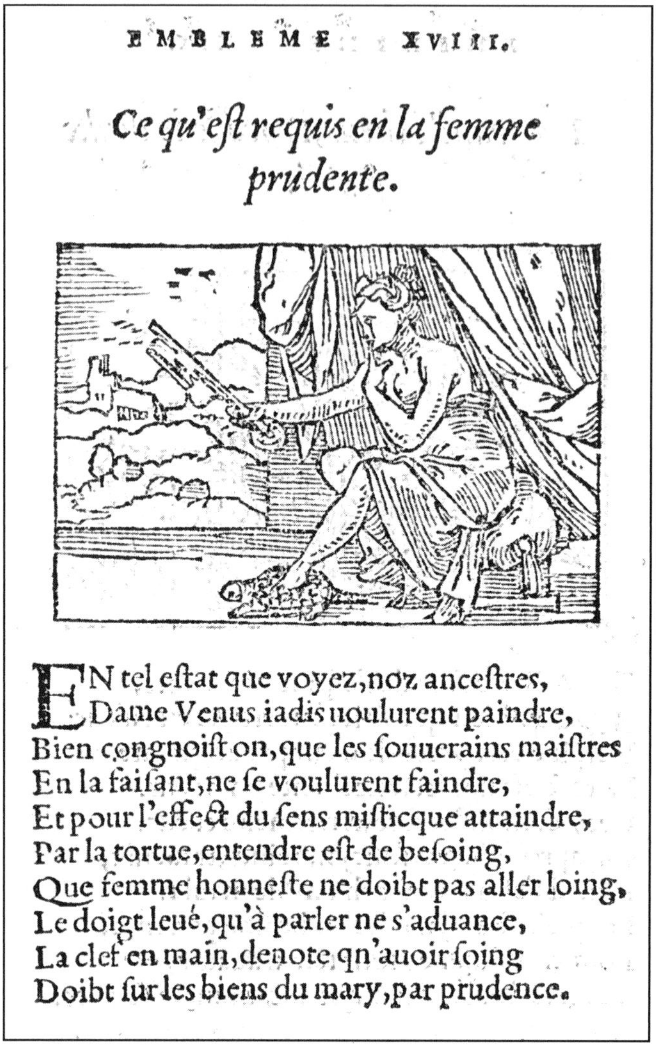

EMBLEME XVIII.

Ce qu'eſt requis en la femme prudente.

EN tel eſtat que voyez, noz anceſtres,
Dame Venus iadis uoulurent paindre,
Bien congnoiſt on, que les ſouuerains maiſtres
En la faiſant, ne ſe voulurent faindre,
Et pour l'effect du ſens miſticque attaindre,
Par la tortue, entendre eſt de beſoing,
Que femme honneſte ne doibt pas aller loing,
Le doigt leué, qu'à parler ne s'aduance,
La clef en main, denote qu'auoir ſoing
Doibt ſur les biens du mary, par prudence.

Figure 11.3: Emblem XVIII, from Guillaume de La Perrière, *Le Théâtre des bons engins* . . . (Paris: de Tournes, 1545). © Herzog August Bibliothek Wolfenbüttel Sign. A: 162.3 Eth. (b) (PURL: http://diglib.hab.de/drucke/162-3-eth-b/start.htm?image=00031).

century *Anticlaudian*. In the fifteenth and sixteenth centuries, a series of carpets was based on this work, in which the figure of Natura is often displayed, but never in a manner that would supplant her other allegorical renderings.[22] The wall carpet exhibited today in the Electoral Palatinate Museum of Heidelberg provides a particularly impressive example of this iconography. The tapes-

22. Kemp, "Natura," 11.

MVLIERIS FAMAM NON
formam uulgatam eſſe
oportere.

Alma Venus quæ nam hæc facies quid denotat illa,
Teſtudo molli quam pede diua premis?
Me ſic effinxit phidias, ſexumq; referri,
Fœmineum noſtra iuſſit ab effigie,
Quodq; manere domi & tacitas decet eſſe puellas,
Suppoſuit pedibus talia ſigna meis.

Figure 11.4: Emblem F 1, from Andrea Alciato [Andreas Alciatus], *Emblematum liber* (Augsburg, 1531, Emblematisches Cabinet 10. Reprint Hildesheim; New York: Olms, 1977).

try, which was probably a part of the inheritance of the Elector Palatine Otto Henry (1502–59), was executed in 1531. Entitled "The way of earthly prudence to heavenly wisdom" (fig. 11.5), this so-called Prudentia-Teppich has been the subject of detailed studies by art historians.[23] The constellation of the three figures—Natura, Prudentia, and Sapientia—is essential to the whole and is highly interesting. Prudentia is seated on a throne in the center of the carpet and then ascends on the right-hand side to take over the sceptre from Sapientia, who floats down as the goddess of heaven. On the left-hand side,

23. Stemper, "Die Wandteppiche"; and compare Stemper, "Der Prudentia-Teppich."

Figure 11.5: Prudentia-Teppich (Prudentia Carpet). Courtesy of Kurpfälzisches Museum der Stadt Heidelberg: Inv. Nr. Te 25.

Natura stands with her head directed upwards beneath Jove, who, as the counterpart of Sapientia, floats as the king of heaven. Closer to the foreground, Natura directs horses ridden by angels, who point with their shields to the senses. Placella, a crowned female figure in the foreground, cannot be interpreted definitely, but perhaps represents a play on the Latin verb *placere* and symbolizes courtesy and agreeableness.[24] It is unclear whether the composition corresponded to some pedagogic-allegorical program aimed at teaching people how to perfect themselves, but this might be assumed, given the male observer in the foreground on the right-hand side who observes the activities of the female majesties.

24. Stemper, "Die Wandteppiche," 165.

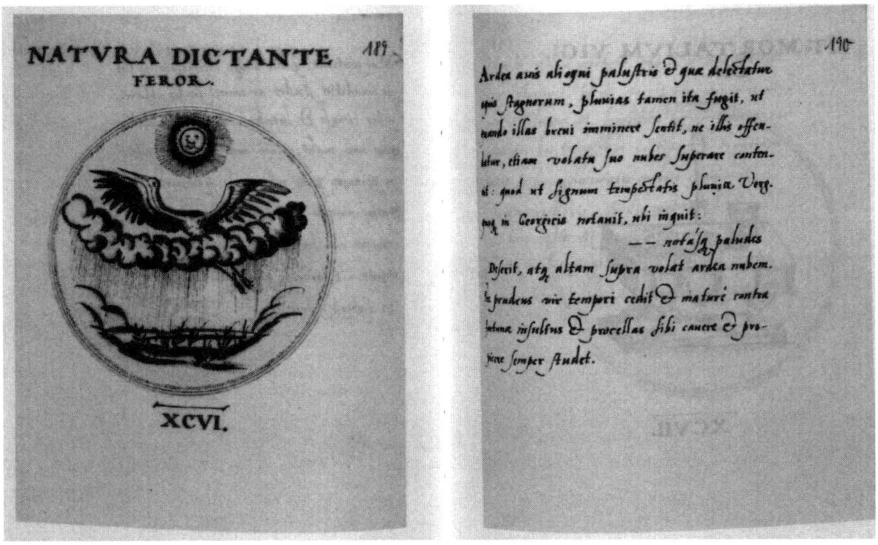

Figure 11.6: Emblem 46 (p. 192), from Joachim Cameraris d. J., *Symbola et emblemata tam moralia quam sycr: Die handschriftlichen Embleme von 1587*, edited by Wolfgang Harms and Gilbert Heß (Tübingen: Niebmeyer, 2009), 512.

The work on symbols and emblems of the German physician and naturist Joachim Camerarius (1534–98) was published in several volumes in 1600.[25] The captions of some emblems refer explicitly to Natura, even though she is not present in the usual representation as a woman. For example, the emblem "Naturae maturavit opus" (The work of Nature has matured) shows an arbor from which the leaves are falling, and the caption reads, "Cur adeò, huic fatum vim properare, doles?" (Why do you suffer so hard that you are thus doomed?).[26] The emblem "Natura potentior ars" (Nature has more power than art), on the other hand, shows a bear licking a fruit or stone in order to polish its surface and bears the caption, "Ars polit, haud fingit, natura utrumque ministrat" (Art polishes, but does not create, Nature supports in either way). Camerarius's collection of handwritten emblems was published in 2009 and is not identical with the above-mentioned edition. It includes the emblem "Naturae dictante feror" (At the behest of Nature I arise) (fig. 11.6), which portrays a heron penetrating raining clouds as they wet the earth, symbolizing a sort of

25. Camerarius, *Symbola et emblemata* (1986).
26. Ibid., pt. 1, no. 65.

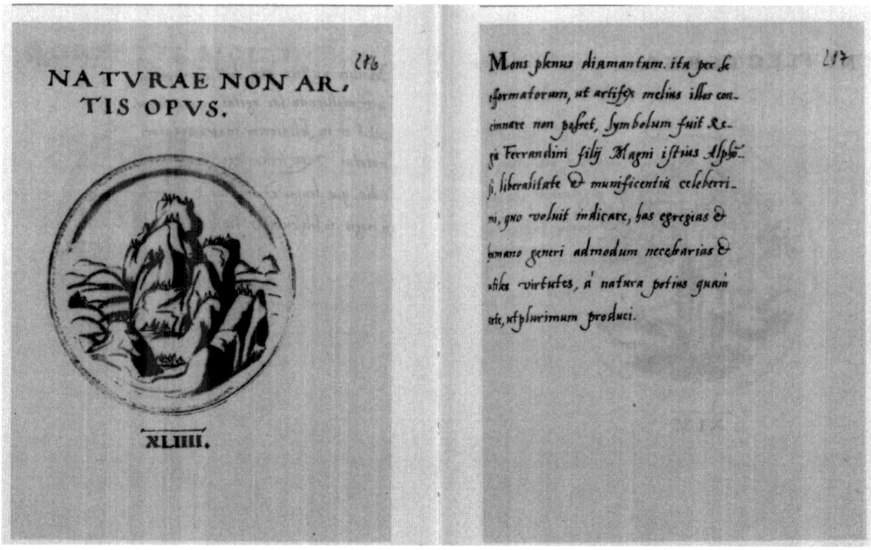

Figure 11.7: Emblem 44 (p. 290), from Joachim Camerarius d. J., *Symbola et emblemata tam moralia quam sacra: Die handschriftlichen Embleme von 1587*, edited by Wolfgang Harms and Gilbert Heß (Tübingen: Niemeyer, 2009), 561.

border of the sublunary world as the beaming sun shines from above. The legs of the heron protrude through the clouds, thus connecting the worlds above and below. In the printed version, the emblem was furnished with a new and more elaborate emblem that depicts the heron flying above the rain clouds.[27] Another emblem—omitted in the printed version—bore the title "Naturae non artis opus" (A work of Nature, not created by art) and displays rocky mountains decorated by diamonds (fig. 11.7). It may be interpreted as a demonstration of the virtue endowed by Nature and not acquired through art.[28]

Baroque emblematics such as those just discussed were of special interest to the Freemasons.[29] For example, one copper engraving portrays nine brothers sitting around a table (fig. 11.8) with a beam of light originating from the eye of God. The beam forms a triangle that carries the inscription, "Lux lucet in tenebris" (The light shines in the darkness) and hits a mirror. The beam reflected downward, on the other hand, has the inscription, "Tenebrae eam

27. Camerarius, *Symbola et emblemata* (2009), 513.
28. Ibid., 562.
29. Wüseke, *Freimaurerische Bezüge*.

Figure 11.8: Copper engraving, reproduced from Eduard B. Wüseke, *Freimaurerische Bezüge zur barocken Emblematik: Kommunikationszeichen an der Schwelle zur Neuzeit* (Münster: Bauhütten Verl., 1990).

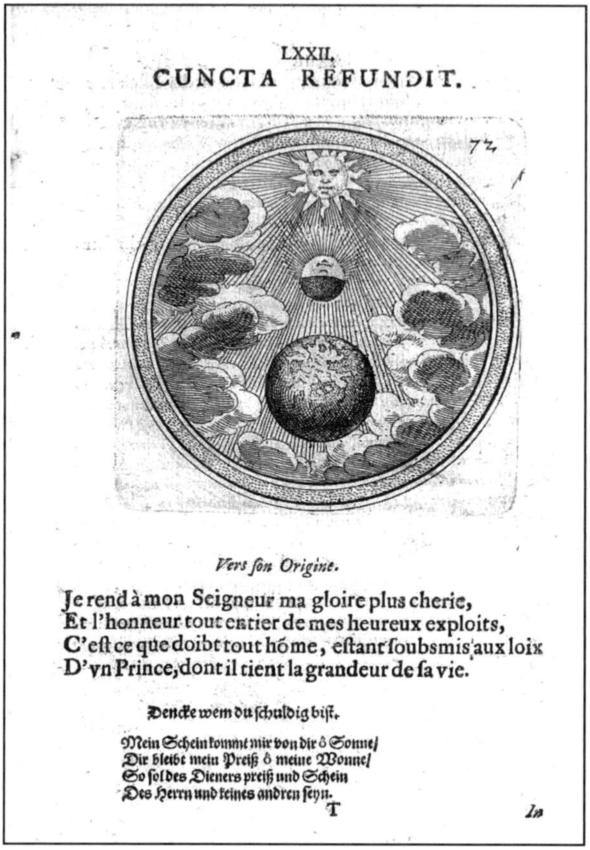

LXXII.
CUNCTA REFUNDIT.

Vers son Origine.

Je rend à mon Seigneur ma gloire plus cherie,
Et l'honneur tout entier de mes heureux exploits,
C'est ce que doibt tout hôme, estant soubsmis aux loix
D'vn Prince, dont il tient la grandeur de sa vie.

Dencke wem du schuldig bist.

Mein Schein kommt mir von dir ô Sonne/
Dir bleibt mein Preiß ô meine Wonne/
So sol des Dieners preiß und Schein
Des Herrn und keines andren seyn.

T In

Figure 11.9: "Cuncta refundit," from Julius Wilhelm Zincgref, *Emblematum Ethico-Politicorum Centuria* (Heidelbergae: Ammonia, 1666), p. LXXII. © Herzog August Bibliothek Wolfenbüttel Sign. M: Li 10083 (PURL: http://diglib.hab.de/drucke/li-10083/start.htm?image=00165).

non comprehenderunt" (The darkness did not comprehend it). Traditionally, the mirror symbolized the act of self-reflection and insight, while the eye of God stood for the source of all light and power. In this illustration, Natura does not appear personified, but rather as the reflection of the beam of light. In other illustrations, however, a female Natura actually holds a mirror, thus herself becoming a mirror of the divine light. This type of trope appears in many variations. For example, another similar emblem, "Cuncta refundit [She transfers all things]," by Julius Wilhelm Zincgref (1591–1635) shows very clearly the hierarchy of God, Nature, and Man in its vertical ordering of sun, moon, and earth (fig. 11.9). Here, the moon is in the position of a mediating Nature, while the sun is identified with Christ, the light of the world. At the same time, the moon becomes a symbol for the transmission of the divine light to man. Accordingly, the subscription explains,

Roll. II Nr. 35 **Zaumzeug** *und Winkelmaß in der Hand einer weiblichen Gestalt*

SERVA MODVM.

Maßhalten im Glück

Mens SERVARE MODUM, rebus sufflata secundis,
Nescit, et affectus fraena tenere sui.

Wahre das Maß!

Das Herz weiß das Maß nicht zu wahren und den Gefühlen Zügel anzulegen, wenn es vom Glückshauch getroffen wird.

Figure 11.10: "Serva Modum," from Gabriel Rollenhaben, *Nucleus Emblematum Selectissimorum* ... (Utrecht: Passaeus; Arnhem: Iasenius, [1611]–1613), 2 Teil, Nr. 53.

Figure 11.11: "Serva Modum," from Jean Baptiste Boudard, *Iconologie tirée de divers auteurs* (Parme: Sebstverl.; [Drucker]: Carmignani, 1759), reproduced in Eduard B. Wüseke, *Freimaurerische Bezüge zur barocken Emblematik: Kommunikationszeichen an der Schwelle zur Neuzeit* (Münster: Bauhütten Verl., 1990), 120.

> The moon as Nature reports
> Got all its [the moon's] light from the sun
> We experience all the gifts of light
> By Jesus Christ from above.[30]

Still another, included in the *Nucleus emblematum selectissimorum* of the Magdeburg poet Gabriel Rollenhagen (1583–1619) was published in 1613 and entitled "Serva Modum [Keep the right mode]." In it, a woman with a goniometer and reins in her hands stands before a cultural landscape between church and castle as symbols of clerical and worldly power (fig. 11.10). Its signature reads,

30. "Der Mond wie die Natur bericht / Hat von der Sonnen all sein Liecht / Uns wiederfehrt all Liechtes Gab / Durch Christum von oben herab"; ibid., 103 (my translation).

"The heart cannot beware the right mode to rein the feelings when it is hit by the puff of fortune." This emblem can be interpreted from rather different points of view. On the one hand, the sociohistorical doctrine of female repression from the perspective of gender studies takes as given the domestication of woman, presumed to be emotionally weak. On the other hand, the natural philosophical view identifies woman with reason (*prudentia*), a characteristic of Natura.

The woodcut "Scienza" from Boudard's *Iconologie*, however, shows a similar but allegorically different scene. It depicts a woman holding a mirror in her right hand and a globe with a goniometer in her left (fig. 11.11) and can be interpreted as an allegory of "science" personified as a goddess with her mirror understood as an instrument of self-exploration rather than as a characteristic of vanity.[31] This personification of Scienza also implies that Natura shows man the right way to investigate her secrets (see below).

Boudard's *Iconologie* includes other, analogous depictions of women as natural philosophy. The illustration "Theorie," for example, presents a woman with a large tome closed on her knees and a circle on the top of her head (fig. 11.12), while the notion of ethics is represented by a woman holding in one hand an angle scale and in the other a rope restraining a domesticated lion, a symbol of balance (fig. 11.13). In yet another image, "Calm consideration" is displayed by a woman looking at a pendulum, which is held by a rope descending from the clouds (fig. 11.14). She represents the virtual golden chain, her eyes concentrating on the divine link. (This image is reminiscent of the method of inducing hypnosis through eye fixation that the Scots physician James Braid [1795–1860] introduced in the nineteenth century.)

Natura as Teacher

The Goddess Natura emerged in the Middle Ages for the first time in the writings of the School of Chartre as literally the "discovery of Nature" that anticipated the Renaissance.[32] In the view of Andreas Speer, a philosopher who investigated the theoretical motivation of those scholars, Plato's explication in *Timaeus* that "all that originates, originates by a necessary cause"[33] was important. The "philosophia

31. Ibid., 61.
32. Speer, *Die Entdeckung der Natur.*
33. Ibid., 290.

Figure 11.12: "Theorie" from Jean Baptiste Boudard, *Iconologie tirée de divers auteurs* (Parme: Sebstverl.; [Drucker]: Carmignani, 1759), reproduced in Eduard B. Wüseke, *Freimaurerische Bezüge zur barocken Emblematik: Kommunikationszeichen an der Schwelle zur Neuzeit* (Münster: Bauhütten Verl., 1990), 121.

mundi" (philosophy of the world) would search for the "causae rerum" (the causes of things) and try to reduce them to their basic principles.[34] Physics (*physica*) would become "a science of the mathematical nature of the sensible world [*mundus sensibilis*]."[35] When common education flourished in the late Middle Ages, the "golden age of magic" flourished as well, with magic being performed not just by a few specialists.[36]

Once again, Alanus ab Insulis in the *Anticlaudian* provides key insights. In his treatise, he described Natura as a "great artist," a "sollers" (from the Latin *sol-*

34. Ibid., 291.
35. Ibid., 293.
36. Kieckhefer, *Magie im Mittelalter*, 78.

Figure 11.13: "Éthique/ Etica," from Jean Baptiste Boudard, *Iconologie tirée de divers auteurs.* (Parme: Sebst-verl.; [Drucker]: Carmignani, 1759), reproduced in Eduard B. Wüseke, *Freimaurerische Bezüge zur barocken Emblematik: Kommunikationszeichen an der Schwelle zur Neuzeit* (Münster: Bauhütten Verl., 1990), 120 and 122.

lus or "whole," and *ars* or "art"), a "whole art" and a virgin "mother of all things."[37] In contrast to the ideology of the Cathars, Nature was not demonized. She did not represent evil, but appeared as a wise leader of man giving him reason.[38] Benign, wise Nature would paint blossoms similar to a royal message. Bernard Silvestris (ca. 1085–ca. 1178), another representative of the School of Chartres, identified Urania with the "Queen of the astral wisdom." This queen in the form of the starred heaven was reflected in the earthly world of blossoms,[39] thus establishing a topos of natural philosophy that would have an important impact

37. Rath, "Einleitung," 51.

38. Ibid., 54.

39. Ibid., 56.

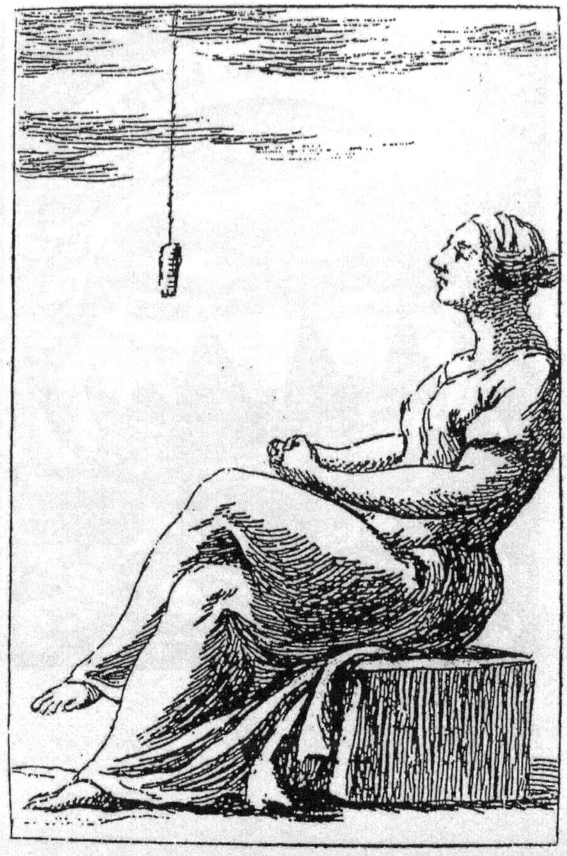

Figure 11.14: "Calm consideration" from Jean Baptiste Boudard, *Iconologie tirée de divers auteurs* (Parme: Sebst-verl.; [Drucker]: Carmignani, 1759), reproduced in Eduard B. Wüseke, *Freimaurerische Bezüge zur barocken Emblematik: Kommunikationszeichen an der Schwelle zur Neuzeit* (Münster: Bauhütten Verl., 1990), 123.

on scientific discourse until the nineteenth century. Lorenz Bausch (1605–65), physician and founder of the Academia naturae curiosorum (later called the Leopoldina), stated, "In heaven there is a glamor from glimmering stars like little blossoms, on earth there is an excellent radiation of the rarest flowers like little stars. . . . That is the work of the architect Nature."[40] He advised his readers in a Hermetic manner reminiscent of the *Tabula smaragdina*: "Believe that below, what you see above. No miracle."[41] The physicist and natural philosopher Gustav Theodor Fechner (1801–87) gave another example for magically com-

40. "In cœlo nitidissimarum stellarum quasi quorundam flosculorum coruscatio est, in terris elegantissima rariorum florum quasi stellularum radiatio. . . . Architectricis scilicet Naturæ . . . hoc opus est." Bausch, *Schediasma de unicornu fossili*, in Fehr, *Anchora sacra*, 202–3 (my translation).

41. "crede infra, quidquid vides supra. Nimirum"; ibid., 203.

bining the stars with blossoms—the heavenly light and with earthly flora—in the middle of the nineteenth century. In his very stimulating and speculative work *Nanna oder über das Seelenleben der Pflanzen,* he explained that "Nanna, the wife of Baldur (the God of Light), is the blossom, the world of flowers, whose most beautiful time coincides with Baldur's lordship of light. . . . [T]he poetics of antiquity imagines the most subtle glamor of flowers none other than bathed in the dew [from the heaven]."[42]

Alanus also characterized Natura as an *alma mater* creating the microcosm similar to the macrocosm in his *Complaint of Nature.* There, Nature asks man, "How could my appearing be so frightening for you? . . . How could you forget that I, your familiar nourishing mother, created the human nature so similar to the big world, that in it the scripture of the world appears like in a mirror?"[43] The metaphor of the disrupted veil of Nature and the violent unveiling of her signify the vice of men groping her and violating her integrity. "The vice of men misleads them to an irreverent attitude toward their own mother, Nature, so that he wants to rob her violently of her veils making the greatest chaos of dissent [*maximum chaos dissensionis*] between him and her into a permanent state. This I [Nature] have only to suffer from man!"[44]

Renaissance humanists took over the new appreciation and admiration of Nature. In his treatise *The Praise of Folly,* Erasmus of Rotterdam (1466–1536) declared Nature to be the scale for human life.[45] There, Erasmus praised the "poor simple people" living without any arts and "only from natural inspiration and suggestion."[46] Yet where there is much wisdom, there are also many needless things since, according to Erasmus, "who much experiences has also much

42. "Nanna, Baldurs (des Lichtgottes) Gattin, ist die Blüte, die Blumenwelt, deren schönste Zeit mit Baldurs Lichtherrschaft zusammentrifft. . . . die Poesie des Alterthums denkt sich den zartesten Blumenglanz nie anders als vom Thau [Tau des Himmels] gebadet"; Fechner, *Nanna oder über das Seelenleben,* iv (my translation).

43. "Wie konnte dich mein Erscheinen so erschrecken? . . . Wie konnte deinem Gedächtnis entschwinden, daß ich [*Natura*] es bin, deine dir so innig vertraute Nährmutter, die die menschliche Natur der großen Welt so ähnlich gestaltete, daß in ihr wie in einem Spiegel die Weltenschrift erscheint?"; Rath, "Einleitung," 59 (my translation).

44. "Es sind die Laster, die den Menschen zu seiner Ehrfurchtlosigkeit gegen die eigene Mutter Natur verführen, so daß er sie mit gewalttätiger Hand ihrer Hüllen berauben will, indem er das größte Chaos der Mißhelligkeit (maximum chaos dissensionis) zwischen sich und ihr zu einem Dauerzustand macht. Es ist einzig und allein der Mensch, von dem ich [*Natura*] solches erdulden muß!"; ibid., 65 (my translation).

45. Erasmus, *Das Theür vnd künstlich.*

46. "schlecht einfeltig volck . . . allein auß anregen vnd eingeben der natur"; ibid., sheet 28/p. 3 (my translation).

to suffer."[47] Moreover, "Nature hates confounding, concealing, boasting and receives much more blissfully what is not adulterated and weakened."[48] Generally, Nature would be the "master of the art, teacher, inventor, advisor"—not vice versa—a treasure, which was wasted if not used.[49] Erasmus attacked the "erring scholars," among them physicians and lawyers, as "evil Christians" who fail to heal their own souls and who disregard divine orders. As a result, no doctor would live well and no lawyer would die well.[50] Erasmus also referred to the *Praise of the Donkey* by Cornelius Agrippa (1486–1535) in offering his view that an adept of wisdom would need the character of a donkey, namely, an innocent, pure heart, an even temper, a peaceful relation to all animals, and a patient bearing of all burdens put on him.[51] Erasmus further reminded his readers to adopt a Christian countenance and to become a fool and a donkey when they wanted to become an angel and to unite with God.[52]

About seventy years after Agrippa, the physician and alchemist Michael Maier (1568–1622) treated Natura as a guide for researching scholars. His *Atalanta fugiens* (1618), illustrated by the well-known engraver Matthäus Merian the Elder (1593–1650), contains a very meaningful emblem (fig. 11.15). Natura, as a knowing woman, precedes a naturalist following her footsteps (quasi direction) equipped with a stick (quasi reason), glasses (quasi experience), and a lantern (quasi light to study her signatures). The message is that the scholar must follow Natura. In the German translation of the *Atalanta fugiens* (1708), this emblem, the "Zwey und viertzigstes Sinnbild von Geheimnuß der Natur [Forty-second Emblem of the Secrets of Nature]" bore the motto, "May Nature, Reason, Exercise and Literature be the guide, staff, spectacles and lamp for him who participates in chemistry."[53] The accompanying epigram read,

> Nature be your guide; follow her with your art willingly, closely,
> You err, if she is not your companion on your way.

47. "vnd wer vil erfert der müß vil leiden"; ibid., sheet 29/p. 1 (my translation).

48. "Die natur haßt das fitzen / vermenteln / gleissen / vnd bekompt vil glücksäliger das mit keiner kunst ist gefälscht vnd geschwecht)"; ibid., sheet 11/p. 2 (my translation).

49. "der kunst meisterin / lererin / erfinderin / angeberin"; ibid., sheet 11/p. 2 (my translation).

50. "Gelehrten / die verkerten," and "böse Christen"; ibid., sheet 116/p. 1 (front page) (my translation).

51. Ibid., sheet 86/verso.

52. Ibid., sheet 90/p. 1.

53. "Dem Sucher der Chymischen Kunst muß die Natur / Vernunfft / Erfahrenheit und das fleissige Lesen / Leiten / und an statt eines Führers / Stabs / ja einer Leuchte und Lampe dienen"; Maier, *Chymisches Cabinet*, 124, as translated in de Jong, *Michael Maier's Atalanta Fugiens*, 266.

Emblema XLII. *De ſecretis Naturæ.* 177
In Chymicis verſanti Natura, Ratio, Experientia & lectio, ſint Dux, ſcipio, perſpicilia & lampas.

EPIGRAMMA XLII.

Dux Natura tibi, túque arte pediſſequus illi
Eſto lubens, erras, ni comes ipſa viæ eſt.
Det ratio ſcipionis opem, Experientia firmet
Lumina, quò poſſit cernere poſta procul.
Lectio ſit lampas tenebris dilucida, rerum
Verborúmque ſtrues providus ut caveas. Z Cas-

Figure 11.15: Emblem 34 (p. 177), in Michael Maier, *Atalanta fugiens, hoc est, Emblemata nova de secretis naturae chymica....* (Oppenheim: de Bry, 1618). © Herzog August Bibliothek Wolfenbüttel (PURL: http://diglib.hab.de/drucke/196-quod-1s/start.htm).

> Reason be your staff, Exercise may strengthen your sight,
> On account of which the things that are far away can be discerned,
> Literature be your lamp, shining in the darkness,
> In order to guard you against an accumulation of things and words.[54]

Maier explained that this illustrated the four wheels of the philosophical: Nature, reason, experience, and the philosophical scriptures. The "chemists" should thus pay attention to Nature; she would be the guide as one followed in her footsteps.[55] In the original Latin edition, Nature was called "dux natura tibi [your (female) leader)]."

54. "Dich leitet die Natur / drum folge ihren Wegen / Sonst tritt'st du aus dem Pfad der rechten Wahrheits Bahn:/ Dein Staab sey die Vernunfft / das Licht muß dir zulegê / Die edle Wissenschafft / wañ du das Werck fängst an./ Das Lesen ist die Lamp so in dem Finstern scheinet / Doch überleg dabey was auch der Weiß recht meinet"; Maier, *Chymisches Cabinet*, 124, as translated in de Jong, *Michael Maier's Atalanta Fugiens*, 266–67.

55. Maier, *Chymisches Cabinet*, 125.

The title page of the *Musæum hermeticum* (1625) displays a Hermetic modification of Maier's emblem (fig. 11.16), however. Within an oval frame, the picture shows Natura bearing a hexagonal star (a hexagram or Star of David) that symbolizes the bond between heaven and earth. She is double-breasted and unclothed to the waist belt, reminiscent of Isis. A scholar with glasses, stick, and lantern is followed by another with the same attributes. Nature thus appears here even more clearly as a goddess in possession of an old, secret knowledge and in whose footsteps her naturalists must follow.

In the early eighteenth century, the imagery of the necessary succession was still presented in natural historical tracts, as this quotation ascribed to the physician Anton Joseph Kirchweger (d. 1764) indicates: "If such an artist does not know the way of Nature, he does not know Nature's work, either. It is impossible for him to understand Nature, but he must stick to the prescribed processes, within which to work and to fail in many ways, because he understands neither rule nor reason."[56] Moreover, man's internal ill nature would need help from an erudite doctor, "Because the learned medicus knows well, that Nature does not need more than a consolidation, by which she can get power to help herself; such a consolidation could not better be reached than by a regeneration through the quintessence, because all is pure and a fixed and spiritualistic medicine."[57]

Art, the "Ape of Nature"

In his treatise *De incertitudine et vanitate scientiarum declamatio invectiva Declamation* (or *Attacking the Uncertainty and Vanity of the Sciences and the Arts*), Agrippa von Nettesheim warned against the arrogance of scholars who overestimate their art.[58] The theologian and freethinker Sebastian Franck (1499–ca. 1543), who dealt intensively with this work, concurred.[59] In his view, such

56. "Denn so ein Künstler den Weg der Natur nicht weiß, weiß auch nicht, wie die Natur arbeitet, derselbe kann ohnmöglich die Natur verstehen, sondern er muß blind auf die vorgeschriebene *Processe* fallen, solche arbeiten und vielfältig drinnen fehlen, dieweil er weder *regulam* noch *rationem* verstehet"; Kirchweger [alleged author], *Aurea catena homeri*, preface (my translation).

57. "Denn ein gelehrter Medicus weiß wohl, daß die Natur nicht mehr als eine confortation vonnöthen, durch welches sie schon selbst wieder potens wird sich zu helffen; solche confortation aber könne besser nicht erlanget werden, als durch solche regeneration in Quintam Essentiam, da alles rein und pur und eine fixe und spiritualische Medicin ist"; ibid., 283 (my translation).

58. Agrippa von Nettesheim, *Incertidudine et vanitate scientiarum*.

59. Franck, *Ausführlicher Bericht*, 3.

Figure 11.16: Title page, *Musæum Hermeticum, Omnes Sopho-Spagyricæ Artis Discipulos Fidelissime Erudiensn* (Frankfurt: Jennis, 1625). © SLUB Dresden; Sign. Chem.359, misc. 1 (PURL: http://digital.slub-dresden.de/werkansicht/dlf/11683/7/cache.off?Seite=&cHash=a69b932d870b9c2a98ca4178d360d826).

scholars could be evil and corrupt the whole of humankind.[60] They could become obstacles to mercy and faith, whom believers in God would drop in favor of God's word and wisdom.[61] In Franck's view, Christ did not choose scribes for his apostles, but rather "simple people, even children, fools lacking art, and donkeys." According to Agrippa's *Praise of the Donkey*, this animal would have been elevated to a symbol for all Christians.[62]

Not that art would teach, illuminate, or enhance Nature, however. Quite the contrary, Nature would invent, teach, and improve the arts.[63] Within this context, Franck, like Robert Fludd (1574–1637) later, personified art as a (female) ape and called her "der Natur äffin" (Nature's ape). Franck explained the circumstances of the case: "Art is an ape domineering Nature, pretending whenever possible to express the nature of a thing, for example, by painting a man that she can never achieve. Because Nature is life and being; art only scratches external things from the outside."[64] It would be even better to do some soul-searching and to stay "inside" rather than to learn all of the outward arts.[65] Franck's advice was that learning should be accomplished from the "Light of Nature," which was, however, shadowed and faded owing to the fall of man. Since this light was common to all men, everyone would have the faculty of judgment in his or her own heart. Franck called this the "implanted word" of God and argued that when Nature is obscured within us, God is also "faded within us."[66] The task was therefore to illuminate darkened Nature, "Because all Nature comes from God and God Himself is Nature, God helps our darkened Nature (which is Himself faded within us) with His mercy, so that Nature can see more clearly."[67]

The topos of the ape had been used as a metaphor of imitation as early as the twelfth and thirteenth centuries.[68] Around 1200, "simian" became a slogan

60. Ibid., 6.

61. Ibid., 16.

62. "einfältig leut / Ja kinder / Kunstloß Narren vnnd Esel"; ibid., 18 (my translation).

63. Ibid., 52.

64. "Die Kunst it ein äffin vnd anmasserin der natur / Dann wie sie kann / anmast sie sich / die Natur eins dings außzutrücken / als das malen einen Menschen / das sie doch nimmer erlangt. Dann die Natur ist leben und wesen / die Kunst schabt nuhr von aussen an eusserlichen dingen"; ibid., 66 (my translation).

65. Ibid., 67.

66. "das eingepflantzt Wort" and "in uns verblichen"; ibid., 70 (my translation).

67. "Nuhn wie alle Natur von Gott ist / vnd Gott selbs die Natur / so hilfft Gott vnserer verfinsterten Natur (das ist jhm selbs in vns verblichen) mit seiner gnadt / dass die natur klärer siehet"; ibid., 71 (my translation).

68. Curtius, *Europäische Literatur*, 522–23.

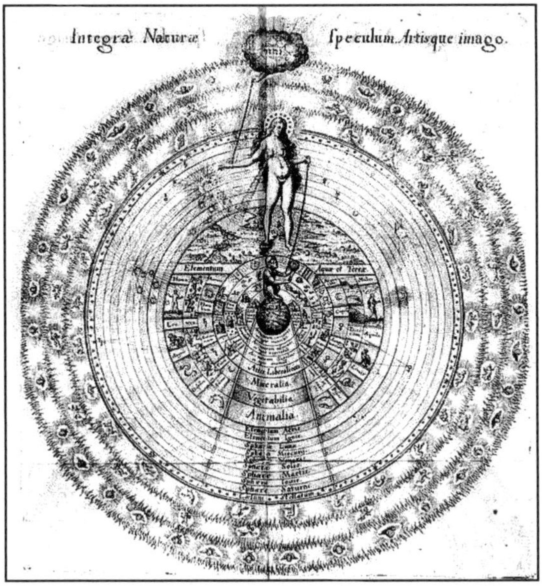

Figure 11.17: Illustration from tract 1, Robert Fludd, *Utriusque cosmi maioris scilicet et minoris metaphysica, physica atque technica historia*, vol. 1, *De microcosmi historia . . . .*, (Oppenheim: de Bry, 1617). © Herzog August Bibliothek Wolfenbüttel: Sign. M: Na 4° 41.

in Latin school poetry and was presumably introduced by Alanus. "The real ape [*simius*] turns to a 'simia,' when he imitates man."[69] Man himself could turn into a "simia naturae" (aper of nature) in imitating Nature through his art. Similarly, art could appear as a *simia*, that is, as a *scimmia della natura* in the Italian literature of the sixteenth and seventeenth centuries. As a metaphor in natural philosophy, the metaphor of the ape—the docile disciple of Nature who imitates her positively—symbolized the leading function of Nature for man. Man could only recognize and use her secrets by imitating her like an ape, particularly her magic arts.

Two allegories are noteworthy in this regard, both represented in large plates in Robert Fludd's *Utriusque cosmi* of 1617. At the beginning, there is the allegory "Integrae Naturae speculum Artisque imago" (Mirror of the whole nature and image of the art). Nature is visually interposed as a medium between God and ape (that is, man) (fig. 11.17). A chain connects the hand of God to the virgin Natura and from her to an ape, personifying art and science imitating Nature above him to improve her products. In this constellation, Nature combines the fiery heaven (God), the ethereal heaven (stars), and the sublunar, earthly world

69. "Der reale Affe (*simius*) wird zur *simia*, wenn er den Menschen nachahmt"; ibid., 522.

(elements). Natura is depicted as a powerful and energy-dispensing woman who nourishes the world. Her heart is a sun lighting the stars (Milky Way), and her crescent-shaped abdomen (uterus) is a medium for transmitting astral influences to the earth. The chain connects the left hand of God that reached out of the fiery cloud to the right hand of Nature, and then from the left hand of Nature to the left forearm of the great ape. Nature is thus a substantial length of chain resembling the golden chain in Greek mythology. The physical operations of the human arts are also described: Nature "imitates, helps, and perfects," and while "Nature is not a goddess, she is God's nearest servant."[70]

Art historian Wolfgang Kemp extensively treated Robert Fludd's Natura as the World Soul,[71] and particularly Fludd's concept of "explication," that is, the two producers of the macrocosm: "The remaining parts being less pure, the realms of the ethers and elements, are subordinated to Nature . . . and therefore are correctly called natural. But there are two producers of all of this part of the macrocosm: Nature and her ape, which we call art."[72] According to Fludd, who referred to Zoroaster and Heraclitus, Nature seemed to be an "invisible fire," the world soul, "But we illustrated it in the way of some more recent philosophers so that the reader can better understand its essence. We depicted it therefore as a naked woman of a tender and prosperous age attributing to her a star crown and golden resplendent hair."[73] Fludd conceived the Isis vision in the dream of Apuleius (2nd century CE) as a pristine model for Nature standing on water and earth.[74] Fludd's sources were likely copper engravings from the *Imagini* of Vincenzo Cartari (ca. 1531–69), which was published in several editions from 1581 into the seventeenth century.[75]

In the second tract of Fludd's *Utriusque cosmi*, an impressive allegory of the erudite-docile ape appears on the title page (fig. 11.18). He sits in the middle of a disk on which the different arts are depicted, and he points to an arithmetical

70. "Imitatur, Adjuvat, Perfecit" and "Natura non Dea, sed proxima Dei ministra"; ibid., 8 (my translation).

71. Kemp, "Natura," 88–101.

72. "Reliquæ verò impurioris massæ partes, ætherea nempe & elementaris, quoniam naturæ . . . subjiciuntur, meritò inde naturales appellantur. Duæ autem harum partium in hac Macrocosmi fabrica contentarum causæ effectrices enumerantur: *Natura* scilicet, & ejus *Simia*, quam Artem appellamus"; Fludd, *Utriusque cosmi*, 11, as quoted in Kemp, "Natura," 89 (my translation).

73. "Finximus itaque virginem nudam ætate tenera & florente, comâ stellifera & pulcherimè deauratâ"; Fludd, *Utriusque cosmi*, 11, as quoted in Kemp, "Natura," 89 (my translation).

74. Ibid., 92.

75. Cartari, *Imagini delli dei de gl'Antichi*, 65–72.

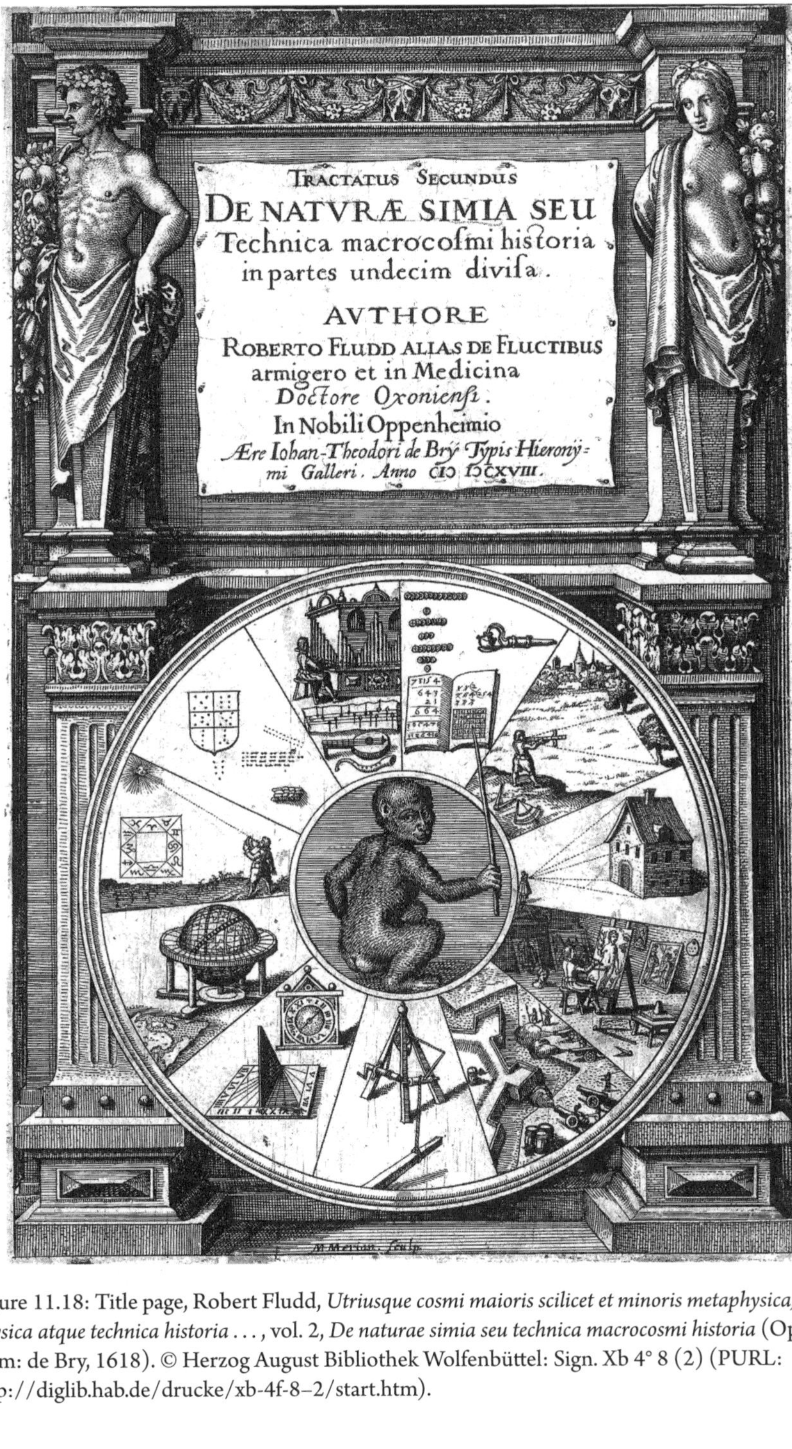

Figure 11.18: Title page, Robert Fludd, *Utriusque cosmi maioris scilicet et minoris metaphysica, physica atque technica historia . . .*, vol. 2, *De naturae simia seu technica macrocosmi historia* (Oppenheim: de Bry, 1618). © Herzog August Bibliothek Wolfenbüttel: Sign. Xb 4° 8 (2) (PURL: http://diglib.hab.de/drucke/xb-4f-8-2/start.htm).

book. This stresses the author's conviction that the arts depicted are based on numbers and relations between numbers. In the second half of the second tract, entitled "The Temple of Music," the importance of music and its harmonies for mathematics and mechanics becomes obvious. Fludd's influence on Athanasius Kircher (ca. 1601–80) here is evident, as he took over some of Fludd's diagrams for his own illustrations.[76]

The scholar of English literature Hans-Joachim Zimmermann demonstrated how the motif of the "academic ape" was passed down in different variations into the nineteenth century, using the example of the allegory "Accademia" from the *Iconologia* of Cesare Ripa (ca. 1560–ca. 1645).[77] In the Middle Ages, the ape was a symbol for the devil, the "ape of God" according to the theologians.[78] By the sixteenth century, three species of apes—following Aristotle and Pliny—were delineated: "Apes without a tail (*simia*, ape or, more precisely, Barbary macaque), apes with a tail longer than the body (*cercopithecus*, a tail ape or guenon), and apes with a tail shorter than the body (*cynocephalus*, a dog-headed ape or baboon)."[79] The article entitled "Accademia" in Ripa's *Iconologia* mentioned the academic ape in these various terms in describing the corresponding woodcut, entitled "Tra quali [libri] risieda vn Cinocefalo, o vero Babuino [In between the books sits a baboon]" (fig. 11.19). According to Egyptian mythology, the dog-headed baboon could paint hieroglyphs; he was worshipped as the divine incarnation of Thot.[80] The article in the *Iconologia* reinterpreted the creatively writing baboon as a receptively reading one. In this light, he could be seen as a prototype of imitation.[81] Around 1800, however, the academic ape was devalued and discredited as the illustration from Filippo Pistrucci's edition of the *Iconologia* (1819) suggests (fig. 11.20). There, the ape sits slumped, looking at the floor away from the books, which are being trod upon by Accademia.

76. Gouk, *Music, Science and Natural Magic*, 101.

77. Zimmermann, *Der akademische Affe*. It is remarkable that Zimmermann did not mention Fludd's illustrations and their possible impact on Ripa's *Iconologia* in his analysis, however.

78. Ibid., 47.

79. "Affen ohne Schwanz (*simia*, d. h. Affe: Berberaffen), Affen mit Schwanz länger als der Körper (*cercopithecus*, d. h. Schwanzaffe: Meerkatze) und Affen mit S'chwanz kürzer als der Körper (cynocephalus, d. h. Hundskopf: Pavian)"; ibid., 46 (my translation).

80. Ibid., 82.

81. Ibid., 85.

Figure 11.19: Illustration from Cesare Ripa, *Iconologia* (ca. 1600), edited by Filippo Pistrucci (1819), reproduced in Hans-Joachim Zimmermann, *Der akademische Affe: Die Geschichte einer Allegorie aus Cesare Ripas 'Iconologia.'* Supplement of the Sitzungsberichte der Heidelberger Akademie der Wissenschaften; Philosophisch-Historische Klasse 6 (Wiesbaden: Reichert, 1991), 80.

The discourse on the relation between Nature and art, *Natura* and *Ars*, started in the High Middle Ages and was especially intense in the context of the natural magic of the early modern period, yet this discourse continued into the Enlightenment of the eighteenth century. In his eighteenth-century monograph *Unterredungen über die Schönheit der Natur* [Discussion on the Beauty of Nature], the Swiss philosopher and mathematician Johann Georg Sulzer (1720–79) dealt with all possible realms of nature and natural impressions. From an aesthetic point of view, he imagined the "harmonic chain of the creatures": "Cross the whole of nature through all figures, from man to worm, and you will see with

Figure 11.20: Illustration from Cesare Ripa, *Iconologia* (ca. 1600), edited by Filippo Pistrucci (1819), reproduced in Hans-Joachim Zimmermann, *Der akademische Affe: Die Geschichte einer Allegorie aus Cesare Ripas 'Iconologia.'* Supplement of the Sitzungsberichte der Heidelberger Akademie der Wissenschaften; Philosophisch-Historische Klasse 6 (Wiesbaden: Reichert, 1991), 438.

astonishment that the snake and the fly as well as the elephant are links of an excellent chain; which would be imperfect, were the least of these links absent.... They compose a perspective chain where one link after the other diminishes by insensible transformations."[82] Man to ape would be a paradigm for such a chain.[83] Sulzer also imagined nature as a chamber of marvels. One could thus admire the beauty of a natural history collection because "[a]lthough the winter puts a veil over the treasures of nature, I still have her with me. The realm of the plants remains forever green in my rooms.... I have also a continuous summer at home, and every day I discover new beauties of the harmonic order of nature, every day new delights."[84] From this point of view, the world is similar to an "art chamber

82. "durchgehe die ganze Natur durch alle Gestalten, vom Menschen bis zum Wurm, so wirst du mit Erstaunen sehen, daß die Schlange und die Fliege, so gut als der Elefant, Glieder einer fürtreflichen Kette sind; die unvollkommen seyn würde, wenn der geringste dieser Ringe fehlte.... Sie machen ein perspektivische Kette aus, daran ein Ring nach dem andern durch unmerkliche Veränderungen abnimmt"; Sulzer, *Unterredungen über die Schönheit der Natur*, 26–30 (my translation).

83. Ibid., 27.

84. "Wenn auch der Winter einen Schleyer über die Schäze der Natur ziehet, so habe ich sie bey mir. Das Reich der Pflanzen bleibt mir in meinen Zimmern immer grün.... Ich habe also einen beständigen Sommer

that includes so many different peculiarities that everyone will find something to enjoy."[85] Divine wisdom constructed the world in such a manner that it would please different minds: "that everyone perceives almost just that [aspect] of the world, that pleases him, the rest, that displeases him, is hidden to him."[86]

The title copper engraving of Sulzer's book *Allgemeine Theorie der Schönen Künste*, which was created by Daniel Chodowiecki, is impressive (fig. 11.21). It depicts an ensemble of women reminiscent of muses and obviously symbolizing the fine arts. One of them stands exalted among the others demonstrating the classical position of Natura mediating between the divine and the earthly domain. Above her head, the sun, as the central divine celestial body, shines, and she transfers its beams to the earthly world. In the background, an archaic group of humans, standing before a tent, is engaged in everyday activities. The entries to the keywords "Natur (Schöne Künste)" and "Natürlich (Schöne Künste)" with their explanations are instructive with regard to the concept of art.[87] Nature as an effective cause would be the "leader and teacher of the artist." The more the artist imitated Nature with his methods or the selection of his materials, the more perfect the result of his work.[88] Consequently, Sulzer defined Nature as "the supreme wisdom per se" and "the real school of the artist."[89]

What was the special art of Nature? At first, Sulzer underlined her art of designing in the sense of the traditional doctrine of signatures. She would set all external appearance in accordance with the inner character of the things, "so that the form, colors, the rough and smooth, the soft and stiff always coincide totally with the inner qualities of the things."[90] The artist should accept Nature as his only teacher. Sulzer also came back to the central project of alchemy: imitating, supporting, and perfecting Nature. Man—respectively, the artist—had both to follow the way that Nature had shown him and to support her in her intentions.

---

bey mir, und ich entdekke täglich neue Schönheiten der harmonischen Einrichtung der Natur: täglich neues Vergnügen"; ibid., 31–32 (my translation).

85. "Kunstkammer, da die Merkwürdigkeiten in solcher Verschiedenheit sind, daß ein jeder Mensch etwas zu seiner Belustigung findet"; ibid., 182 (my translation).

86. "daß ein jeder von der Welt fast nur das siehet, was ihm gefällt; das übrige, so ihm misfallen würde ist ihm verborgen"; ibid., 183 (my translation).

87. Sulzer, *Allgemeine Theorie der Schönen*, 506–14.

88. "Führein und Lehrerin des Künstlers"; ibid., 507 (my translation).

89. "die höchste Weisheit selbst" and "die eigentliche Schule des Künstlers"; ibid. (my translation).

90. "alles Aeußerlichen mit dem innern Charakter der Dinge, daß die Gestalt, die Farben, das Rauhe und Glatte, das Weiche und Harte, immer mit den innern Eigenschaften der Dinge gänzlich übereinkommen"; ibid., 508 (my translation).

Figure 11.21: Title page, Johann Georg Sulzer, *Unterredungen über die Schönheit der Natur nebst desselben moralischen Betrachtungen über besondere Gegenstände der Naturlehre.* Von neuem aufgelegt (Berlin: Haude und Spener, 1770). Image from Wikimedia Commons (URL: http://de.wikipedia.org/w/index.php?title=Datei:Sulzer_Titelkupfer_Chodowiecki_1771.jpg&filetimestamp=20081029211440).

From this, genius could result: "All excellent works of the fine arts are in their essential parts the fruits of Nature, which have ripened through experience and closer reflection of what Nature gives the genius for his disposal."[91] The artist should thus be faithful to the "voice of Nature [Stimme der Natur]" speaking within him. He should strain his external and internal senses so as not to miss a single one of the works of Nature. The geniuses and inventors in the realm of the fine arts would be just "the most hardworking and sagacious observers of Nature."[92] The artist should create his object from Nature so as to make it appear like a natural object.[93] Therefore, artificially produced objects should look as if they existed without art and only by the impact of Nature.

Natural Philosophy and Gynephilia: A Crucial Link

Nature was generally personified as a noble woman, and so her depiction had specific gender implications. There was always a more or less subtle erotic tension when naturalists, physicians, artists, or poets viewed Nature as the ideal guiding their work and as a source for their inspiration and power. This erotic tension became a particular topic of occultism and natural magic and was expressed more or less openly in both medicine and alchemy. Moreover, since Nature was identified with a divine woman, the image of earthly women was exalted. Women thus seemed nearer to divine wisdom.

The treatise of Agrippa von Nettesheim *De nobilitate et praecellentia foeminei sexus*, published in 1529, is an extraordinary example of the highlighting of women in the natural philosophical context.[94] Agrippa used a series of arguments—drawn mainly from philosophy and theology—to demonstrate this nobility and preeminence. His treatise was frequently reprinted and translated into other languages; for example, a German translation appeared in 1720, two hundred years after the first edition, although the work is largely forgotten today.[95]

91. "Alle vorzügliche Werke der schönen Künste sind in ihren wesentlichen Theilen Früchte der Natur, die durch Erfahrung und nähere Überlegung dessen, was die Natur dem Genie an die Hand giebt, reif geworden"; ibid., 509 (my translation).

92. "die fleissigsten und scharfsinnigsten Beobachter der Natur"; ibid., 510 (my translation).

93. Ibid., 511.

94. Agrippa von Nettesheim, *De nobilitate et praecellentia foeminei sexus*.

95. Agrippa von Nettesheim, *Anmuthiges und curieuses Tractätgen*.

At first Agrippa pointed out that, in regard to reason and mind, there was no difference between the sexes. Both were equally gifted. Relative to other things, however, the female sex clearly took precedence. Agrippa enumerated a series of qualities to prove his thesis.

(1) The *etymological* argument stressed that "Eve" meant "life," whereas "Adam" derived from "earth." The name of Eve, moreover, was more closely related to the Tetragrammaton JHWH (Jehovah) than the name of Adam.

(2) The *order of the creation* recorded in Genesis argued for the pre-eminence of woman. She was created by God as the last, and therefore the most complete, creature.[96]

(3) Eve also differed from Adam in regard to the *matter* from which she was created. Whereas Adam was created from "dead paste or feces,"[97] Eve was made of "a purified, lively Matter gifted with a reasonable soul participating in the spirit of God."[98] Here, Agrippa reversed the traditional gender difference whereby man represented the spiritual principle and the divine light, but he went even further. In his view, man had been fashioned by heavenly influences from an earthen lump; he was thus a creation of Nature. Woman, on the other hand, "is created only by God without the support from the stars," thereby making her more able than man to understand divine secrets.[99] This idea of a wise woman, a female *sophia,* fit very well within the contemporary natural philosophy of alchemy and occultism.

(4) Because of the cosmological location of Natura as a sort of medium between God and man, women take over the role of Nature in the human realm. Their position and function imply analogies with cosmic Nature. Their beauty seems to be a reflection of the divine splendor that originates from the light of God and emanates over all creatures. Since the female body is more beautiful than the male body, God elected it for residence.[100]

96. Ibid., 18.

97. "ex inanimato quopiam aut vili luto"; ibid., 23 (my translation).

98. "ex materia purificata, vivifica et animata, anima inquam rationali mentem participante divinam"; ibid., 23 (my translation).

99. "mulier autem supra omnem coeli influxum ac naturae promptiduninem absque ulla virtute cooperante a solo deo creata est"; ibid., 23–24 (my translation).

100. Ibid., 24.

(5) Agrippa also stressed a physiological argument. Nature *prefers* women since they could bear children and give milk as a life-sustaining substance not only for children but also for the nourishment of the old and weak. In general, a woman had more healing power, so she could give off heat as a life force by pressing her breast on the chest of old exhausted men.[101]

(6) The biblical female figures of Eve and Mary also played an important role in Agrippa's arguments. The fall of mankind was not committed by Eve but by Adam because God had forbidden *him* to eat the fruit.[102] So Adam committed original sin, not Eve. The most striking argument for female superiority, however, was that the "noblest of all pure creatures" was a woman—Mary, the Virgin Mother.[103] Female superiority was also obvious because of the fact that many arts and sciences (the "liberal arts") kept the female names of those who invented them and because in geography, female names were given to prominent parts of the world: Asia (nymph), Europa (daughter of Agenois), and Africa (daughter of Epaphi).[104]

(7) Since, according to Agrippa, women are prone to divination and were the first to serve as prophets and sybils,[105] an ordinary woman was wiser than an erudite man and an old countrywoman could have more experience than a physician, who was perceived as an adept.[106] This statement would be re-echoed in romantic natural philosophy and medicine around 1800.

(8) In the end, Agrippa pled emphatically for the emancipation of women, deploring traditional misogynistic disrespect. Women had to give way to men "as if they would have been conquered in a war."[107] Their suppression resulted not from the divine order but "from custom, education, fortune, and tyrannic opportunity."[108]

101. Ibid., 35–39.
102. Ibid., 47.
103. Ibid., 63.
104. Ibid., 73.
105. Ibid., 74.
106. Ibid., 87.
107. "gleich als ob sie im Krige [*sic*] überwunden wären"; Agrippa von Nettesheim, *Abigail*, 204 (my translation).
108. "durch die Gewohnheit, durch die Erziehung, durch das Looß, und durch die tyrannische Gelegenheit"; ibid., 110 (my translation).

Agrippa's tract was well known to "feminist" contemporaries like the erudite Marguerite de Navarre (1492–1549), but was not the first of its kind.

There were actually allusions to the superiority of women as early as the late thirteenth century. Heinrich von Meißen, who died in 1318 in Mainz and was dubbed "Frauenlob," compiled an encyclopedia of learned women.[109] There, he held that women often had more talent than men and, with an adequate education, they could easily become as erudite as men. As he noted, Minerva was adored by the pagans as a goddess because of her skills and knowledge in the arts, which were supposed to have been invented by her.[110] He concluded that feminine ingenuity often surpassed that of men.[111] The *querelle des femmes*, however—that is, the broad discussion of gender in the Renaissance and early modern period—was mainly initiated by Giovanni Boccaccio in his tract *De mulieribus claris*, written in the early 1360s.[112] The most comprehensive text about famous women in the history of culture, it was used by many authors. In it, Boccaccio sketched an ambiguous image, at once subversive and conservative, directing his social analysis particularly against ecclesiastical doctrine.[113]

Even earlier, in the last novella of the *Decamerone*, Boccaccio had highlighted the plight of women in his telling of the legend of Griselda, a story that had an especially great impact on other authors. The daughter of a poor farmer, Griselda becomes the wife of a nobleman named Gualtieri, the Marquis of Saluzzo. After having his two children, Griselda is tricked by Gualtieri into believing that he has killed them both. He then pretends to have tired of her, announces that he has taken a new wife, and turns her out of their home with only the clothes on her back as he brings in a young girl whom he pretends is his new wife. Through all of this, Griselda is ever patient and uncomplaining, and recognizing this, Gualtieri brings her back, revealing that his "new wife" was actually their daughter and that neither of their children was dead. Griselda is rewarded for her patience not only by being reunited with her children and husband but also with the title of a noblewoman. Boccaccio's telling of this story so impressed Petrarch (1304–74) that he translated his friend's work from Italian into Latin and wrote his own

109. Frauenlob [Frawenlob], *Die Lobwürdige Gesellschafft.*

110. Ibid., 25.

111. Ibid., 33.

112. Kolsky, *Ghost of Boccaccio,* 2.

113. Ibid., 177.

version of the story in the form of a letter. This was one of the first works of Petrarch to be translated into German and illustrated with woodcuts.[114]

That Agrippa knew and was inspired by Boccaccio's writings seems fairly certain, since he took over some of the Italian's major arguments. Boccaccio, for example, had assumed that woman was created by God out of purified rational matter, whereas man was created by nature out of feces or soil.[115] Women were thus much more susceptible to divine enlightenment than men.[116] One of Boccaccio's physiological arguments is not to be found in Agrippa's treatise, however. According to Boccaccio, menstruation reflected female cleanliness because her intemperate humors could properly and regularly be expelled (as opposed to the less natural technique of nose-bleeding).[117]

From the fifteenth to the eighteenth century—over a period of some four hundred years—there were more writings on the superiority of women than there were explicitly misogynic ones. In addition to the works by Boccaccio and Agrippa, Marc Angenot has pointed out that some eighty works on the superiority of women can be traced in the French literature alone.[118]

The reason for the adoration of the female sex in the Renaissance and humanistic period was the new approach that held Nature as tantamount to a holy scripture. Within this context, Nature was no longer—as in the traditional humoral pathology—associated with the deficient female qualities of the earth (dark, wet, and cold), but rather with the divine qualities of heaven (light, ethereal, and warming). These qualities were personified by certain female figures that symbolized the mediating function between the divine and the earthly. Nature became the medium that linked human beings and all natural things with God, the original creating power. Certain images and metaphors were coined: queen of heavens, virgin, wet nurse, wise woman, guide for naturalists. Consequently, Nature was imagined as the real teacher of philosophers and alchemists. They had to follow in her footsteps as depicted in the emblem of Michael Maier discussed above.

114. Petrarca, *Petrarchas Griseldis.*

115. Boccaccio, *Historien von allen den fürnembsten Weibern*, 7.

116. Ibid., 8.

117. Ibid., 14.

118. Angenot, *Champions des femmes.*

Those imaginings showed Nature personified as a female. She was a divine magician beaming the splendor of God down to the microcosm. All women shared her characteristic traits, especially the ability to produce living creatures. There was, however, a clear hierarchy. Nature was inferior to God, but superior to man. Analogously, Agrippa's theory presented woman as superior to man but inferior to God, Christ, and the Holy Spirit. The crucial point is that in the Renaissance and early modern period, the idea of the superiority of the female sex was stimulated by two factors that interplayed intriguingly: first, natural philosophy and its personifications and, second, the image of the Virgin Mary the mother of God, as a religious mediator of the splendor of God the father.

Works Cited

Agrippa von Nettesheim, Heinrich Cornelius. *Abigail; das ist, Des lob-würdigen Frauen-Zimmers Adel und Forträfligkeit; for mer dan hundert Jahren von Heinrich Kornel Agrippen latinisch beschriben* . . . [Lübeck:] Schernwebel; Jäger, 1650.

———. *Anmuthiges und curieuses Tractätgen von dem Vorzug des Weiblichen vor dem Männlichen Geschlecht.* Ehemals aus dem Lateinischen ins Frantzösische, anietzo aus dem Frantzösischen ins Teutsche übersetzt von I. K. L. N.p., 1720.

———. *De incertidudine et vanitate scientiarum declamatio invective.* [Köln]: Cervicornus, 1532. Translated as *Anmuthiges und curieuses Tractätgen von dem Vorzug des Weiblichen vor dem Männlichen Geschlecht* (see the previous entry).

———. *De nobilitate et praecellentia foeminei sexus: Von Adel und Vorrang des weilichen Geschlechts.* Edited and translated by Otto Schönberger. Würzburg: Königshausen und Neumann, 1997.

Angenot, Marc. *Les champions des femmes: Examen du discours sur la supériorité des femmes 1400–1800.* Montréal: Les Presses de l'Université du Québec, 1977.

Bausch, Johann Lorenz. *Schediasma de unicornu fossili.* Added to Johann Michael Fehr, *Anchora sacra; vel scorzonera* . . . Jena: Trescher, 1666.

Boccaccio, Giovanni. *Historien von allen den fürnembsten Weibern* . . . Frankfurt am Main: Feirabends, Hüter, Lechler, 1566. *De claris mulieribus*, dt. 2nd vol., translated into German as *Des ander Theil / Vom herkommen des Adelichen Fürtrefflichen Weiblichen gesclechtes* . . . Frankfurt am Main, 1566.

Boudard, Jean Baptiste. *Iconologie tirée de divers auteurs: Ouvrage utile aux gens de lettres, aux poëtes, aux artistes, et généralement à tous les amateurs des Beaux arts.* 3 vols. Parme: Carmignani, 1759. Reprint, Vienna: de Trattnern, 1766; New York: Garland Publishing, 1976.

Camerarius, Joachim [d.J.]. *Symbola et emblemata (Nürnberg 1590 bis 1604)*. With an introduction and [Registern] by Wolfgang Harms and Ulla-Britta Kuechen. 2 vols. Naturalis Historia Bibliae 2/1–2/2. Graz: Akademische Druck- u. Verlagsanstalt, 1986–88.

———. *Symbola et emblemata tam moralia quam sacra: Die handschriftlichen Embleme von 1587*. Edited by Wolfgang Harms and Gilbert Heβ. Tübingen: Niemeyer, 2009.

Cartari, Vincenzo. *Imagini delli dei de gl'Antichi* . . . Venetia: Tomasini, 1647.

Frauenlob [Frawenlob], Johann. *Die Lobwürdige Gesellschafft der Gelehrten Weiber*. N.p., 1531.

Curtius, Ernst Robert. *Europäische Literatur und lateinisches Mittelalter*. 2nd ed. Bern: Francke, 1954.

———. "Zur Literaturästhetik des Mittelalters II." *Zeitschrift für romanische Philologie* 58 (1938): 129–232.

Debus, Allen G. *The Chemical Philosophy: Paracelsian Science and Medicine in the Sixteenth and Seventeenth Centuries*. 2 vols. New York: Science History Publications, 1977.

de Jong, H. M. E. *Michael Maier's Atalanta Fugiens: Sources of an Alchemical Book of Emblems*. Janus Supplements 8. Leiden: Brill, 1969.

Eamon, William. *Science and the Secrets of Nature: Books of Secrets in Medieval and Early Modern Culture*. Princeton, NJ: Princeton University Press, 1994.

Eliade, Mircea, and Ioan P. Culianu. *Handbuch der Religionen*. Edited by H. S. Wieser. Frankfurt: Suhrkamp Taschenbuch, 1995.

Erasmus, Desiderius, von Rotterdam. *Das theür vnd künstlich Bůchlin Morie encomion, das ist, Ein Lob der Thorhait*. Ulm: Varnier, ca. 1534.

Fechner, Gustav Theodor. *Nanna oder über das Seelenleben der Pflanzen*. Leipzig: Voß, 1848.

Franck, Sebastian. *Ausführlicher Bericht / Was von Dünsten vund menschlicher Weißheit zu halten sey / etwas aus der Declamation Henrici Cornelij Aggripe* . . . Frankfurt am Main: Jennis, 1619.

Frauenlob [Frawenlob], Johann. *Die Lobwürdige Gesellschafft der Gelehrten Weiber*. N.p., 1531.

Gouk, Penelope. *Music, Science and Natural Magic in Seventeenth-Century England*. New Haven, CT: Yale University Press, 1999.

Gronemeyer, Nicole. *Optische Magie: Zur Geschichte der visuellen Medien in der Frühen Neuzeit*. Bielefeld: transcript, 2004.

Kemp, Wolfgang. "Natura: Ikonographische Studien zur Geschichte und Verbreitung einer Allegorie." PhD diss., Tübingen University, 1973.

Kieckhefer, Richard. *Magie im Mittelalter*. Translated by Peter Knecht. München: Beck, 1992.

Kirchweger, Anton Joseph [alleged author]. *Aurea catena homeri, oder Eine Beschreibung von dem Ursprung der Natur und natürlichen Dingen* . . . Frankfurt: Böhme, 1723.

Kolsky, Stephen. *The Ghost of Boccaccio: Writings on Famous Women in Renaissance Italy*. Late Medieval and Early Modern Studies 7. Turnhout, Belgium: Brepols, 2005.

Maier, Michael. *Chymisches Cabinet / Drer grossen Geheimnussen der Natur / Durch wohl ersonnene sinnreiche Kupfferstiche und Emblemata . . . : Der Chymischen Republic Und Dero Liebhabern / Zur Speculation, Betracht- und Unersuchung aus wohlmeinender Veneration und Liebe zum zweyten mahl in der Lateinischen Sprach ausgefertiget / vor jetzo aber zum ersten Mahl in das Hochteutsche übersetzet ist* . . . Frankfurt: Oehrling, 1708.

Möbius, Paul Julius. *Ueber den physiologischen Schwachsinn des Weibes*. Sammlung zwangloser Abhandlungen aus dem Gebiete der Nerven- und Geisteskrankheiten 3. Halle: Marhold, 1900.

Modersohn, Mechthild. *Natura als Göttin im Mittelalter: Ikonographische Studien zu Darstellungen der personifizierten Natur*. Berlin: Akademie Verlag, 1997.

Petrarca, Francesco. *Petrarchas Griseldis* (Ulm: Johann Zainer, 1473/74). Facsimile edition. Edited by Henirich Steinhöwel. Potsdam: Müller, 1921.

Rath, Wilhelm. "Einleitung." In Alanus ab Insulis, *Der Anticlaudian oder Die Bücher von der Himmlischen Erschaffung des neuen Menschen*, edited by Wilhelm Rath, 15–92. 2nd ed. Stuttgart: Mellinger, 1983.

Schäfer, Peter. *Mirror of His Beauty: Feminine Images of God from the Bible to the Early Kabbalah*. Princeton, NJ: Princeton University Press, 2002.

Schott, Heinz. *Magie der Natur: Historische variationen zu einem Motiv der Heilkunst*. Aachen: Shaker, 2014.

Speer, Andreas. *Die Entdeckung der Natur: Untersuchungen zur Begründungsversuchen einer "Scientia naturalis" im 12. Jahrhundert*. Studien und Texte zur Geistesgeschichte des mittelalters 45. Leiden: Brill, 1995.

Stemper, Anneliese. "Der Prudentia-Teppich des Pfalzgrafen Ottheinrich im Kurpfälzischen Museum zu Heidelberg." *Heidelberger Jahrbücher* 2 (1958): 68–95.

———. "Die Wandteppiche." In *Ottheinrich: Gedenkschrift zur vierhundertjährigen Wiederkehr seiner Kurfürstenzeit in der Pfalz (1556–1559)*, edited by Georg Poensgen, 141–71. Heidelberg: Verlag der Studentenschaft der Universität Heidelberg, 1956.

Sulzer, Johann Georg. *Allgemeine Theorie der Schönen Künste . . .* Leipzig: in der Weidmannischen Buchhandlung, 1793.

———. *Unterredungen über die Schönheit der Natur nebst desselben moralischen Betrachtungen über besondere Gegenstände der Naturlehre.* Berlin: Haude und Spener, 1770. Reprint, Frankfurt: Athenäum, 1971.

Würzburg, Konrad [Conrad] von. *Die goldene Schmiede: Aus Gothaischen Handschriften herausgegeben und erklärt von W. C. Grimm.* Frankfurt: Körner, 1816.

Wüseke, Eduard B. *Freimaurerische Bezüge zur barocken Emblematik: Kommunikationszeichen an der Schwelle zur Neuzeit.* Münster: Bauhütten Verlag, 1990.

Zimmermann, Hans-Joachim. *Der akademische Affe: Die Geschichte einer Allegorie aus Cesare Ripas 'Iconologia.'* Supplemente der Sitzungsberichte der Heidelberger Akademie der Wissenschaften, Philosophisch-Historische Klasse. Vol. 6. Wiesbaden: Reichert, 1991.

About the Contributors

KU-MING (KEVIN) CHANG is associate professor at the Institute of History and Philology of the Academia Sinica, Taipei, Taiwan. He works on early modern science, especially the chymical and medical work of Georg Ernst Stahl. He is a coeditor (with Sheldon Pollock and Benjamin A. Elman) of *World Philology* (2014), a collection of essays that compare philological traditions in major civilizations. He is also revising a book manuscript on the history of the dissertation as a genre of academic writing and publication.

NICHOLAS H. CLULEE is emeritus professor of history at Frostburg State University, where he taught for forty-two years and was department chair for eleven. He is author of *John Dee's Natural Philosophy: Between Science and Religion* (1988, 2012), and numerous articles on John Dee and late sixteenth-century natural philosophy, alchemy, and magic. He was a National Endowment for the Humanities Fellow in 1984/85.

DANE T. DANIEL is associate professor of history at Wright State University, where he teaches courses in European history and the history and philosophy of science. His research focuses on early modern science and religion, especially the natural philosophical and theological writings of Theophrastus Bombastus von Hohenheim, or Paracelsus (1493/94–1541). He has published over a dozen articles on Paracelsus and the Paracelsians, including the Partington Prize-winning "Invisible Wombs: Rethinking Paracelsus's Concept of Body and Matter" (*Ambix*, 2006) and "Paracelsus on the 'New Creation' and Demonic Magic: Misunderstandings, Oversights, and False Accusations in His Early Reception," in *World-Building and the Early Modern Imagination* (2010).

MARGARET D. GARBER is associate professor of history of science at California State University, Fullerton. She has published articles on intersections of optics, medicine, and alchemy (*chymia*) and is currently working on a manuscript of the medico-chymical correspondences of physician members of the Academia Naturae Curiosorum (also known as the Leopoldina Academy). She was a Dibner Research Fellow at the Huntington Library in 2010/11.

Bruce Moran is professor of history at the University of Nevada, Reno, where he teaches courses in the history of science and early medicine. His general research interest is in the intersection of cultures, learned and lay, scribal and artisanal, Latinate and vernacular as they relate to the investigation of nature and the body in early modern Europe. Among many articles and books are *Distilling Knowledge: Alchemy, Chemistry, and the Scientific Revolution* (2005) and *Andreas Libavius and the Transformation of Alchemy: Separating Chemical Cultures with Polemical Fire* (2007). He has been a Dibner Distinguished Fellow in the history of science and technology at the Huntington Library (2010/11), and, most recently, a Gorden Cain Distinguished Fellow at the Chemical Heritage Foundation (2014).

Karen Hunger Parshall is professor of history and mathematics at the University of Virginia. In addition to numerous articles and chapters, she is the author, among other books and editions, of *James Joseph Sylvester: Jewish Mathematician in a Victorian World* (2006), *Taming the Unknown: A History of Algebra from Antiquity to the Early Twentieth Century* (with Victor J. Katz, 2014), and *Experiencing Nature: Proceedings of a Conference in Honor of Allen G. Debus* (coedited with Paul H. Theerman, 1997). She was a John Simon Guggenheim Fellow in 1996/97 and served from 1996 to 1999 as the editor-in-chief of *Historia Mathematica*.

Mar Rey Bueno is coeditor of *Azogue*, the journal for the historical-critical study of alchemy (www.revistaazogue.com). She is author of, among other books and editions, *El Hechizado: Medicina, alquimia y superstición en la corte de Carlos II (1661–1700)* (1998); *Los señores del fuego: Destiladores y espagíricos en la Corte de los Austrias* (2002); *Lastanosa: Art and Science in Baroque* (coedited with Miguel López Pérez, 2008); and *Chymia: Science and Nature in Medieval and Early Modern Europe* (coedited with Miguel López Pérez and Didier Kahn, 2010).

Heinz Schott is emeritus professor of the history of medicine at the University of Bonn (Germany). He is author of, among other books and editions, *Zauberspiegel der Seele: Sigmund Freud und die Geschichte der Selbstanalyse* [Magic Mirror of the Mind: Sigmund Freud and the History of Self-analysis] (1985);

Die Chronik der Medizin [The Chronicle of Medicine] (1993); *Geschichte der Psychiatrie: Krankheitslehren—Irrwege—Behandlungsformen* [History of Psychiatry: Nosology—Meanders—Forms of Treatment] (with Rainer Tölle, 2006); and *Magie der Natur: Historische Variationen über ein Motiv der Heilkunst* [Magic of Nature: Historical Variations on a Motif of the Healing Art] (2 vols., 2014). From 1987 to 2014, he was director of the Institute of the History of Medicine at the University of Bonn.

JOLE SHACKELFORD is assistant professor in the Program for the History of Science, Technology, and Medicine at the University of Minnesota. Recent scholarship includes the 2013 monograph *Northern Light and Northern Times: Swedish Leadership in the Foundation of Biological Rhythms Research.* His book *A Philosophical Path for Paracelsian Medicine: The Ideas, Intellectual Context, and Influence of Petrus Severinus* (2004) received the George Urdang Medal from the American Institute for the History of Pharmacy in 2007.

ANKE TIMMERMANN was EU Marie Curie postdoctoral fellow at the Medical University of Vienna, Austria (2011–13), and subsequently Munby Fellow in bibliography at the University of Cambridge, England. She is the author of *Verse and Transmutation: A Corpus of Middle English Alchemical Poetry (Critical Editions and Studies)* (2013); "Scientific and Encyclopaedic Verse" in A. S. G. Edwards and Julia Boffey, eds., *Companion to Fifteenth-Century English Poetry* (2013); "Doctor's Order: An Early Modern Doctor's Alchemical Notebooks," *Early Science and Medicine* (2008), and other articles on manuscript studies and the history of the book, the history of science, and especially the history of alchemy and medicine.

MICHAEL THOMSON WALTON took his PhD at the University of Chicago in 1979. He coedited, with Allen G. Debus, *Reading the Book of Nature: The Other Side of the Scientific Revolution* (1998). He is the author of *Medical Practitioners and Law in Fifteenth Century London* (with Phyllis J. Walton, 2003); *Genesis and the Chemical Philosophy: True Christian Science in the Sixteenth and Seventeenth Centuries* (2011); and *Anthonius Margaritha and the Jewish Faith: Jewish Life and Conversion in Sixteenth Century Germany* (2012). Two of his articles, "John Dee's *Monas Hieroglyphica:* Geometrical Cabala" and "Boyle and Newton on the Transmutation of Water and Air," were reprinted in *Alchemy and Early Modern Chemistry: Papers from Ambix* (edited by Allen G. Debus, 2004). Michael Walton died in August 2013.

Index

Page references in *italics* indicate illustrations.

A

C

P

Q

R

T